Prediction and Control of Dynamical Instabilities in Engineering Systems

In this practical guide that discusses both theoretical and real-world results, Paul A. Meehan draws on his extensive experience to provide a vital compilation of the methods and applications for predicting and controlling instabilities in engineering systems in a digestible and implementable form.

Through *Prediction and Control of Dynamical Instabilities in Engineering Systems*, readers will gain an in-depth understanding of classical and nonlinear instabilities in engineering systems and the mathematical tools and other techniques used to identify, analyse and control them. After first introducing instability phenomena and their growing importance, the author then goes on to describe general predictive modelling tools and analysis methods. Subsequently, recent real applications across a wide range of engineering systems are explained, including prediction of squealing noise, aeroelastic flutter, rail corrugation, rolling mill chatter, mining dragline swing instabilities, intractable railway contact wear instability, human body system instabilities and spacecraft attitude instabilities. Analytical and simulation predictions of instability phenomena are presented and compared with experimental or field results. Various applied techniques are outlined, successful example applications are described and mathematical tools are presented in a form that enables readers to apply them to the control of instability phenomena in their own fields.

This book is an invaluable reference for academic researchers and industry figures in a variety of fields, particularly because it highlights the commonalities and generalisation in approaches toward the prediction and control of instabilities. The insights within are also a useful resource for students from intermediate undergraduate level and up to better understand the topic and its applications.

Paul A. Meehan is a professor in the School of Mechanical and Mining Engineering at The University of Queensland. He earned his Bachelor of Engineering (Hons I) and PhD from The University of Queensland in 1992 and 1997, respectively.

Prediction and Control of Dynamical Instabilities in Engineering Systems

Paul A. Meehan

CRC Press
Taylor & Francis Group
Boca Raton London New York

CRC Press is an imprint of the
Taylor & Francis Group, an **informa** business

Designed cover image: Paul A. Meehan and Hannah K. Meehan

First edition published 2026
by CRC Press
2385 NW Executive Center Drive, Suite 320, Boca Raton FL 33431

and by CRC Press
4 Park Square, Milton Park, Abingdon, Oxon, OX14 4RN

CRC Press is an imprint of Taylor & Francis Group, LLC

ISBN: 9781032803449 (hbk)
ISBN: 9781032803463 (pbk)
ISBN: 9781003496533 (ebk)

DOI: 10.1201/9781003496533

Typeset in Times
by codeMantra

Contents

Preface

Engineering has always advanced by stretching the limits of what is possible – lighter, faster, more powerful machines that deliver unprecedented performance. Yet, as we press harder against these frontiers, we inevitably step into relatively newer regimes of dynamical behaviour, i.e., nonlinear instabilities and their control. Here, the once-linear and predictable gives way to the nonlinear and unstable. These behaviours, long regarded as undesirable and elusive, now emerge as both the greatest challenges and richest opportunities in modern engineering. Nonlinear instabilities and control are not rare curiosities – they are widespread, recurring across mechanical systems of every scale. We see them in the squeal of a brake or a railway wheel, in the flutter of aircraft wings, in the chaotic oscillations of satellites, in the corrugation of railways or in the chatter of machine tools and rolling mills. What was once dismissed as nuisance noise or destructive instability is now recognised as a profound manifestation of nonlinear dynamics. Intriguingly, such "instabilities" are not alien to nature. They appear central to some biological systems – the irregular rhythm of a healthy heart or the instability of the human inner ear – phenomena that grant remarkable adaptability and sensitivity. The frontier of nonlinear dynamics thus lies not only in control but also in understanding what can be learned, and even harnessed, from instability itself.

This book explores that frontier as I have attempted to do over the past 30 years. It begins with a guided introduction to instability phenomena and why they matter more than ever. From there, the predictive modelling tools and analytical methods that underpin modern instability analysis are revealed to make them accessible to both engineers and scientists. With this foundation, the discussion turns to real-world applications: the prediction of squealing noise in brakes, the onset of aeroelastic flutter, the growth of rail corrugation, the seemingly intractable problem of railway wear instabilities, rolling mill and machining chatter, the swinging of mining draglines, spacecraft attitude instabilities and even human body systems. These examples are drawn together into two categories – classical instabilities and nonlinear instabilities – revealing patterns, common mechanisms and unifying insights. I have attempted not to leave theory and simulation invalidated by reality predictions set alongside experimental data and field results. But prediction is only half the story. The ability to control instability – to suppress it, to mitigate it or, in some cases, to exploit it – is the ultimate goal. Practical techniques are presented, tested and shown in action across the same range of systems. What I hope emerges is a body of methods that not only diagnose instability but also offer pathways to master it to make our lives better.

This book is intended as both a reference and a companion for engineers and scientists. I hope it simplifies and makes efficient my learning discovery from a university student. It bridges advanced mathematics and applied practice, translating powerful tools into a form that can be readily used. It is written for a broad audience, from undergraduate engineers beginning to encounter instability to seasoned researchers tackling nonlinear problems in cutting-edge applications. Across many fields, instability is no longer an anomaly – it is the terrain upon which future advances will be won. This book seeks to equip the reader not only with knowledge but also with the tools to push the boundaries of engineering further still.

Acknowledgements

More than 30 years ago, I sat at the crossroads of research, pondering whether I should do my honour's thesis on a strange pendulum that defied gravity and uprighted itself under certain high-frequency pivot vibrations. I was dismayed as I had missed out on what seemed like an exciting applied topic of calculating new orbits to Mars. I then spent that seminal year of discovery first building a demonstration of the strange pendulum and then understanding its surprising behaviour. Since then, the concept of the Kapitza pendulum has been found in many applications such as the stability of insect flight and attosecond dynamics of electrons. That decision led to the long winding journey culminating in this book, with much help along the way.

Hence, there are many contributors I wish to acknowledge – those who have supported, encouraged and believed in me throughout this journey. Much of the material in this book draws on three decades of research within the Nonlinear Mechanics Group at The University of Queensland. Many of the examples are taken from this body of work, though this is by no means suggestive of the fact that there are no equivalent or even stronger examples in the broader literature. I apologise to those authors whose contributions are not cited here; this omission reflects the fact that the book also represents a personal journey of learning, rather than a measure of relative importance to the field.

This work has been made possible through the support of many contributors, including the Australian government and a wide range of industry partners. Numerous postdoctoral researchers and PhD students have also played an essential role in the work presented. I am indebted to them, and it has been both a privilege and a pleasure to work alongside such talented individuals. I hope they see their influence throughout these pages and take pride in their contributions.

I am also grateful to The University of Queensland, which has given me the opportunity to "play with the sand on the beach of knowledge." In particular, the university's support of sabbaticals provided the chance to collaborate with colleagues and friends around the world, broaden my horizons and dedicate the necessary time to complete this book.

There are too many academic colleagues to name who have contributed indirectly but in meaningful ways. I acknowledge you all, although I refrain from naming individuals for fear of omission. I must, however, mention a few inspirations: my late uncle, Jens Andreasen, and Pra Murthy, both pioneers in their fields, authors of seminal books and sharers of many enlightening conversations, which continue to motivate. Special thanks go to my thesis supervisor Sam Asokanthan, who inspired me and sent me out initially on this journey.

Finally, I turn to my family: to my Mum and late Dad, for the support and learning that laid the foundation for all that followed, and to my wife and family, Trisha-Kate, William, Joseph, Genevieve, Patrick and Hannah, without whose love and patience this book would not have been possible. You have given me the highs and lows of chaos, which I have – fortunately – been unable to control. A special extra note of thanks goes to Hannah, who predominantly drew the figures for Chapter 1 and contributed to the front cover design. I hope you enjoy it as much as I do.

Nomenclature

MATHEMATICAL SYMBOLS

$A^*_{1\text{-}4}$	Pitching moment flutter derivatives (Section 3.3)
A_w	Vibration amplitude (Section 3.1)
b	Damper distance from system centre of mass when $y, z = 0$ (Section 4.5)
b, B	Half, total chord length (Section 3.3)
B	Bucket/load position parallel to the boom axis (Section 4.4)
c	Damping constant (i.e., modal; Sections 3.1, 3.2, 3.3)
c_t	Torsional damping constant (Section 4.4)
$\hat{c}$	Nondimensional damping constant
$C(\tau)$	Control signal
C_1, C_2	Viscous damping coefficients of input and output shafts (Section 3.6)
$\mathbf{C}$	Control matrix associated with the OGY method
$\mathbf{C}$	Damping matrix (Sections 3.1, 3.2)
C_ξ	Sensitivity of creep force to normal force variations (Section 3.5)
$d(t)$	Divergence of trajectories
D	Fractal dimension
D_C	Correlation dimension
D_L	Lyapunov dimension
D	Rayleigh dissipation function
$\underset{\sim}{D}_1$	Horizontal distance from the slew axis to the bucket image in the vertical boom plane (Section 4.4)
$\underset{\sim}{D}_2$	Perpendicular horizontal distance from the bucket swing velocity to the slew axis (Section 4.4)
E	Elastic modulus (Section 3.4)
E	Total system energy (Section 3.4)
$\hat{E}$	Nondimensional total system energy
f	Forward slip (Section 3.4)
$f(\mu,\theta)$	Spragging criterion function (Sections 3.1,3.2)
F_f	Dynamic friction force (Sections 3.1, 3.2)
F_N	Dynamic normal force (Sections 3.1, 3.2)
$\mathbf{F}$	Force matrix (Sections 3.1, 3.2)
g	Gravity acceleration
$\mathbf{g}(\mathbf{x},t)$	Periodic perturbation vector function
G_r	Exponential growth rate of instability (Section 3.5)
h	Plunge or heave (Section 3.3)
h_G	System angular momentum
h_Y	Vertical distance of the bucket below the X-axis at zero swing angle (Section 4.4)
$\hat{h}_G$	Nondimensional system angular momentum (Section 4.5)
$\hat{h}_G$	Constant nondimensional system angular momentum
H	Hamiltonian function

H	Angular momentum (Section 4.4)
h_i, H_i	Entry and exit thickness for stand i (Section 3.4)
$H^*_{1\text{-}4}$	Plunging force flutter derivatives (Section 3.3)
I	Moment of inertia
$\hat{I}$	Nondimensional moment of inertia (Section 4.5)
$\underset{\sim}{I}$	Instantaneous moment of inertia of the system about the Y-axis (Section 4.4)
I_x, I_y, I_z	Principal moments of inertia about body fixed axes
I_1, I_2	Moments of inertia associated with the input and output drive systems (Section 3.6)
J	Mass moment of inertia (Section 3.3)
$\mathbf{J}$	Jacobian matrix
k	Spring constant
k	Modal stiffness (Sections 3.1, 3.2, 3.3)
k_c, K_H	Hertzian contact stiffness (Sections 3.1, 3.2, 3.5)
k_{stall}	Ratio of the stall slope to lift slope in the bilinear lift curve (Section 3.3)
k_w	Work roll stiffness (Section 3.4)
$k_{\varsigma,\mu 1}$, $k_{\varsigma,\mu 2}$, $k_{\varsigma,\mu 3}$	Bilinear friction/creep curve parameters (Sections 3.1, 3.2)
$\hat{k}$	Nondimensional spring constant (Section 4.5)
K, K_1, K_2	Control feedback gains
$K_{b,c}$	Sensitivity of wear variations to wheel/rail contact deflection variations and modal sensitivity of the wheel/rail relative displacement to a change in the input longitudinal profile (Section 3.5)
K	Reduced frequency (Section 3.3)
$\mathbf{K}$	Proportionality vector associated with the OGY method
$\mathbf{K}$	Stiffness matrix (Sections 3.1, 3.2)
L	Lagrangian function
L	Aerodynamic lift (Section 3.3)
L	Interstand length (Section 3.4)
m	Mass (i.e., modal, damper, bucket and airfoil; Sections 3.1, 3.2, 3.3, 4.4, 4.5)
m_T	Total mass
$\hat{M}$	Nondimensional external time varying torque
M	External time varying torque amplitude
M	Aerodynamic moment (Section 3.3)
$\hat{M}$	Nondimensional external time varying torque amplitude
$\hat{M}_C$	Nondimensional control torque
$\hat{M}_E$	Nondimensional excitational time varying torque
$\hat{M}_E$	Nondimensional excitational torque amplitude
$\boldsymbol{M}$	Vector of external moments applied about the body fixed axis ($=\left[M_x\ M_y\ M_z\right]^T$)
$M_{\max}$	Control torque limitation
$M(t_0)$	Melnikov function

$\mathbf{M}$	Map matrix of partial derivatives ($=\partial\mathbf{F}/\partial\mathbf{X}$) associated with the OGY method
$\mathbf{M}$	Mass matrix (Sections 3.1, 3.2)
N	Static normal force (Sections 3.1, 3.2)
p	Control parameter
pf	Exponential coefficient associated with the contact roughness fractal distribution (Section 3.2)
p_0	Unperturbed control parameter value
$\delta p_{\max}$	Control parameter perturbation limitation
$\hat{p}_y$	Generalised momentum associated with generalised coordinate $\hat{y}$
$\mathbf{p}$	Vector of generalised momenta
$\mathbf{p}_L$, $\mathbf{p}$	Left and right undamped eigenvectors (Sections 3.1, 3.2)
P	Roll, normal force (Sections 3.4, 3.5)
P	Bucket/load position perpendicular to the boom axis (Section 4.4)
q_i	Generalised coordinate
$\dot{q}_i$	Generalised velocity
Q_i	Generalised force associated with q_i
$\mathbf{q}_0(t)$	Set of homoclinic or heteroclinic orbits in the unperturbed system
R^n	$n \times 1$ vector set of real values
$\mathrm{r}_{\mathrm{CoP}}$	Position of centre of pressure (Section 3.2)
S_1, S_2	Torsional stiffnesses (Section 3.6)
S_x	Inertial coupling parameter between the pitch and plunge (Section 3.3)
$\mathbf{s}$	Shift vector ($=\partial\mathbf{F}/\partial p$) associated with the OGY method
SPL, $\mathrm{SPL}_{\mathrm{const}}$	Sound pressure level, constant (Section 3.1)
t	Time
t_e	Time to reach the equilibrium state
T	Kinetic energy
T_{body}	Rotational kinetic energy assuming the system is a rigid body
T_{damp}	Kinetic energy associated with the damper mass movement
t_f, T_f	Entry and exit tension force (Section 3.4)
T_{rotor}	Kinetic energy associated with the rotation of the rotor
T_r	Time varying torque applied to the rotor bearing
T_r	Rotor torque amplitude
$\hat{T}_r$	Nondimensional rotor torque
$\hat{T}_E$	Nondimensional excitational time varying rotor torque
$\hat{T}_E$	Nondimensional amplitude of the excitational rotor torque
$\hat{T}_C$	Nondimensional control rotor torque
$T_{\max}$	Control torque limitation
$\mathbf{u}(t)$	General Lyapunov forcing control input
$U(\mathbf{x})$	Lyapunov function
U, W	Local aerodynamic flow speed (Section 3.3)
v_i, V_i	Entry and exit velocity for stand i (Section 3.4)
v_{cl}	Chatter lower critical speed (Section 3.4)
V	Potential energy

V_c	Lateral crabbing velocity (Sections 3.1, 3.2)
V_o	Rolling, sliding speed (Sections 3.1, 3.2)
V_p	Work roll peripheral velocity (Section 3.4)
$\hat{V}$	Lyapunov function of the rigid body system
W	Strip width (Section 3.4)
W_{frict}	Frictional power dissipated in the rolling contact (Section 3.5)
$W_o()$	Positive solution to the Lambert W function (Section 3.2)
x	Input shaft twist variable (Section 3.6)
x	Roll stack displacement (Section 3.4)
x	Rail longitudinal profile position (Section 3.5)
x_α	Displacement at CG due to the angle of attack (Section 3.3)
x_f	Fixed point
x_{CoP}	Position of centre of pressure (Section 3.2)
x_n	nth iterate or Poincare map point
$\mathbf{x},\mathbf{x}(.),\mathbf{X}$	System state vector
$\mathbf{x}_e$	Equilibrium state vector
$\dot{\mathbf{x}} = \mathbf{f}(\mathbf{x},t)$	First-order state space equations as an explicit function of t
$\dot{\mathbf{X}} = \mathbf{F}(\mathbf{X},p)$	First-order state space equations as a function of p
$\mathbf{X}_f$	Fixed point
y	Damper mass displacement in the rigid body system
y	Output shaft twist variable (Section 3.6)
y	Modal displacement (Sections 3.1, 3.2, 3.5)
$\hat{y}$	Nondimensional damper mass displacement in the rigid body system
$\overline{y}$	Infinite set of equilibrium values damper mass displacement
$\overline{y}'$	Critical point value of damper mass displacement
$\hat{y}$	Nondimensional damper mass displacement in the rigid body system
z	Displacement of the damper mass in dual-spin spacecraft
$z_{\text{in,out}}$	Modal rail profile variation entering, exiting the rolling contact region (Section 3.5)
$\overline{z}$	Infinite set of equilibrium values damper mass displacement
$\mathbf{Z}$	Denote the set of states for which $\dot{U}(\mathbf{x}) = 0$
α	Angle of attack or pitching (Section 3.3)
$\alpha_{1,2}$	Input and output angles of Hooke's joint (Section 3.6)
α_μ, β_μ	Friction curve shape parameters (Section 3.2)
α_c	Critical angle of attack (Section 3.3)
β	Output angle of Hooke's joint (Section 3.6)
β	Dragline boom angle (Section 4.4)
Γ	Vector region containing $\mathbf{x}_e$
δ	Constant
$\delta_{w,a}$	Unit sliding direction (Sections 3.1, 3.2)
Δ	Delay time
Δ	Discriminant that determines the complexity of the eigenvectors and hence the type of mode coupling (Sections 3.1, 3.2)
Δ	Variation from the nominal value (Section 3.4)
Δz_o	Nominal steady change in the profile (wear) per pass (Section 3.5)

Δt	Time between successive wheel passes (Section 3.5)
ε	Perturbational amplitude
ς	Lateral creep (Sections 3.1, 3.2)
ς	Damping ratio (Section 3.5)
ς_o	Modal damping ratio (Section 3.4)
θ	Angular rotation of the rigid body system
θ	Angular misalignment of shafts (Section 3.6)
θ	Sprag angle (Sections 3.1, 3.2)
θ	Bucket swing angle (Section 4.4)
θ_A	The quasistatic angle of attack (Sections 3.1, 3.2)
θ_c	Contact angle (Section 3.1)
$\theta_x, \theta_y, \theta_z$	Angular rotations about the body fixed frame of reference
θ_r	Angular rotation of the rotor relative to the platform
λ	Maximal Lyapunov exponent
λ	Eigenvector (Sections 3.1, 3.2)
λ_n	nth largest Lyapunov exponent
λ_i	Input and output shaft twist amplitude ratio (Section 3.6)
Λ	Amplitude of a periodic function using Melnikov's method (Section 4.5)
μ	Mass ratio ($= m/m_T$; Section 4.5)
μ	Rolling or sliding friction coefficient (Sections 3.1, 3.2)
μ_s	Static friction coefficient (Sections 3.1, 3.2)
μ_k	Kinetic friction coefficient (Section 3.2)
ξ	Creep or microslip (Section 3.5)
ρ	Fluid density (Section 3.3)
τ	Nondimensional time
ϕ	Slew angle (Section 4.4)
ω	Angular velocity/natural frequency
$\bar{\omega}$	Infinite set of equilibrium values for the angular velocity
$\bar{\omega}_x, \bar{\omega}_y, \bar{\omega}_z$	Infinite sets of equilibrium values for angular velocity about the body fixed frame of reference
$\bar{\omega}_r$	Infinite set of equilibrium values for the angular velocity of the rotor relative to the platform
$\bar{\omega}', \bar{\omega}^*$	Critical point values of angular velocity
$\hat{\omega}$	Nondimensional angular velocity
$\boldsymbol{\omega}$	Angular velocity vector ($= \left[\omega_x \ \omega_y \ \omega_z\right]^T$)
$\omega_0, \omega_1, \omega_2$	Natural frequencies
ω_c	Primary combination resonance frequency (Section 3.6)
ω_{ref}	Desired reference angular velocity
Ω	Perturbational (forcing) frequency
Ω	Angular velocity of the input shaft (Section 3.6)
$\Delta\Omega_c$	Change in the angular velocity of the input shaft
$\Delta\Omega_{\max}$	Maximum change in the input shaft angular velocity

SUBSCRIPTS

$(\ )_{\langle n \rangle}$	nth map iterate
a,b	Dominant modes (Section 3.2)
ae	Aerodynamic (Section 3.3)
c	Crabbing, critical (Section 3.1)
eff	Effective (Sections 3.2, 3.3)
d	Damped (Section 3.5)
f	Flutter onset (Section 3.3)
h	Plunge or heave (Section 3.3)
h	House (Section 4.4)
i	Mode number (Sections 3.1, 3.2)
i,j	Matrix element (Sections 3.1, 3.2, 3.3)
o	Nominal value (Sections 3.1, 3.2)
min, max	Minimum value, maximum value
r	Rail mode (Section 3.1)
ref	Reference input value
s	System (Sections 3.1, 3.2)
w	Wheel mode (Section 3.1)
α	Angle of attack (Section 3.3)

ACRONYMS

FFT	Fast Fourier Transform
INTELSAT	The International Telecommunications Satellite
KAM	Kolmogorov (1954); Arnold (1963); Moser (1962)
MATLAB®	Matrix and Simulation Analysis Software Package
NASA	National Aeronautics and Space Administration
OGY	Ott, Grebogi and Yorke (1990)
OPF	Occasional Proportional Feedback
RPF	Recursive Proportional Feedback
UPO	Unstable periodic orbit

1 Introduction

Many major events in the world's history have been driven, unseen, by instability processes. As far as we know, all major extinctions were caused by an instability in climatic conditions, causing a rapid change in the temporary stability under which species had optimally evolved. Global pandemics are caused by an unstable spread and growth of micro-organisms that infect humans. Similarly, human innovation results in unstable growth in new technologies that cause revolutions in the way we live. Recognising and understanding instability in these processes may be critical to our future success as a species. The prediction and control of this behaviour are critical for continued expansion of the capability of engineering systems. For mechanical systems, in the recent past, such behaviour was typically undesirable and difficult to predict. Some examples include the occurrence of wheel and brake squeal, aerodynamic flutter, growth of chaotic attitude oscillations in spacecraft, rail corrugations, rolling mill and tool chatter. Interestingly, such behaviour is arguably the norm in biological and human body processes and appears to provide useful advantages. This chapter introduces instability along with the answers to the important questions including the following: what is instability, what engineering processes does it act in and can we control it.

1.1 WHAT IS INSTABILITY?

Instability is generally defined as the lack of stable behaviour. In dynamical systems, it refers to the inherent tendency of systems to exhibit growing fluctuations and/or unpredictable behaviour over time. Generally, this means that at least one or more of the system variables (degrees of freedom) move away from a stable state with time. Even a slight change in the starting state or input parameters of a system can lead to significantly different outcomes. This sensitivity to initial conditions is a fundamental characteristic of dynamical instability in many natural and man-made systems throughout the world and the universe. Familiar examples include the behaviour of the weather, financial markets, the Big Bang and human societies. Hence, it pervades our everyday life and history of mankind. Let's take a look at some of these further.

Dynamical instability plays a crucial role in weather systems. The atmosphere is a complex, nonlinear system that exhibits a type of nonlinear behaviour following dynamic instability known as chaos. Small changes in temperature, pressure or humidity cause exponential divergence from stable conditions, which can have dramatic effects on the formation and intensity of weather patterns. This sensitivity to initial conditions (instability) combined with nonlinearity makes it challenging to predict weather accurately beyond a few days. Lorenz showed that this complex behaviour can come from a very simplified mathematical model of the weather to cause unpredictability of atmospheric dynamics. He graphed this behaviour to identify a famous Lorenz attractor, which is like a special indicator of chaotic behaviour.

DOI: 10.1201/9781003496533-1

This dynamic behaviour of weather drives one of the most important global challenges that humans face today, i.e., another climate instability that may cause another major extinction of life on Earth. Will humans be able to control the unstable growth of carbon dioxide emissions to avoid extreme climatic change?

Financial markets also demonstrate dynamical instability. Market dynamics are influenced by numerous factors, such as investor sentiment, economic conditions, political events and even climate change. The interplay of these factors creates a highly complex and nonlinear system, making it challenging to predict the behaviour of financial markets accurately. The concept of market crashes and speculative bubbles arises from the inherent instability of financial systems. For instance, small fluctuations in investor confidence, e-trading or market conditions can trigger vicious feedback, causing significant price movements and volatility.

The weather and financial markets are global systems of dynamic instability. There are many other examples of dynamic instabilities in our daily lives if you look carefully and notice the mechanisms behind the instability. One of these mechanisms is a positive feedback cycle or more colloquially known as a "vicious cycle." For example, my children identified one of these recently at home. The family had discussed and implemented a washing-up schedule. Rules which seemed to work well for a few weeks then gradually broke down, i.e., became unstable. The rules prescribed that each person must put their own dirty dishes into the dishwasher during the day. Then, at night, the person on the wash-up duty had to stack the dishwasher with any remaining dinner preparation utensils and run it. After cycle completion, they would empty it before breakfast so that the cycle could continue again. In theory (under no perturbations), it was a perfect plan (stable equilibrium point), but in practice, it was unstable. The fatal flaw was that if the person on duty forgot to unstack the dishwasher before morning or others neglected to stack their plates during the day, dishes would pile up (see Figure 1.1). This led to dissatisfaction and loss of motivation for the next person on duty – and the vicious cycle would repeat. Hence, they would be more likely to also neglect a part of their duties in the next cycle. In such a way, the annoyance and neglect of duties (perturbations) would grow over each cycle, i.e., a vicious cycle. This is a simple little example of an unstable system that had a significant impact on our lives, i.e., led to annoyance and family disharmony! I would say we have solved this problem completely, but we are by no means perfect parents or a perfect family. However, the family insight gained by identifying the vicious cycle has modified behaviours and expectations and therefore lowered the impact (or growth rate) of the washing-up instability.

1.2 WHY STUDY INSTABILITY?

The washing-up instability identified in the last section, albeit simplified, shows why we should study and identify instability. In essence, if we want to solve a problem, then we need to understand it. More specifically, if we can identify an instability and its driving mechanism, then with this insight, we will be able to understand it and maybe avoid or control it. Here are some examples.

Celestial dynamics based on Newton's and Einstein's laws play a crucial role in the stability of planetary orbits. This modelling and prediction have given us important understanding of the history of the solar system and life on Earth.

FIGURE 1.1 Washing-up instability.

A small deviation in the initial conditions of a planet's orbit can sometimes lead to its eventual expulsion from the solar system or collision with another celestial body. It is now clear that such an orbital mechanics instability is likely to have led to an ejected meteor, from the outer main part of the asteroid belt, over 60 million years ago (see Figure 1.2). This then impacted Earth, and the following devastation and climate disruption caused the mass extinction of 75% of plant and animal species on Earth. This insight gives us great understanding of the history of life on Earth, i.e., why (non-avian) dinosaurs became extinct and what led to the present human domination of the planet. Apart from this, we can use this insight to predict and maybe avoid similar planetary impacts that cause significant effects on life on Earth.

In a similar way, understanding dynamical instability is essential to us for many applications. In weather forecasting, meteorologists use sophisticated models to predict the instability behaviour of atmospheric conditions that we all rely on, daily. Financial analysts employ various mathematical models and statistical techniques to assess the stability and predictability of financial markets. By gaining insights into the causes and patterns of dynamical instability, researchers and practitioners can predict dynamic behaviour and develop strategies to mitigate its adverse effects in many engineering systems, like those discussed subsequently.

1.3 WHERE HAS INSTABILITY BEEN OBSERVED IN ENGINEERING SYSTEMS?

Instability in engineering systems can have significant consequences, typically leading to reduced performance, increased noise and/or wear and tear and even failure. In this book, we look in detail at several examples of instability observed in various

FIGURE 1.2 Orbital mechanics instability leading to mass extinction of 75% of life on Earth.

engineering disciplines, from mechanical systems to transportation and even biomedical applications. Some of the notable instances of instability include railway wheel squeal, brake squeal, aerodynamic flutter, rolling mill chatter, tool chatter, rail corrugation, transmission driveline instability and inner ear cochlear amplification.

Railway wheel squeal (or curve squeal) is an instability that sometimes occurs in railway systems, when a train negotiates a curve (see Figure 1.3). The frictional interaction between the wheel and the rail can result in a loud tonal high-pitched squeal noise that is very annoying. A typical occurrence can be clearly heard for miles in the wee hours of the night, due to a freight train passing through the track curves of a city. This phenomenon occurs due to the dynamic interaction between mainly wheel vibrations (behaving like flexible rolling drums) and friction between the wheel and the rail. This causes self-excited oscillations and instability. Wheel squeal causes noise pollution and can lead to accelerated wear of the wheel and rail surfaces.

Brake squeal is another type of instability commonly encountered in automotive and industrial braking systems (i.e., including trains). It occurs when the vehicle or system is slowed such that the friction between the brake pad and disc or drum causes vibrations. These propagate pressure fluctuations through the air, which we hear as an annoying high-frequency tonal noise (see Figure 1.4). Like wheel squeal, there is a dynamic interaction between the structure (i.e., brake components) and friction which becomes unstable due to various mechanisms that we will identify

FIGURE 1.3 Railway wheel squeal instability.

FIGURE 1.4 Brake squeal instability.

and discuss in Chapter 3. The instability of the brake system is influenced by numerous factors, including material properties, friction conditions, geometry and applied loads. The difficulty in identifying these and their interactions has made brake squeal a "fugitive" phenomenon to solve. Hence, although it has been with us for at least a

FIGURE 1.5 Aircraft wing flutter instability.

century, it remains one of the most important issues for the automotive industry. This is because the annoying noise and vibration result in high levels of customer concern and large warranty costs [37].

Aerodynamic flutter is an instability phenomenon observed in aircraft (see Figure 1.5) and a wide range of other aerodynamic systems like wind turbines. Flutter is an aeroelastic vibration phenomenon of an elastic structure in a fluid that results from an unstable interaction between the fluid and structural dynamics. Flutter vibrations occur typically above a lower critical wind speed and increase to an amplitude determined by the nonlinearities in the aerodynamics and/or structural dynamics. Flutter appears to have first been identified in the tail of a World War 1 aircraft and since then has galvanised the focus of many researchers to solve the problem. Still now, it remains one of the most important issues for aircraft and structural engineering industries, motivating careful design to avoid failures such as the Tacoma Narrows Bridge disaster. The mechanism driving the instability is similar to brake squeal, due to a dynamic interaction between aerodynamic and structural forces. Here, the instability is powered by the aerodynamic energy, as opposed to the frictional energy in wheel and brake squeal.

Rolling mill chatter refers to different types of unstable vibrations that can occur in rolling mills used in metal processing. For instance, third octave mode chatter occurs above a critical mill speed due to an unstable interaction between the oscillations of the mill stand and inter-stand tension. This creates excessive processed strip thickness fluctuations that compromise the quality of the final product. The fifth octave mode chatter increases slowly, over a much longer time period, due to microscopic oscillating wear or banding of mill rolls. This interacts with excited millstand vibrations and transfers the undesirable banding to the processed strip. Hence, rolling mill chatter can lead to surface defects, reduced mill throughput and increased maintenance costs and therefore is a significant challenge in the metal manufacturing industry.

Tool chatter is a common form of instability encountered in machining operations. When the cutting tool and the workpiece interact, vibrations can increase, resulting in poor surface finish, reduced tool life and decreased machining accuracy. The typical instability mechanism is due to tool structure vibrations interacting with cutting force dynamics that regenerate as the tool passes over the previously cut and corrugated pass, i.e., a vicious cycle. Tool chatter can be influenced by various factors, including cutting parameters, tool geometry and structural dynamics. Proper selection of cutting conditions and tooling design are critical in minimising the impact of tool chatter in machining operations.

Rail corrugation refers to the formation of irregular patterns on railway tracks, typically characterised by periodic undulations. It is an international phenomenon that has remained persistent and grown in prevalence in its multiple forms over many decades. The phenomenon is manifested as variations in the longitudinal rail profile that induce excessive vibrations in trains as they pass over them (see Figure 1.6). This causes undue noise, restricting vehicle speeds and in some cases causing fatigue defects. Its removal by grinding places substantial costs for maintenance on the railway industry. It occurs due to an unstable dynamic interaction between the train and track vibrations and the frictional power and wear occurring in the rolling contact. The corresponding growth process is a bit like the fifth octave mode rolling mill chatter. Rail corrugation leads to increased maintenance costs, reduced ride comfort and potentially derailments without removal via track regrinding. Understanding the

FIGURE 1.6 Rail corrugations (rail wear instability) on inner rail exciting train vibrations.

underlying mechanisms and implementing effective maintenance strategies are crucial for managing rail corrugation.

Transmission driveline instability refers to the dynamic interactions occurring within the various components of a vehicle's driveline, such as the engine, transmission and driveshaft. These interactions can result in undesirable growth of oscillations and vibrations that reduce the performance and efficiency of the driveline. A typical example is a driveline connected by a Hooke's (or universal) joint used to transmit power between rotating shafts with an angular misalignment (i.e., for rear drive vehicles like trucks). This system may exhibit severe twisting motion primarily due to the fluctuating angular velocity ratio inherent in such systems. This motion causes noticeable noise, severe mechanical shakings and premature fatigue failures in shafts, dog clutches, gear teeth, etc. The occurrence as well as severity of these detrimental torsional vibrations have been found to depend on the range of engine or drive shaft speeds. Here, the instability is due to a dynamic interaction and self-excitation between inherent joint oscillations with the twisting dynamics of the transmission shafts. Effective design and control strategies are required to suppress the driveline instability, which effects many heavy road vehicles, and therefore ensure smooth power transmission.

Crack growth is a well-known instability phenomenon in engineering, playing a pivotal role in the structural integrity of many components we use daily, such as cars, bridges, aircraft, pipelines and other load-bearing structures. Understanding crack growth as an instability process is essential for designing engineering systems to withstand the challenges of real-world applications. For instance, without the prediction and control of crack growth, we could not reliably fly in airplanes without a high risk of component failures that may lead to crashes. Crack growth involves the propagation of a crack or a fracture within a structure, often initiated by factors such as stress concentrations, material defects and/or environmental conditions. This process can be characterised as an instability because once a crack initiates, its growth rate tends to increase uncontrollably over time, eventually leading to catastrophic failure if not properly managed. The vicious cycle behind crack growth is due to a positive feedback loop between crack growth and strain energy release. When a material is subjected to external cyclic loads, it undergoes deformation, and energy is stored within the structure. As a crack initiates and propagates, this stored strain energy is released, promoting further crack growth. This causes an exponential growth in the size of the crack over many load cycles until the strain energy released exceeds a critical value, causing catastrophic failure. Recognising the interplay of factors that contribute to crack initiation and propagation is essential for designing engineering components with enhanced durability and structural integrity. Researchers continue to advance our understanding and prediction of crack growth mechanisms, paving the way for the development of more resilient and reliable components in many engineering applications.

Lastly, inner ear cochlear amplification is a useful biomedical engineering application of instability. The cochlear amplifier that provides acute sensitivity and amplification is a positive feedback mechanism within the inner ear of mammals, giving us a distinct advantage over other vertebrates. This means we can hear very soft sounds, i.e., whispers, higher-frequency sounds and distinguish pitch/frequency very accurately. The mechanism most likely evolved so that we can hear important environmental cues, such as the sound of an approaching predator as well as nuances in speech so that we can communicate at a very high level of complexity. In this case, the tiny outer hair cells in the inner ear (which act as sensors and actuators) use electromechanical feedback to create large amplifications of vibrations via an instability. If we are healthy, the cochlear amplifier actually outputs very low sound called otoacoustic emissions, which we can use to diagnose hearing functionality in babies (see Figure 1.7). Hence, our human bodies use instability deliberately to greatly enhance our hearing functionality, i.e., range. Understanding the causes and mechanisms of inner ear instability is crucial for diagnosing and treating hearing disfunction in many people.

In conclusion, instability can be observed in a range of engineering systems, ranging from transportation, manufacturing and even biomedical engineering applications. The examples mentioned highlight the importance of understanding and addressing instability to ensure the safe and reliable operation of engineering systems as well as understanding increased functionality, like in the human ear. Effective control strategies are crucial in minimising the adverse effects of instability and enhancing the performance and longevity of these systems. This will be introduced in the following section.

FIGURE 1.7 Healthy instability: otoacoustic emissions for neonatal testing as a symptom of sensitive hearing in mammals.

1.4 CAN WE CONTROL INSTABILITY?

If we can understand the causes, the next step of solving the problem is usually more straight-forward. However, this complex and relevant question has been the focus of much research and development in various fields. While complete control over instability may be elusive in some cases, significant progress has been made in understanding, mitigating and managing instability in different systems. This is primarily due to the use of efficient validated modelling which provides important insights. This modelling allows detailed and wide investigation of phenomena over many scenarios without relying upon time-consuming numerical, experimental or field work.

The ability to control instability depends on several factors, including the nature of the system, the underlying causes of instability and the available resources and technologies. In some cases, instability can be minimised or even eliminated through appropriate design, control strategies and/or maintenance practices. For engineers and in life, in general, the critical step in solving a problem is understanding it. An efficient validated model can be developed and used to investigate the phenomenon and confirm the mechanism of instability. With this insight, the method of control can nearly always be developed. Sometimes, control may even directly incorporate the model itself like in model predictive control. However, complete control over all forms of instability may not always be achievable due to practicalities.

For instance, for avoiding the next cataclysmic meteor impact on Earth, we can use our understanding of the cause, orbital mechanics instability, to control it. The insight gleaned can be used to provide small perturbations in the meteor trajectory early on to avoid collision with the Earth. Perhaps, this could be done by ballistic impact or using another object/satellite as a gravity tractor (Figure 1.8). But is this feasible, so far away from the Earth?

FIGURE 1.8 Controlling orbital mechanics instability to avoid meteor impact.

Similarly, in the realm of financial markets, controlling instability poses significant challenges. However, regulatory bodies and market participants have implemented various measures aimed at reducing the likelihood and impact of instability. These include market circuit breakers, stricter risk management practices and improved surveillance and monitoring systems. While these measures may not entirely eliminate instability, they serve to enhance market resilience and reduce the potential for systemic events.

In engineering systems, extensive efforts have been directed towards controlling various types of instability. Advances in control theory and techniques have allowed engineers to develop sophisticated control systems that can stabilise unstable systems and suppress unwanted oscillations. These control systems rely on feedback mechanisms that continuously monitor and adjust system parameters to maintain stability. Examples include active vibration control systems for structural/building applications and stability control systems for vehicle suspension and braking. A useful approach to controlling instability in engineering systems is to break the positive feedback mechanism or "vicious" cycle that is driving it. This approach will guide a number of methods detailed in this book.

In the field of fluid dynamics, instability control is crucial for optimising performance and preventing undesirable phenomena such as flutter and turbulence. Active flow control techniques, such as the use of actuators and sensors, can be employed to manipulate fluid flows and stabilise unstable flow regimes. By injecting energy into the flow or modifying the boundary conditions, researchers have been able to mitigate flow instabilities and enhance system performance. An approach for controlling flutter is to use insights gained from identifying its "vicious" cycle. In particular, flutter grows from reinforcing feedback between flow-excited pitching and lift oscillations of an aerofoil due to a dynamic interaction between the aerodynamic and structural forces. This reinforcement depends on the position of the elastic centre (centre of pitching) relative to the centre of mass. Therefore, one method of control is to shift the centre of mass of the aerofoil to the elastic centre to eliminate the vicious cycle. Another method is to mitigate the structural vibrations by introducing damping. Due to various design practicalities, these methods may not be able to be performed. Also, as detailed in subsequent chapters, we need to be careful where we introduce the damping as it may actually enhance flutter depending on the type of the mechanism involved.

Similarly, in railway engineering, the vicious cycles driving wheel squeal and rail corrugations provide many insights into their control. For wheel squeal and rail corrugation, we can try and break the dynamic interaction between the contact friction and structural dynamics. For instance, this can be achieved by changing the frictional mechanics between the wheel and rail using specialised friction modifiers (i.e., lubricants). Alternatively, hardening the rail will reduce the extent of positive feedback growth of the corrugations from differential wear. Changing the curving behaviour of the wheels can avoid excessive lateral sliding that drives the wheel drum-like vibrations of wheel squeal. Vibrational damping should also theoretically mitigate both these instabilities depending on the type of the mechanism. But which method is best? An efficient predictive model of the mechanism allows us to quantify the effects of these control measures so that multiple options can be evaluated effectively and efficiently.

In the field of medicine and healthcare, efforts are made to control instability associated with certain health conditions. For instance, medications and therapies can be utilised to manage unstable heart conditions, i.e., stabilise blood pressure, heart rhythm and other physiological parameters. Alternatively, with cochlear amplification, the instability is desired normal behaviour that enhances our range of hearing (i.e., allows us to hear whispers). Hence, we monitor and detect it in babies and try and mimic it with man-made cochlear implants to restore hearing of the inner ear.

These examples highlight the successful control of instability and also instances where it may not be completely solved. In complex systems with multiple interacting components and dynamic behaviours, such as the environment and socioeconomic systems, complete control over instability may be unrealistic. Instead, the focus shifts towards managing and adapting to instability by implementing robust and flexible strategies based on the understanding of the driving mechanism. This includes measures like disaster preparedness, risk management and resilience planning.

While complete control over instability may not always be attainable, significant progress has been made in understanding, managing and mitigating instability in various systems. We can avoid instability through understanding its mechanisms, development of efficient modelling, innovations in control systems and/or improved design and maintenance practices. That way, we can minimise the adverse effects of instability in engineering systems, financial markets, healthcare and other domains. Hence, dynamical instability is fundamental to the behaviour of many systems around us. Its sensitivity to initial conditions and small perturbations can lead to unpredictable and chaotic behaviour over time. This is typically bad (i.e., undesirable) but sometimes good for us (i.e., our hearing). Hence, understanding the causes and implications of dynamical instability is crucial for improving predictions and decision-making in a wide range of fields. By embracing the inherent uncertainty and complexity of dynamical systems, we can enhance our understanding and develop strategies to navigate the uncertain world we live in. While complete control may remain elusive in some cases, our ability to manage and adapt to instability continues to advance, fostering safer and more efficient systems as technology advances. The rest of this book will hopefully provide a guide for readers to achieve this through giving detailed practical examples of predicting and controlling instabilities in engineering systems.

2 Tools of Analysis and Prediction of Instabilities

Despite safety warnings, most of us have at some point in our lives reclined in a chair to relax, stretch or simply have fun. For small angles of leaning, the chair returns to a stable position with four legs on the floor after we release our feet. However, for larger angles of leaning, we reach a critical tipping point (with highest gravitational potential). Past this point, the chair becomes unstable and tips rearward towards another stable condition (of lowest gravitational potential energy) and lies backwards on the floor! Apart from avoiding injury, we can learn a lot from this simple example of nonlinear instability. Firstly, by understanding and modelling the physics using gravity, we can predict the equations of motion of the system. Based on that, we can then identify the critical points where forces are in equilibrium (i.e., the tipping point when the centre of gravity is directly above the back two chair legs) and determine their stability. This predicts the instability without having to experiment and injure ourselves. Climatologists attempt the same thing so that life on Earth doesn't need to experience another extinction period to prove their analysis. Addressing these challenges may require interdisciplinary knowledge, but the analysis is somewhat universal as it relies upon the worldwide language of mathematics.

To predict and control important instability processes in engineering, we need to be able to understand the phenomenon and predict when it occurs. This enables us to ultimately avoid, suppress or mitigate it. During the outbreak of COVID-19, the first step was to understand the mechanism of initial unstable exponential spread. Mathematical models were developed based on this understanding to predict what effect various conditions or parameters had on its growth. This provided information to the world on implementing various control measures including those to inhibit transmission and infection, such as lockdowns, wearing masks and/or vaccinations or alternatively to predict the development of herd immunity. This chapter is dedicated to providing the reader with an understanding of the tools needed to analyse and predict instabilities. Methods of modelling instability are introduced, leading to identifying different types of instabilities using mathematical tools. Then, we review the use of appropriate stability analysis to predict when the phenomenon occurs.

2.1 MODELLING INSTABILITY

The first and perhaps arguably the most important step in analysing instability is to develop a suitably accurate and efficient model that captures the important behaviour and mechanism of instability. It's very easy to draw incorrect conclusions of dynamic behaviour with an inaccurate or inappropriate model. However, there are many texts and undergraduate subjects available on modelling, so we only provide an overview here. The primary aim of modelling instability is to enable numerical and

DOI: 10.1201/9781003496533-2

analytical techniques to be employed, which then accurately identify and characterise the full nonlinear behaviour and mechanism of growth of the system. Modelling is also used to motivate and assist in the development of suitable control methods for the elimination of undesirable nonlinear or unstable phenomena. For these purposes, it is essential that each model must be able to accurately account for both local and global dynamic behaviour of each system. In particular, the inherent nonlinearity and/or instability mechanism, such as coupling, arising from the relative motion of sub-bodies, must be accurately modelled. Hence, it is typically necessary to identify the physical attributes that are important to the global nonlinear behaviour of each system and develop the full model which captures this. There are a few broadly different types of modelling, as summarised below.

Mathematical Models

Modelling instability often begins with identifying the key variables and parameters that influence the system's behaviour. These can include physical quantities like temperature and pressure in a fluid flow, market variables like supply and demand in economics or biological factors like population dynamics in ecological systems. Once these variables are identified, mathematical models are developed to describe their interactions and dynamics. A mathematical model is very powerful as it is the universal language for modelling dynamical behaviour across all domains from physics to even arts. Through dynamical systems theory, a mathematical model can provide an understanding of the evolution of system equations of motion over time. By representing the system's dynamics using differential equations or difference equations, researchers can analyse its stability properties and predict the onset of instability. The development of these equations of motion may be performed using Newton's Laws, Lagrange's Equations, Hamilton's equations or by many other means that a typical science, maths and/or engineering undergraduate degree provides training for. There are also many books on the topic across each domain, so we won't attempt to replicate what is available here. The engineering dynamical systems, detailed in this book, typically are nonlinear multiple-degree-of-freedom dynamic systems with dissipation. Hence, we prefer Lagrange's equations to provide the most efficient and accurate method for developing the equations of motion compared to Newton's (prone to user errors in multiple-degree-of-freedom systems) or Hamilton's methods (mainly conservative systems). However, this is only a preference, and researchers may use whatever method provides them with a set of accurate and appropriate equations of motion of the system of interest.

Computer Models

Mathematical models may also be codified in the form of computer models, such as by using finite-element analysis and/or computational fluid dynamics. Essentially, these are means of codifying mathematical modelling equations of motion and applying numerical solvers to solve them computationally. Complex systems often defy analytical solutions, especially when they involve nonlinear dynamics, complex geometry or stochastic processes. In such cases, researchers may need to turn to computational models to simulate the behaviour of the system and study its stability properties. These models can range from simple agent-based models to sophisticated simulations using techniques like finite-element analysis or Monte Carlo methods. The disadvantage of numerical models is that they are much more computationally inefficient than analytical solutions and therefore greatly restrict the scope of parameter analysis that can be performed, i.e., they are typically restricted to one-off case studies that may be useful for validating assumptions of a more efficient solution.

Experimental Models

A smaller and or more convenient physical model of the real instability process can be developed experimentally. This allows researchers to investigate and prove certain fundamentals of the mechanism under more controlled and easily measured conditions. However, experimental models do not allow the infinite range of measurement, analysis and control that a mathematical model does. Experimental modelling typically involves using a physical apparatus to simulate the fundamental physics, actuators to provide external forces if needed, sensors, signal processing and data recording and analysis, i.e., using a computer. It is very important that these systems are accurate and precise enough to provide scientific rigour. Therefore, it is important to perform an error analysis of the experimental system to determine the extent of noise (unwanted signals) and statistical significance of the results. Many great scientific discoveries have been made using experimental research and scientific methods. Although mathematical analytical solutions are more powerful, experimental and or field models, combined with measurements, are desirable to validate the results of mathematical models so as to confirm theory, assumptions and provide confidence.

2.2 TYPES OF INSTABILITY

In dynamical systems, instability means that some of the outputs or internal states increase typically exponentially with time, without bounds and without external forcing. Not all systems that are not stable are unstable; systems can also be marginally stable or exhibit limit cycle behaviour. Although instability in dynamic systems has this universal definition, it manifests in various forms, each with its own characteristics and implications. Understanding these types of instability is important for understanding the cause behind its occurrence. This also can provide insights into its avoidance or control. Here, we explore some common types:

Not Instability – Resonance

Sudden increases in amplitude due to forced resonance can be mistaken for instability. This is not an instability as it is externally forced and the resultant motion is bounded (for real systems with damping). Resonance is a classical linear phenomenon occurring under harmonic excitation. For a forced linear oscillator (such as small motions in a pendulum), resonance occurs at the natural frequency of the system (ω_o). This may be shown graphically as classical resonance curves relating the output sinusoidal amplitude (A) and the excitation frequency (Ω) for varying degrees of damping (δ), as seen in Figure 2.1.

Engineers apply this theory to predict fatigue failures in structures and machines caused by resonant vibrations. In a nonlinear system, the resonance curve is distorted by the nonlinearities such that there may exist a range of forcing frequencies for which there are three possible output amplitudes (two of which are stable). This is illustrated by the resonance curve seen in Figure 2.2.

The dotted part of the amplitude in Figure 2.2 represents an unstable region, so for increasing and decreasing frequencies, different amplitudes are obtained. This hysteretic effect, known as the jump phenomenon, can be experimentally observed in many engineering systems. The curve also describes the phenomenon of *entrainment,* whereby the periodic motion of the system becomes entrained at the driving frequency, as opposed to the natural frequency in linear systems.

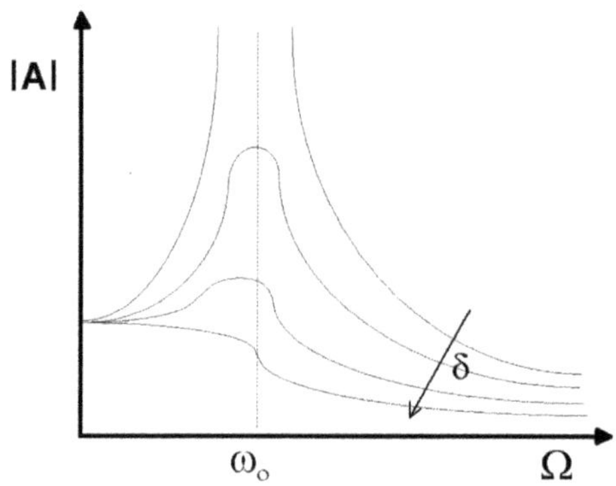

FIGURE 2.1 Linear resonance.

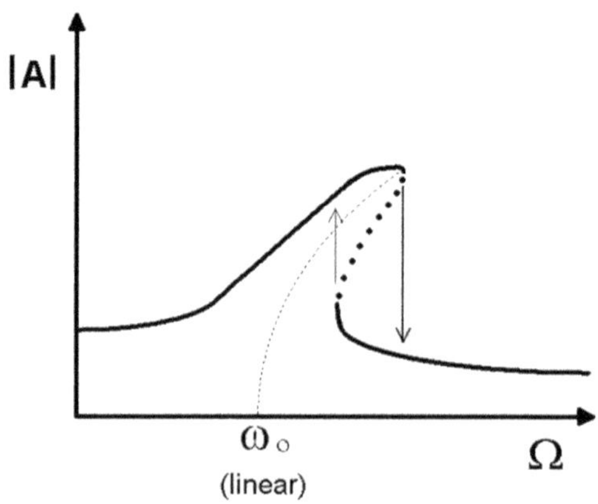

FIGURE 2.2 Nonlinear resonance.

Linear Instability

In linear systems, instability often arises when the system's response grows without bound over time. This typically occurs when the system's response has an exponential growth component. Mathematically, this means that if the equations of motion are expressed in matrix form, their eigenvalues will have positive real parts, indicating that small perturbations amplify exponentially. Often a nonlinear system can be linearised about an equilibrium point (where forces are balanced) to analyse the linear stability. Examples include the linearised perturbed motion of a pendulum or a ball on a hill (or tipping chair) about their highest positions. Without damping, the equations of motion of these linearised systems represent simple harmonic oscillators of the form,

$$\ddot{x} + \lambda^2 x = 0, \tag{2.1}$$

where λ is the eigenvalue and determines the stability according to

$$\lambda^2 > 0 \rightarrow \text{stable oscillatory motion (for lowest point)}, \tag{2.2}$$

$$\lambda^2 < 0 \rightarrow \text{unstable exponential growth (for highest point)}. \tag{2.3}$$

Parametric Instability

A nonlinear system may produce instabilities leading to periodic oscillations above and below the linearised natural frequency known as *subharmonic* or *superharmonic* vibrations, respectively. These are limit cycle oscillations that occur after an exponential growth from stable conditions due to parametric instability. This phenomenon occurs in systems which have parameters that vary as a function of time, as opposed to an external forcing function. These are known as *parametrically excited* systems and have been studied in some detail by authors such as [1] and [2]. Probably, the most general parametrically excited system is described by *Hill's equation*:

$$\frac{d^2w}{d\tau^2} + q(\tau)\cdot w = 0. \tag{2.4}$$

The coefficient $q(\tau)$ is a regular periodic function of time. A special case of this equation known as the Mathieu equation is obtained when

$$q(\tau) = \delta + \varepsilon\cos(\tau).$$

Therefore yielding:

$$\frac{d^2w}{d\tau^2} + [\delta + \varepsilon\cos(\tau)]w = 0, \tag{2.5}$$

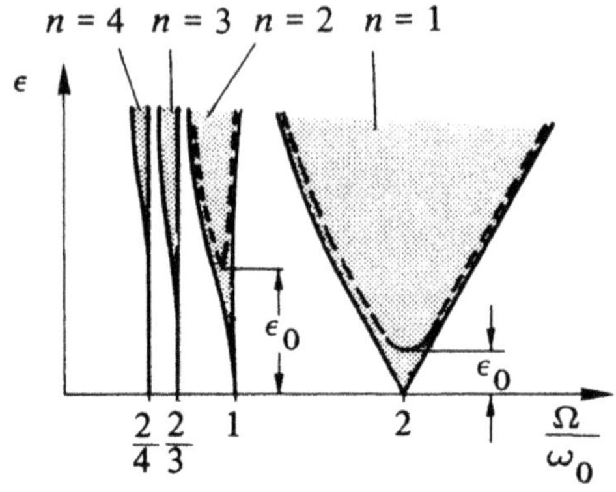

FIGURE 2.3 Instability regions of the Mathieu differential equation [1, p. 102].

where δ and ε are constants. This equation has been studied very extensively and has been shown to yield very interesting stability results (see [3] and [1]). It has been found that the system does not only become unstable at the natural frequency but also at twice the value. Instabilities also occurred at the subharmonics described in Figure 2.3.

In this case:

$$\tau = \Omega t, \quad \frac{\omega_0}{\Omega} = \sqrt{\delta}, \quad \in = \frac{\varepsilon}{\delta},$$

and $\in_0$ is a threshold value which must be exceeded before instability can occur in the damped system. It should also be noted that for a nonlinear system, the instability shown in Figure 2.3 denotes a limit cycle of finite amplitude.

The existence of *subharmonic* and *superharmonic vibrations* in such mechanical systems is of concern to engineers. This is because a multiple-degree-of-freedom linear system is usually modelled as a coupled set of linear oscillators which produces multiple resonant frequencies (one for each degree-of-freedom) when the system is forced. Thus, engineers usually assume that each peak in a vibration frequency spectrum corresponds to a mode or degree-of-freedom within a system. As has been described above, this is not the case in a nonlinear system that can produce multiple peaks on a vibration frequency spectrum for just one degree-of-freedom.

The *Hill* and *Mathieu* differential equations have applications to a number of engineering systems whose equations of motion have the same form. Some of these are mentioned:

- A circuit with periodically varying resistance [1]
- A link–mass–spring mechanism [3]
- A periodically buckled beam [4]
- A pendulum with harmonically varying length or vertically excited pivot point [3].
- A transmission driveline with a universal joint [5].
- Wheel squeal [6].

Mode Coupling Instability

Mode coupling instability occurs when different oscillation modes interact nonlinearly, leading to instability. Physically, it occurs when one mode of vibration in a system affects another mode, and vice versa, causing a feedback loop that amplifies disturbances. The mode coupling could be oscillations, waves or any other independent characteristic patterns or resonances of motion within the system. When these modes interact, energy can be transferred between them, typically leading to the growth of certain modes at the expense of others. This interaction can become self-reinforcing, ultimately leading to instability. Mathematically, this means that when the equations of motion are expressed in matrix form, typically the coupling causes a nonsymmetric stiffness matrix. This can produce complex eigenvalues with positive real parts, indicating that small perturbations are amplified exponentially. Therefore, complex eigenvalue analysis is typically used to determine the instability analysis, which is essentially reducing the coupled system to a group of linear unstable and/or stable single degree of systems. Physically, the energy driving the instability comes from a non-conservative force (path-dependent) like friction in brake squeal or aerodynamic forces in flutter.

Chaotic Instability

Chaotic instability is a type of instability that leads to chaotic motion. This particular term should be used carefully as chaotic motion in itself is strictly not an instability because it is bounded. Chaotic motion refers to bounded motion in which a system exhibits sensitive, unpredictable behaviour, even though it follows deterministic rules. In such systems, small changes in initial conditions can lead to drastically different outcomes over time, making long-term predictions nearly impossible. This phenomenon is often observed in complex dynamical systems like weather patterns, fluid dynamics and the behaviour of many engineering systems like the ones detailed in this book. Chaotic instability refers to the process leading to a chaotic system's sensitivity to initial conditions, where minor variations can amplify and lead to divergent trajectories and thus is a relatively new term. Dynamics is the oldest branch of physics, with Newton's *Principia* [7] being published over 340 years ago. Relatively recently, scientists and engineers have discovered that nonlinear dynamical systems may exhibit nonperiodic or chaotic vibrations, and this field has experienced a rebirth of research activity. However, many of the roots of chaotic dynamics stem back to classical nonlinear dynamics. The following is a brief description of some of these ideas.

A commonly used tool in nonlinear dynamics and stability analysis is the phase plane. This is a plot of position versus velocity $(x, \dot{x})$ over time and displays more global information concerning the dynamic behaviour of a system than time history (t, x). For nonlinear problems, the equilibrium points of the system are usually first examined so that the motion around these points may be characterised, and the phase plane portrait is used for this purpose. A set of dynamical system equations identified from mathematical modelling may be linearised to obtain the local motion about its equilibrium points in the following manner.

For a first-order system, the differential equations may be represented as

$$\dot{\mathbf{x}} = \mathbf{f}(\mathbf{x}),$$

where $\mathbf{x}$ is the state vector (of system variables). The equilibrium points occur when

$$\dot{\mathbf{x}} = \mathbf{f}(\mathbf{x}_e) = 0.$$

Linearisation is then performed about these points by introducing small variables:

$$\mathbf{x} = \mathbf{x}_e + \xi.$$

Then, the linearised system may be written in the following form:

$$\dot{\xi} = \nabla\mathbf{f}(\mathbf{x}_e)\bullet\xi,$$

where ∇ denotes the Jacobian. This has the eigensolutions described by

$$\xi(t) = \xi_0 e^{st},$$

where ξ_0 is a vector of arbitrary constants which depend on the initial conditions. The motion close to the equilibrium points is then characterised by the eigenvalues (i.e., the characteristics of s). The phase plane portraits depicting different solutions and the corresponding nature of stability are shown in Figure 2.4. The nature of the stability of the equilibrium points is determined by the sign of Real(s). From the equation, it can be seen that if Real(s) is positive, the equilibrium point is unstable, and conversely, if it is negative, it is stable. Pure oscillatory motion results from a complex conjugate solution of $s = \pm i\omega$, and is depicted as a closed curve or limit cycle indicating periodic motion. For a nonlinear system, further analysis would be required to determine the stability of the point. The phase space diagrams depicted in Figure 2.4 generally only describe two classic types of motion (after transients have decayed). The motion of the system will either terminate at an equilibrium point or may be attracted onto a limit cycle performing periodic motion.

For nonlinear systems, another final state of motion can exist which is not periodic, yet predictable, known as *quasiperiodic* motion. This is sometimes called combination oscillations because it results from the sum of two oscillating solutions of two different frequencies that are incommensurate, as in the form

$$x = a\cos(\omega_1 t) + b\cos(\omega_2 t),$$

where ω_1/ω_2 is irrational. Because of the incommensurate frequencies, any state of the motion will not be repeated on a phase space representation $(x, \dot{x})$, and thus a closed loop will not be obtained. This motion could be mistaken as chaotic since it is not periodic (but it is predictable). Therefore, other methods (described in Section 2.3)

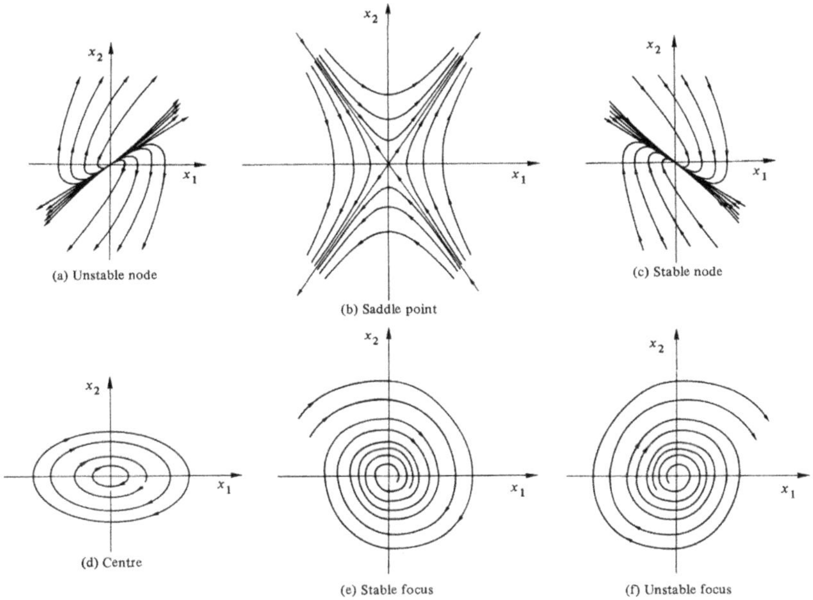

FIGURE 2.4 Phase space plots of six different equilibrium point types [1, p. 85].

need to be used to qualify the motion definitively. In summary, a dynamic system with some form of damping tends to be "attracted" to one of three types of predictable states of motion which may be visualised easily in phase space, as described by

1. An equilibrium point	Point
2. A limit cycle (periodic motion)	Closed curve
3. Quasiperiodic motion	Surface in three-dimensional phase space

These states are appropriately known as attractors.

Chaotic motion is the recently discovered addition to this list. It is characterised by its exponential sensitivity to initial conditions so that a system's dynamics very quickly becomes unpredictable. The random nature of the motion appears in phase space as a *strange attractor.* This attractor is not visualised in phase space the same way as the other equilibrium states which are characterised by the geometric objects that are described above. In the three-dimensional phase space, it is usually depicted by an infinite set of warped parallel surfaces with differing separations, some of which approach the infinitesimal. This shape is associated with a fractal set, one of the new mathematical and geometrical concepts devised to permit the description and quantification of this phenomena.

Understanding these diverse forms of instability is essential for designing robust and reliable systems across numerous disciplines. Engineers and scientists employ mathematical modelling, simulation and experimental methods to analyse and mitigate instability, ensuring the safety, efficiency and performance of complex systems in various applications.

2.3 HOW TO IDENTIFY AND PREDICT INSTABILITY

After developing a model for instability, we need to detect its type and analyse the dynamical outputs to characterise and quantify the behaviour. Stability theory is the fundamental tool used to identify instability. It is a foundational concept across various disciplines, including mathematics, physics, engineering, biology and economics. At its core, stability theory investigates the behaviour of systems in response to small and/or large perturbations. It can predict and provide insights into whether a system's equilibrium or steady state is maintained or disrupted over time. In mathematics, stability theory often deals with differential equations and dynamical systems, examining how small changes in initial conditions or parameters influence the long-term behaviour of solutions. This branch of mathematics has profound implications in fields like chaos theory and control theory. In physics and engineering, stability analysis is crucial for designing robust and reliable systems. Whether it is determining the stability of a physical structure under different loading conditions or analysing the stability of an electrical circuit, understanding the nature of stability ensures the safety and efficiency of engineered systems.

The purpose of this section is to introduce the methods of analysis used to understand, identify and quantify instability in engineering systems. In particular, methods used on the many examples covered in more detail in this book will be described. A brief review of methods for analysing classical instability is first given. Methods used for identifying and quantifying chaotic instability are then presented.

Classical Instability

Classical instabilities can be identified using a range of stability theories developed over the past two centuries. In essence, stability theory provides a powerful framework for analysing the behaviour of diverse systems and understanding the conditions under which they remain stable or undergo significant changes.

Local/Linear (Eigenvalue) Stability Theory

The stability of fixed points of a system of constant coefficient linear first-order differential equations can be analysed using the eigenvalues of the corresponding matrix of equations of motion.

An autonomous system

$$x' = Ax,$$

where $x(t) \in \mathbf{R}^n$ and A is an $n \times n$ matrix with real entries, has a constant solution

$$x(t) = 0.$$

In other words, the origin $0 \in \mathbf{R}^n$ is an equilibrium point of the corresponding dynamical system. This solution is asymptotically stable as $t \to \infty$ ("in the future") if and only if for all eigenvalues λ of A, $\mathrm{Re}(\lambda) < 0$. Similarly, it is asymptotically stable as $t \to -\infty$ ("in the past") if and only if for all eigenvalues λ of A, $\mathrm{Re}(\lambda) > 0$. If there exists an eigenvalue λ of A with $\mathrm{Re}(\lambda) > 0$, then the solution is unstable for $t \to \infty$.

The Routh–Hurwitz stability criterion can be used to determine this criteria without solving the eigenvalues which are the roots of A's characteristic polynomial. In particular, it may be shown that a polynomial in one variable with real coefficients is stable if the real parts of all roots are strictly negative. Alternatively, the eigenvalues can be determined via a complex eigenvalue analysis using matrix analysis, either numerically or analytically. Local asymptotic stability of fixed points of a nonlinear system can be established by linearising about a point p of interest. This can be achieved using a Jacobian matrix. Suppose that v is a C^1-vector field in $\mathbf{R}^n$ which vanishes at a point p, $v(p) = 0$. Then, the corresponding autonomous system

$$x' = v(x),$$

has a constant solution

$$x(t) = p.$$

Let $J_p(v)$ be the $n \times n$ Jacobian matrix of the vector field v at the point p. If all eigenvalues of J have strictly negative real parts, then the solution is asymptotically stable. This condition can also be tested using the Routh–Hurwitz criterion.

Lyapunov Stability Theory

Since analytical solutions of nonlinear systems usually cannot be obtained, the use of stability methods is of major importance in the understanding of nonlinear dynamics. Lyapunov's methods are the most useful and general approach for studying the stability of nonlinear systems. There are two primary methods for stability analysis. The linearisation method is concerned with small motion of nonlinear systems about equilibrium points. Therefore, it justifies the use of linear methods for the design and analysis of weakly nonlinear physical systems, as discussed in the previous section. In contrast, the direct method, based on so-called Lyapunov functions, is not restricted to small motions. In principle, this method is applicable to all dynamical systems, linear, nonlinear, finite order or infinite order in all types of motion, small and large. A brief summary of the major results of Lyapunov's direct method will be given here; however, a more comprehensive review can be found in Slotine and Li [8] and Junkins and Kim [9].

The basic philosophy of Lyapunov's direct method stems from a fundamental physical observation that if the total energy of a system is continuously dissipated, then it must eventually settle down to an equilibrium point. Using this heuristic argument, it may be concluded that the stability of a system may be determined by examination of the behaviour of a single scalar function known as a Lyapunov function. If a suitable Lyapunov function can be found for a system, then the stability of the motion can be analysed. The following brief outline is the basic stability concept underlying Lyapunov's direct method.

Firstly, consider a continuous, finite-dimensional dynamic system described by the vector differential state equation

$$\dot{\mathbf{x}} = \mathbf{f}(\mathbf{x},t), \quad \mathbf{x} \in R^n,$$

where $\mathbf{x}(t)$ is the $n \times 1$ state vector at time t. An equilibrium state $\mathbf{x}_e$ may be defined for this system as

$$\dot{\mathbf{x}} = \mathbf{f}(\mathbf{x}_e,t) = 0, \quad \forall t \geq t_e,$$

where t_e is the time at which the equilibrium state is encountered by the system. The equilibrium state $\mathbf{x}_e$ is said to be locally stable if there exists a positive definite single-valued function $U(\mathbf{x})$ which is continuous and has continuous partial derivatives with respect to the components of the vector $\mathbf{x}$ such that its time derivative along any state trajectory of the system is negative semi-definite. These conditions may be more rigorously expressed as

$$\begin{aligned}
&(1)\ \mathrm{U}(\mathbf{x}_e) = 0, \\
&(2)\ \mathrm{U}(\mathbf{x}) > 0 \qquad \text{for all } \mathbf{x} \neq \mathbf{x}_e,\ \mathrm{x} \in \Gamma, \\
&(3)\ \dot{\mathrm{U}}(\mathbf{x}) \leq 0 \qquad \text{for all } \mathbf{x} \neq \mathbf{x}_e,\ \mathrm{x} \in \Gamma,
\end{aligned}$$

where Γ is some region containing $\mathbf{x}_e$ and $U(\mathbf{x})$ is called a Lyapunov function for the system. It is of note that the "energy rate" $\dot{U}(\mathbf{x})$ is evaluated along a typical trajectory $\mathbf{x}(t)$ and the conditions (2) and (3) must hold for the infinite set of trajectories of the dynamical system arising from all the initial states in Γ.

For local asymptotic stability, condition (3) is restricted further to

$$(3^*)\ \dot{U}(\mathbf{x}) < 0 \quad \text{for all } \mathbf{x} \neq \mathbf{x}_e,\ \mathrm{x} \in \Gamma$$

These two theorems provide local stability conditions for motion in the vicinity of the equilibrium state. If the stable region Γ is extended to infinity and $\mathrm{U}(\mathbf{x})$ is radially unbounded in the form

$$(4)\ U(\mathbf{x}) \to \infty \quad \text{as} \quad \|\mathbf{x}\| \to \infty,$$

the equilibrium state $\mathbf{x}_e$ is said to be globally asymptotically stable. It is of note that there is no one unique Lyapunov function for a given system, and some may be more conservative than others in the size of the stable region Γ. Therefore, the conditions required to prove stability, based on a particular choice of Lyapunov function, may be very conservative. A result by Mukherjee and Chen [10] may also be used to prove global asymptotic stability in spite of the fact that the function derived for $\dot{U}(\mathbf{x})$ is only negative semidefinite or $\dot{U}(\mathbf{x}) \leq 0$. A good review of this theorem is given in [9], and an outline of the results is given here.

For the case described by conditions (1–3), let $\mathbf{Z}$ denote the set of points for which $\dot{U}(\mathbf{x}) = 0$. A sufficient condition for asymptotic stability is that the first $k-1$ derivatives of $U(\mathbf{x})$ vanish on $\mathbf{Z}$, up through some even order $k-1$, as defined by,

$$(5)\quad \frac{d^jU}{dt^j} = 0 \quad \text{for all } \mathbf{x} \in \mathbf{Z}, \quad \text{for} \quad j = 1,2,\ldots,k-1,$$

and the first (the kth) nonzero derivative of $U(\mathbf{x})$, evaluated on $\mathbf{Z}$ is of odd order and is negative definite for all points on $\mathbf{Z}$,

$$(6)\quad \frac{d^kU}{dt^k} < 0 \quad \text{for all } \mathbf{x} \in \mathrm{Z}, \quad \text{for } k \text{ odd.}$$

It should also be noted that if a Lyapunov function cannot be found, nothing can be concluded about the stability of the system since the Lyapunov stability theorem only provides sufficient conditions for stability. However, if there exists a continuously differentiable function $U(\mathbf{x})$ such that

(1) $U(\mathbf{x}_e) = 0$ and $\dot{U}(\mathbf{x}_e) = 0$,

(2) $\dot{\mathrm{U}}(\mathbf{x}) > 0$ for all $\mathbf{x} \neq \mathbf{x}_e$, $\mathrm{x} \in \boldsymbol{\Gamma}$,

(3) There exist points $\mathbf{x}$ arbitrarily close to $\mathbf{x}_e$ such that $U(\mathbf{x}) > 0$,

the equilibrium state $\mathbf{x}_e$ is unstable.

With the use of these theorems, rigorous local or global conclusions about the stability of motion of an equilibrium state may be derived without solving a system of general nonlinear equations. These results for equilibrium points may be extended to invariant sets such as limit cycles; however, the details will not be given in this book. Unfortunately, there is no general systematic procedure for generating Lyapunov functions for arbitrary nonlinear systems. However, for various classes of systems, a variety of methods have been devised for constructing Lyapunov functions, as described in [8].

It is also noteworthy that Lyapunov theory also has wide applications in control design, and since the advent of modern control theory, many refinements to Lyapunov's methods have been developed for this purpose. In Chapter 5, the use of the theory to design stable nonlinear control systems will be discussed.

Chaotic Instability – Chaos Following Instability

In a nonlinear dynamic system, the equilibrium configuration may have multiple states that can change with parameter values. Where a dynamical system has two or more possible final states or attractors, the initial transient motion may hesitate between them for an arbitrarily long period of time before settling to one or the other. Chaotic motion may be described as occurring when the initial transient motion never settles down to these previously defined states but instead seems to wander precariously between the previous attractors for all time. This new final state is in fact a strange attractor, indicating chaotic motion can only exist in systems with two or more equilibrium states. This is a characteristic only exhibited by nonlinear systems. No general predictive criteria for the onset of chaotic motions in nonlinear systems has been devised, but the motion is generally found close to critical conditions

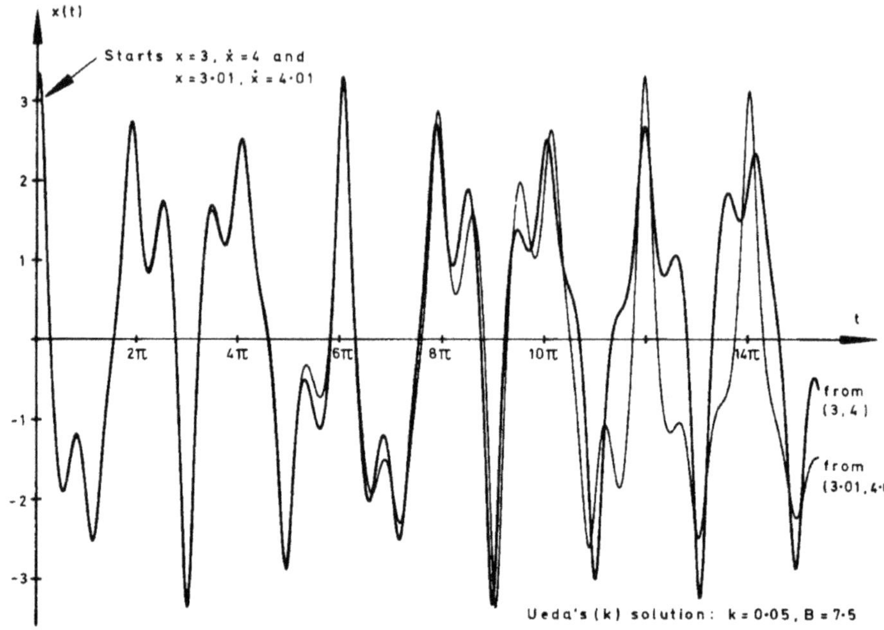

FIGURE 2.5 Time history of Duffing's equation showing sensitivity to initial conditions [11].

for the changes in the nature of attractors (i.e., change in stability). The changes in the number of equilibrium points are often called bifurcations, and the study of this phenomena is known as bifurcation theory. Moon [4] provides a very good introduction and review of tools for nonlinear and chaotic vibrations, which provides a good basis for the following.

The unusual property of a chaotic attractor, apart from its refusal to have periodicity, is that it is found that nearby trajectories in phase space move exponentially apart, i.e., a property of instability. This may be seen more clearly on a time history plot for two nearby initial conditions, as seen in Figure 2.5.

This diagram results from the nonlinear Duffing's equation:

$$\ddot{x} + kx + x^3 = B\cos t,$$

which may be used to model the motion of a sinusoidally forced structure undergoing large elastic deflections. It depicts two numerical integrations that are initialised from sets of almost equal initial states. The two adjacent starts remain close to each other, initially, but then rapidly move apart, and their motions become uncorrelated. This illustrates the underlying unpredictability of chaotic motion, in engineering systems where some uncertainty in the initial conditions of a system is inevitable. The same system and divergence may also be depicted in the phase space seen in Figure 2.6.

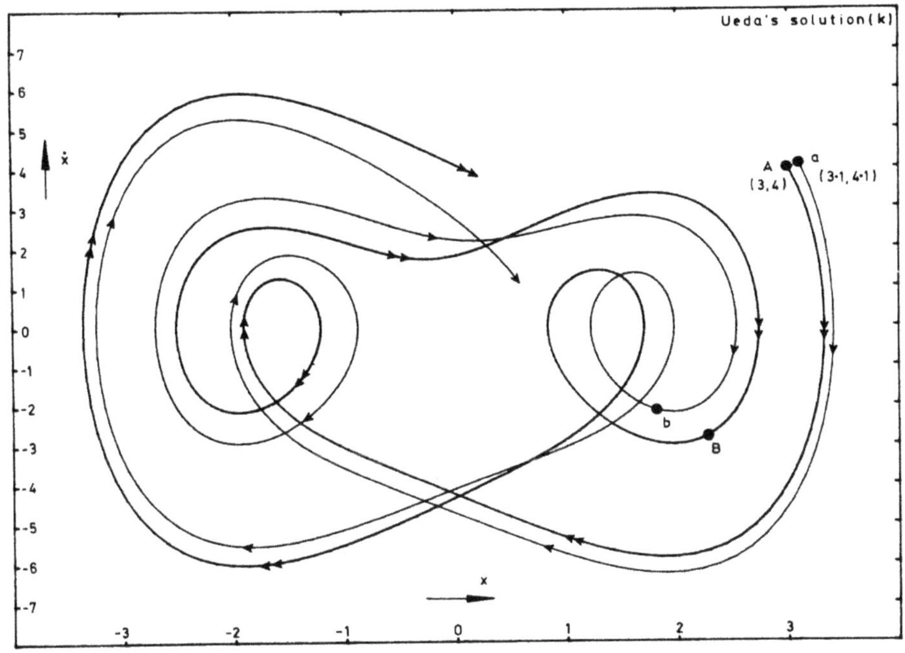

FIGURE 2.6 Phase space of Duffing's equation showing sensitivity to initial conditions [11, p. 5].

Identifying and Predicting Chaotic Vibrations

Engineers in the field of dynamics are often called upon to identify the source of undesirable vibrations in a wide range of systems including vehicles, equipment and structures. Knowledge about the nature and magnitude of these vibrations is often required as well so that effective control or suppression may be performed and the fatigue life of the component may be predicted. For this purpose, the characteristics unique to chaotic vibrations may be identified using a number of experimental and numerical tools. Moon [4] describes the following checklist which is extended here for identification and quantification of nonperiodic or chaotic motion in a nonlinear dynamic system:

1. Identify nonlinear elements
2. Check for random input
3. Observe time history
4. Observe phase plane history
5. Examine the Fourier spectrum of signals
6. Obtain Poincaré map
7. Vary system parameters (routes to chaos)
8. Quantify chaotic behaviour
9. If possible, obtain and compare with analytical predictions of chaos.

In practice, the experimentalist usually employs a number of these methods to verify chaotic vibrations.

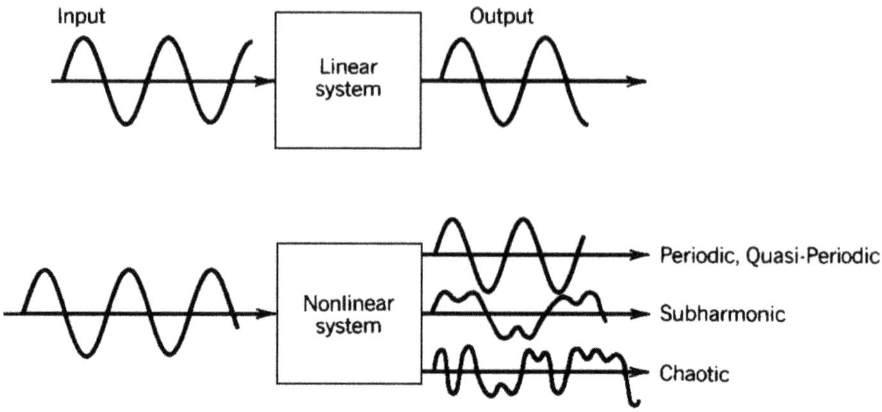

FIGURE 2.7 Possible input–output response for linear and nonlinear systems [4, p. 50].

Nonlinear Elements

As is described in Chapter 4, chaotic motions only occur in dynamical systems which have nonlinear elements. For a linear system, periodic outputs are always obtained (after transients decay) for periodic inputs. This is illustrated in Figure 2.7.

Most physical systems do contain nonlinearities that could cause other possible output motions such as chaotic or quasiperiodic. Some of these nonlinear elements commonly associated with mechanical systems are listed below.

- Nonlinear damping, (i.e., stick-slip friction)
- Backlash or play
- Nonlinear elastic or spring elements
- Fluid forces
- Nonlinear boundary conditions such as fixed stops or deadzone
- Nonlinear body forces such as magnetic or electric forces

The nonlinearities may also arise due to sources resulting from the inherent dynamics of a system such as

- Kinematics, e.g., Coriolis and relative motion accelerations in multibody systems
- Geometric nonlinearities caused by large displacements or rotations.

The identification of these elements may be difficult in some cases as the nonlinearities may arise subtly, such as with play or nonlinear friction. In general, an engineering system with more than one equilibrium position has nonlinearities.

Random Inputs

Nonperiodic motions in the past have usually been attributed to external random inputs or noise. It is only now that scientists and engineers are realising that the motion

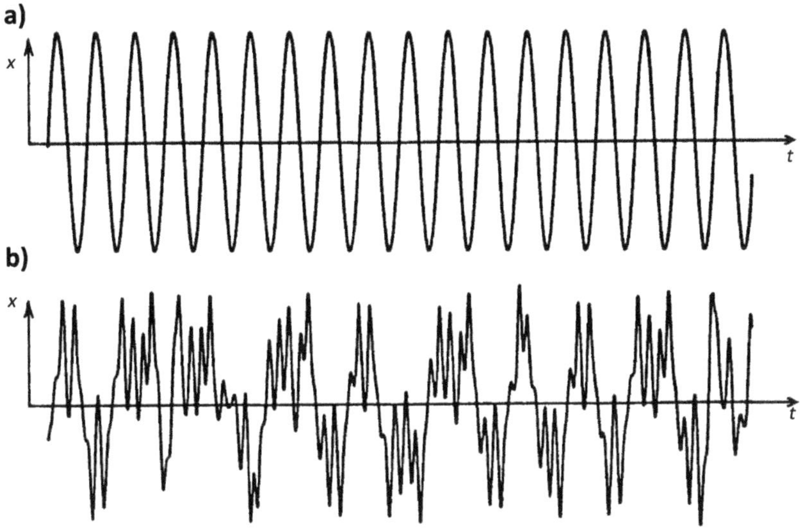

FIGURE 2.8 Time history plot showing (a) periodic motion; (b) chaotic motion [4, p. 160].

was chaotic and arose from the nonlinearities within the system. Experimentally, it is thus important to reduce the amount of external noise to as low as practicable so that it is not confused with chaotic vibrations. Although noise is present in all physical systems, the output signal-to-noise ratio can usually be kept to low values.

Time History

This is usually used as the first clue that a system may have chaotic vibrations. The generalised coordinate corresponding to the degree-of-freedom concerned is plotted against time. If the resulting motion observed has no visible pattern or periodicity, then the system might be chaotic. However, this method is not sufficient to conclude that the motion is chaotic. A system may exhibit long periods of seemingly transient chaos before settling to a periodic attractor or may have long-period behaviour which is difficult to detect. In nonlinear systems, it is also possible to have a nonperiodic (or quasi-periodic) signal caused by the sum of two incommensurate periodic signals. A typical example of a time history plot exhibiting periodic and chaotic motion is depicted below in Figure 2.8.

Fourier Spectrum

The Fourier spectrum may be used to distinguish between chaotic and quasi-periodic motion (where the time history fails). Chaotic motion appears as a broad spectrum of frequencies in the output spectrum even when the input is a single-frequency sinusoid (see Figure 2.9). Quasi-periodic motion would appear in the spectrum as two frequencies. Care should be used, however, because a system with a number of degrees of freedom may be nonchaotic and still appear as a number of frequencies on the spectrum. Random motion would be distinguished as well, appearing as a broad linear band of frequencies.

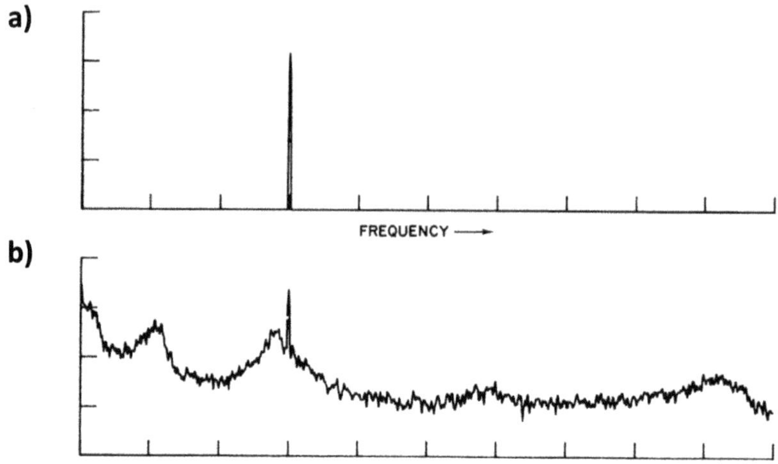

FIGURE 2.9 Frequency spectrum of (a) periodic motion; (b) chaotic motion [4, p. 58].

Phase Plane

The phase plane is the plot of the set of points of a variable and its derivative, i.e., position versus velocity $(x, \dot{x})$. This displays more global information concerning the dynamic behaviour of a system than the time history (t, x). An example of two different phase plane plots is seen in Figure 2.10. For a periodic system, the motion would appear as a closed curve, as depicted in Figure 2.10a. However, for chaotic motion, the resulting curve does not close, as seen in Figure 2.10b. The problem with this method, however, is that it is very difficult to determine whether the orbit has closed or not in practice. This challenge is seen in Figure 2.10b. Quasiperiodic motion also has an unclosed orbit.

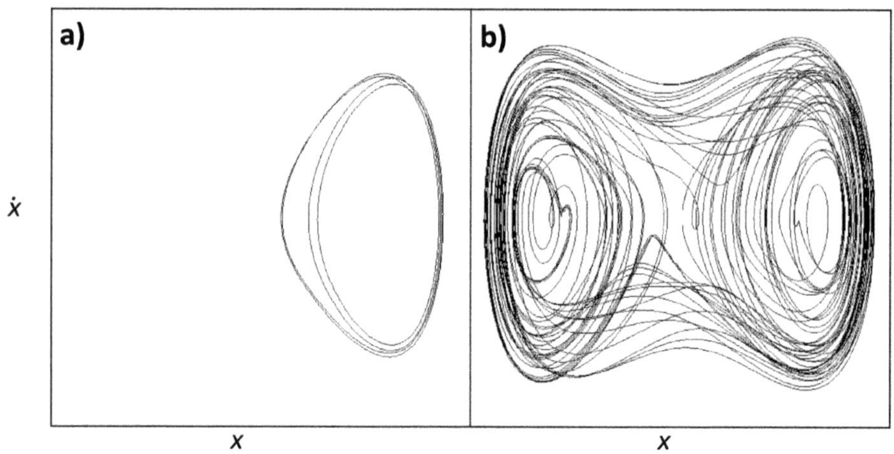

FIGURE 2.10 Phase plane of (a) period-4 motion; (b) chaotic motion.

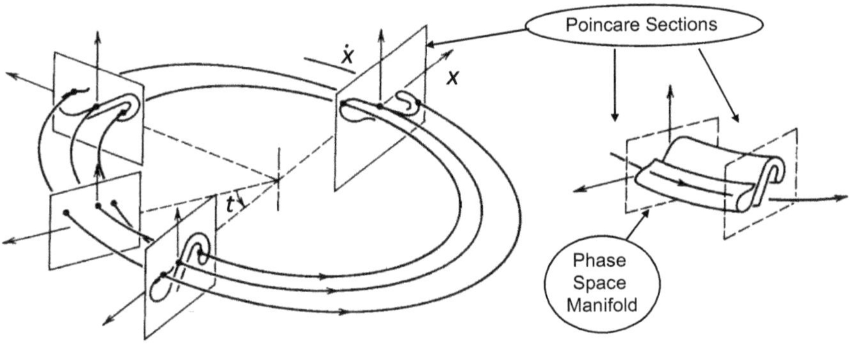

FIGURE 2.11 Several Poincaré sections of trajectory in three-dimensional phase space [4, p. 67].

Poincaré Map

A modified technique which overcomes the problem of identifying chaos in the phase plane is called the Poincaré map. This is a type of phase plane, where the sequence of data is time-sampled. This means that the motion on the phase plane is only recorded at discrete time intervals (or stroboscopically sampled), as depicted in Figure 2.11. If the sampling time is adjusted to be the same as the forcing frequency, the following depictions result.

- For periodic motion of the same frequency as the forcing or a subharmonic frequency, a single point is obtained.
- For periodic motion of an integer and multiple frequency of the forcing, a finite set of points is obtained.
- For quasi-periodic motion, a closed curve is obtained.
- For chaotic motion, an infinite set of highly organised points in the form of an intricate pattern, known as a strange attractor, is obtained.

Some examples of these possible motions are depicted in Figure 2.12.

A characteristic of chaotic motion is the fractal nature exhibited by the strange attractor. This manifests itself such that repeated enlargements of the attractor's pattern still produce further organised patterns as depicted in Figure 2.12c and d. This structure within structure is called a fractal set. The damping has the effect of contracting the structure of a strange attractor such that for low damping, the attractor will tend to uniformly fill up a section of the Poincaré map and thus not reveal the fractal nature. In fact, if the two-dimensional Poincaré map consists of a fuzzy collection of points, this may be due to a number of reasons:

1. the system has a random or noisy input;
2. the system has very small dissipation;
3. the strange attractor has more than three dimensions or the motion is quasiperiodic with three or more dominant incommensurate frequencies.

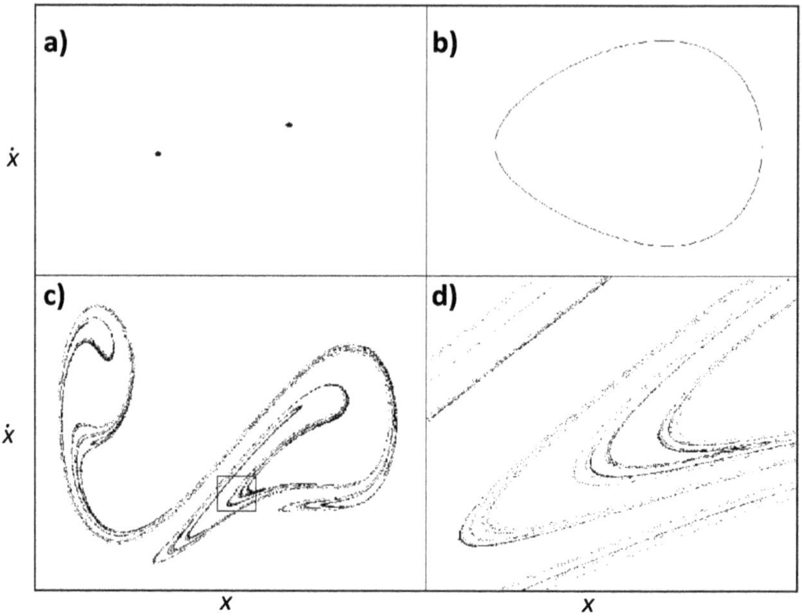

FIGURE 2.12 Poincaré section of (a) period-2 motion; (b) quasiperiodic motion; (c) chaotic motion; (d) blow-up of a strange attractor in (c).

Bifurcation Diagram

A commonly employed method for analysing how a dynamical system evolves with changes in parameters is the bifurcation diagram. This diagram represents a measure of the system's motion (e.g., amplitude) plotted against a system parameter, such as the forcing amplitude or damping coefficient. Usually, the data are time-sampled as with the Poincaré map so that it is easy to observe period doubling and subharmonic bifurcations. Figure 2.13 is an example of a bifurcation diagram for the Henon map (see [12]). Chaotic motion is indicated by a loss of continuity in Figure 2.13. This technique is used to determine the prechaotic and postchaotic dynamical signs as the systems behaviour for a range of control parameter values is described in this diagram. However, other methods must be used in conjunction with this technique for the identification of chaotic motion. This is because loss in continuity in a bifurcation diagram of this type may be due to quasiperiodic motion or high levels of noise.

Routes to Chaos

When identifying chaotic motion, it is common for the experimentalist to vary certain control parameters of the system such as excitation frequency or forcing amplitude. There are a number of benefits in this procedure, the first being that if the system has a range of periodic behaviours for a range of control parameters, then the possibility that nonperiodic motion arises from external random noise may

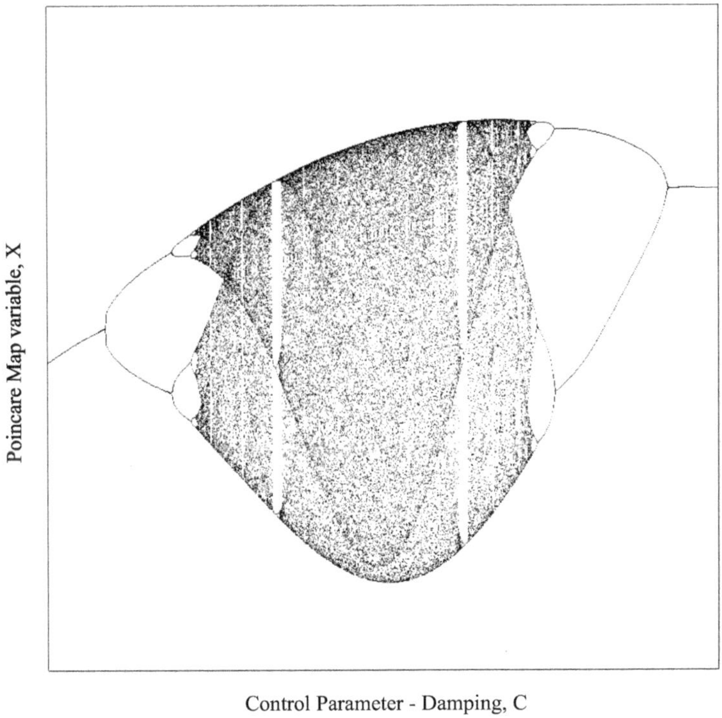

FIGURE 2.13 Bifurcation diagram showing period-doubling for Henon map [13].

be eliminated. The second reason is that chaotic motion is usually preceded by a particular pattern of prechaotic behaviour. Many of these have been identified in research investigations. Hilborn [14] and Moon [4] provide comprehensive overviews of the possible routes to chaos.

Period Doubling

This phenomenon describes the bifurcations from an initial periodic limit cycle. As a control parameter is changed, the motion exhibits a bifurcation to a new limit cycle which has twice the period of the original. Further changes causes further bifurcations to more and more complicated limit cycles with doubling periods. An example of this process is depicted in Figure 2.13 for the Henon map. The remarkable characteristic of this process is that the critical values of the control parameter at which successive bifurcations occur may be predicted by the following scaling law:

$$\lim_{n\to\infty} \frac{\lambda_n - \lambda_{n-1}}{\lambda_{n+1} - \lambda_n} \to 4.6692016,$$

where λ_n is the varied control parameter. This scaling law has been found to be universal in period doubling systems and therefore remarkably holds in a wide variety of unrelated nonlinear systems (and is known as the Feigenbaum number [15]).

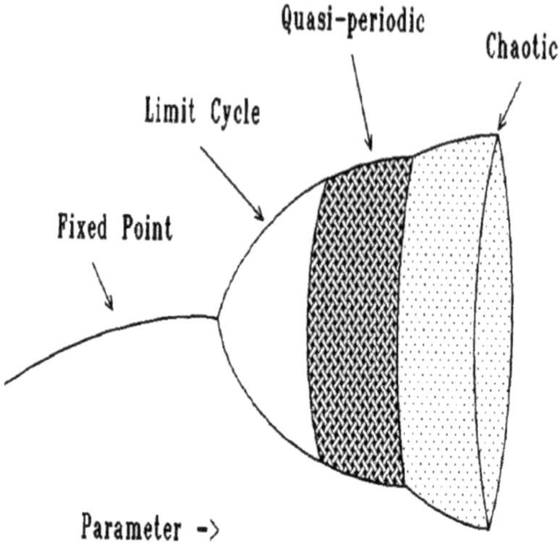

FIGURE 2.14 Quasiperiodic route to chaotic motion [14, p. 257].

This bifurcation process continues geometrically until it accumulates at a critical value after which chaotic motion occurs.

Quasi-Periodicity

Another less renowned route to chaos may be characterised by quasiperiodic motion. Newhouse [16] describes a dynamic system which involves the breakup of the quasiperiodic attractor as a system parameter is varied. This phenomenon has been observed in Taylor–Couette flow before turbulence occurs. A schematic representation of the evolution of the attractors in state space for the quasiperiodic route to chaos is seen in Figure 2.14.

The system may begin with a fixed-point attractor. As a control parameter is changed, a limit cycle trajectory may develop due to a Hopf bifurcation. This term is simply used to describe the birth of a stable limit cycle. As a control parameter is changed further, another Hopf bifurcation may occur, and a second periodic frequency appears in the frequency spectrum of the system. The state-space trajectories of the system then reside on the surface of a torus. If this second frequency is not a rational ratio (or is incommensurate) with the first, then the trajectory will never repeat and will cover the surface of the torus. As described previously, a Poincaré map of this system would reveal a closed loop or drift ring. This is called quasiperiodic motion. Further changes of the system control parameter may cause the system's behaviour to become chaotic. For some systems, more Hopf bifurcations may occur before the appearance of chaotic motion.

Intermittency

Another route to chaos is characterised by long periods of periodic motion interrupted by bursts of chaos. As the control parameter is changed, these chaotic bursts become more frequent and of longer period. Evidence of this phenomenon has been observed in experiments on convection in a cell with a temperature gradient (see [17]).

Transient Chaos

Prechaotic behaviour may also be characterised by the occurrence of initial chaotic motion which eventually settles to a periodic attractor. As the control parameter of the dynamic system is changed, the settling period increases until a critical value is surpassed and fully chaotic motion occurs. The experimentalist must, therefore, ensure that a sufficient period of time has elapsed before pronouncing the state as chaotic.

In physical experiments, it is possible that a number of these prechaotic phenomena may be observed.

Quantification

The above-mentioned techniques are used primarily as predictive criteria for chaos. However, they reveal little quantitative information concerning the nature of the chaotic motion. Two further techniques may be used to diagnose and quantify the chaotic motion in the dynamic system.

Lyapunov Exponents

Chaotic motion indicates a strong sensitivity to initial conditions. This trait can be quantified by measuring how rapidly two nearby trajectories in phase space diverge from one another. The largest Lyapunov exponent provides a measure of this divergence, averaged over multiple points along the trajectory, as illustrated in Figure 2.15.

For chaotic motion, the initially very close trajectories move away from each other, exponentially, such that

$$d(t) = d_0 2^{\lambda t} \tag{2.6}$$

where $d(t)$ is the distance between trajectories and λ is the largest Lyapunov exponent. The criterion for chaotic motion in a system with bounded motion becomes

$$\lambda > 0 \Rightarrow \text{chaotic motion},$$

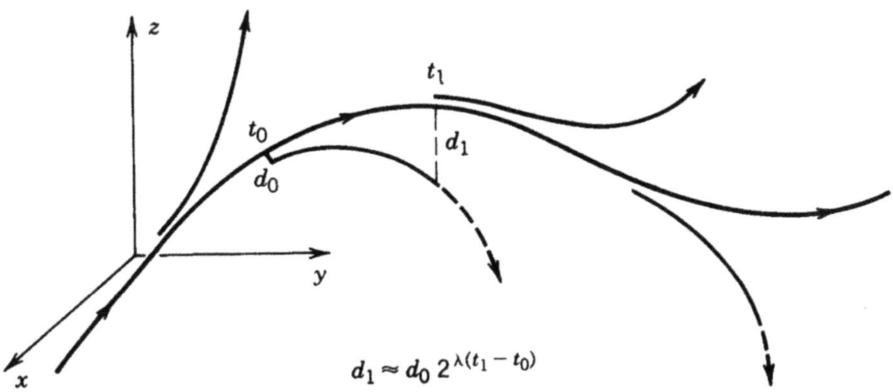

FIGURE 2.15 The largest Lyapunov exponent as a measure of divergence [4, p. 308].

$$\lambda \leq 0 \Rightarrow \text{regular motion}.$$

It is noteworthy that this criterion describes the behaviour of individual trajectories (rather than the bulk of the trajectories) and is thus local in nature. The divergence of these trajectories can only be locally exponential for chaotic motion since the systems' dynamics is bounded (i.e., $d(t)$ cannot go to infinity). In fact, Henon [12] found that chaotic motion involves a combination of exponential stretching, folding and contraction in different directions, which explains why a system can be chaotic yet bounded. Linear systems in contrast do not have this folding mechanism. This creates a problem when trying to measure the Lyapunov exponent of a nonlinear system since only the exponential stretching or divergence of orbits is quantified and not the fold. Therefore, it is unusual to be able to use the above formula and obtain the Lyapunov exponent analytically. However, the formula may be defined discretely and focussed locally, such that the Lyapunov exponent may be calculated from numerical or physical system models by a digital computer. Algorithms for calculating Lyapunov exponents have been devised by a number of authors such as Benettin et al. [18], Shimada and Nagishima [19] and Wolf et al. [20]. A brief description of the method is described in the following.

To overcome the problem of nonlinear folding and contraction, the calculation process must be restarted every time the divergence becomes too large between two initially close orbits. The exponential growth is thus averaged at many points i along the trajectory seen in Figure 2.15 so that equation (2.6) may be rewritten as

$$\lambda = \frac{1}{t_N - t_0} \sum_{i=1}^{N} \log_2 \frac{d(t_i)}{d(t_{i-1})},$$

where $d(t)$ is the distance between trajectories at any given time and λ denotes the Lyapunov exponent. To ensure that the folding motion is not quantified, the equations of motion are linearised for the reference trajectory. This approximation will have little effect on the accuracy of results for small changes relative to the fiducial trajectory. This analysis enables calculation of the largest Lyapunov exponent, and similarly every dynamical system has a spectrum of Lyapunov numbers that describe how lengths, areas and volumes transform in phase space. In particular, dissipative systems exhibit contraction of a volume of initial conditions (in at least one direction). This is seen in Figure 2.16.

Characteristic properties have been identified for the Lyapunov spectrum for particular dynamic behaviour. It is found that for any bounded dynamic motion, the system must have one zero Lyapunov exponent. It is also found that contraction must outweigh expansion ("stretch") so that $\sum_{i=1}^{N} \lambda_i < 0$. For chaotic motion, at least one Lyapunov exponent must be positive. Therefore, it follows that a strange attractor must have at least three Lyapunov exponents. Hence, chaos cannot occur in a first- or second-order continuous time system.

The Lyapunov spectrum may also be used to obtain another quantification of chaotic motion, namely, a type of fractal dimension called the Lyapunov dimension.

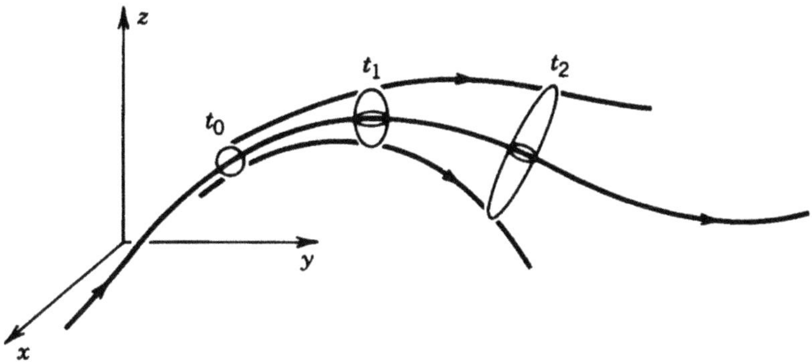

FIGURE 2.16 Divergence measured from a sphere of initial conditions used to define a Lyapunov spectrum [4, p. 315].

This will be discussed in the subsequent section. The number of Lyapunov exponents in the spectrum corresponds to the number of state variables required to describe a dynamical system.

Fractal Dimension

Another characteristic of chaotic motion which may be quantified is the "strangeness" of the chaotic attractor obtained from the Poincaré map of the system. This is called the fractal dimension and strictly is a measure of the extent to which orbits in the phase space fill an arbitrary subspace. The simplest definition is derived from the characteristic, that for a strange attractor

$$N(\varepsilon) \approx \varepsilon^{-D},$$

where $N(\varepsilon)$ is the number of spheres of size ε that are required to cover an orbit in phase space. This can be rearranged in the form

$$D = \lim_{\substack{N \to \infty \\ \varepsilon \to 0}} \frac{\ln N}{\ln\left(1/\varepsilon\right)}.$$

It is found that a non-integer fractal dimension (such as $D=2.3$) implies chaotic motion (Mandelbrot [21]). The fractal dimension also gives an indication as to the minimum number of degrees of freedom required to describe a system's dynamic behaviour. For example, a fractal dimension of $D=4.7$ implies the system requires at least five independent variables to describe its dynamic behaviour. The fractal dimension, thus, gives important information concerning the reduction of multiple-degree-of-freedom systems.

There are in fact several different types of fractal dimensions of which the most renowned has been described above. Four other types are the information dimension, correlation dimension, kth nearest neighbour dimension and Lyapunov dimension. Two of these methods are employed in this book:

The correlation dimension has been successfully employed experimentally by other experimentalists such as described in Baker and Gollub [22] and Moon and Li [23]. Parker and Chua [24] give a detailed review of the technique.

The Lyapunov dimension has been defined as another type of fractal dimension by Kaplan and Yorke [25], which is calculated using the Lyapunov spectrum of a dynamic system. This algorithm is quite simple, and a summary is described below:

Let $\lambda_1 \geq \lambda_2 \cdots \geq \lambda_n$ be the Lyapunov exponents of an attractor.
Let j be the largest integer such that $\lambda_1 + \lambda_2 \cdots + \lambda_j \geq 0$.
Then, the Lyapunov dimension is defined as

$$D_L = j + \frac{\lambda_1 + \lambda_2 \cdots + \lambda_j}{|\lambda_j + 1|}. \tag{2.7}$$

The following notes on this equation may be made:

- For any attractor, $\sum_{i=1}^{N} \lambda_i < 0$, so j is always $< n$.
- For an attracting limit cycle, $\lambda_1 = 0 > \lambda_2 \cdots > \lambda_n$; therefore, $D_L = 1$.

Melnikov's Method

Since chaotic motion behaviour is found to be very complex and highly nonlinear, there have been no generalised techniques found to predict the occurrence of chaotic motion analytically in all nonlinear systems. Various techniques have been derived; however, they may only be applied to simple systems of a particular class. Also, due to numerous assumptions, the methods are usually inaccurate when compared to physical systems, and the results need to be verified numerically or experimentally. Due to these difficulties, the trend has been to numerically investigate the dynamics of an engineering system to identify the occurrence of chaotic motion and then subsequently make appropriate simplifications to attempt to find analytic criteria. Analytical prediction of chaotic motion in engineering systems has been approached by a number of independent routes. Moon [4] provides a good survey of these techniques. May [26] and Feigenbaum [15] have developed criteria based on period-doubling behaviour to predict when a system becomes chaotic by simply observing the prechaotic behaviour for the system. Predictions have also been derived for intermittency and transient route to chaotic motion. These methods cannot be applied to systems without this characteristic behaviour and also usually require experimental or numerical knowledge of the prechaotic behaviour of the system. A number of authors have also investigated various heuristic criteria, of which Szemplinska-Stupnicka [27] gives a brief review. Criteria for chaotic motion arising in conservative systems have been devised as well, including those derived from the Kolmogorov–Arnold–Moser (or KAM) theorem (Kolmogorov [28]; Arnold [29]; Moser [30]) and the overlap criteria. Two criteria that extend this research to nonconservative systems are the Melnikov and Shil'nikov methods. These criteria are derived by considering the nonconservative systems as conservative systems with perturbations so that knowledge of the conservative dynamics of the system may be used. These techniques have led to specific criteria for the occurrence of chaotic motion in nonconservative systems based on the mathematical models of the system. Presently, a brief review of Melnikov's method will be given due to its applicability to the modelling of rotating multibody systems like spacecraft and draglines investigated in this book.

Melnikov's renowned idea of the existence of homoclinic (orbit connected to one saddle equilibrium point) tangling in the phase space of a chaotic system (see Figure 2.17c) provides an exact mathematical proof that the basin boundaries (the boundaries between the basins of attraction of coexisting attractors/orbits) are fractal and hence proves the sensitivity of a bounded system to initial conditions. Strictly speaking, the criterion is only valid for the prediction of transient chaos. In practice, Melnikov's method is used for obtaining a predictive relation for the onset of chaos as a function of parameters in a physical system. The method, however, is restricted to a class of perturbed conservative systems which have homoclinic or heteroclinic orbits (connected to one or two saddle equilibrium points, respectively). More specifically, it is a perturbation technique for proving the existence of transverse homoclinic or heteroclinic orbits to hyperbolic periodic orbits in a class of time-periodic vector fields. This existence of stable and unstable orbits implies the existence of a horseshoe-type Poincaré map in the vicinity of each homoclinic point and consequently predicts chaotic motion via the Smale–Birkhoff Theorem. In essence, the method gives global information about the perturbed system's dynamics using global information about the unperturbed integrable system dynamics. Wiggins [31,32] and Guckenheimer and Holmes [33] provide a detailed derivation and mathematical discussion of the technique. However, for the purposes of this book, a brief outline of the method will be given.

The Melnikov function is used to measure the distance between the unstable and stable orbits in phase space when the distance is small. This approach has been used in problems with minimal dissipation, where the equations for homoclinic and heteroclinic orbits of the conservative system are known. In particular, the method focuses on systems with small damping and forcing, allowing the equations of motion to be expressed as

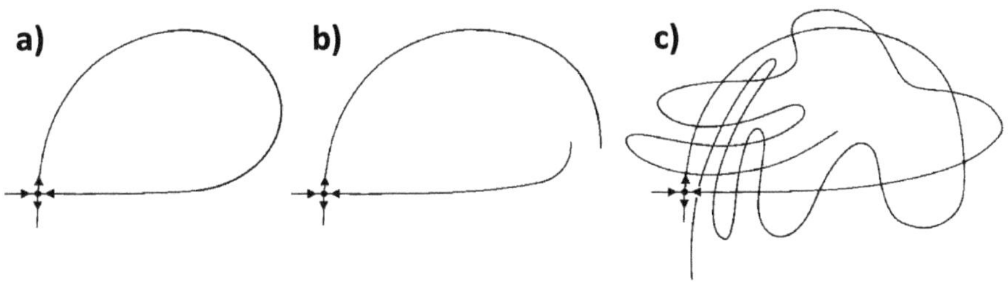

FIGURE 2.17 Sketch of homoclinic orbit under increasing perturbations; (a) small perturbations, (b) larger perturbations causing disconnected branches, (c) large enough perturbations causing branches to intersect infinitely (homoclinic tangle).

$$\dot{\mathbf{x}} = \mathbf{f}(\mathbf{x}) + \varepsilon\mathbf{g}(\mathbf{x},t); \quad \mathbf{x} = \begin{pmatrix} u \\ v \end{pmatrix} \in \Re^2,$$

where $\mathbf{f}(\mathbf{x})$ is a conservative vector field defined on $\Re^2$ for the undamped unforced system ($\varepsilon = 0$) and $\varepsilon\mathbf{g}(\mathbf{x},t)$ is a small periodic perturbation which need not be conservative. It is also assumed that a saddle point exists in the unperturbed conservative problem such that a homoclinic (or heteroclinic) orbit exists, as depicted in Figure 2.17a. For small forcing, the stable and unstable branches of the saddle do not intersect, as shown in Figure 2.17b. However, at a sufficiently large amplitude of forcing, the two branches may intersect. It can be shown that if they intersect once, they will intersect an infinite number of times in the form of a homoclinic tangle, as shown in Figure 2.17c.

The complexity of this homoclinic tangle arises due to the stretching, folding and contraction of manifolds in phase space due to nonlinearity. For a dissipative system, it is noted that the contraction must be greater than exponential stretching. The Melnikov function provides a measure of the separation of the manifolds of these orbits in three-dimensional phase space $\{u, v, t\}$ as a function of time t.

The Melnikov function, denoted $M(t_0)$, is written as the integral:

$$M(t_0) = \int_{-\infty}^{\infty} \mathbf{f}(\mathbf{q}_0(t)) \wedge \mathbf{g}(\mathbf{q}_0(t), t + t_0)\, dt,$$

where the symbol $\wedge$ is the wedge product defined by $a \wedge b = a_1b_2 - a_2b_1$, and $\mathbf{q}_0(t)$ is the solution for the set of homoclinic or heteroclinic orbits in the unperturbed system. Melnikov theory simply states that the condition for chaotic motion is that the Melnikov function has a simple zero, $M(t_0) = 0$. In other words, chaotic motion arises when the unstable and stable manifolds intersect.

By evaluation of the Melnikov function under this condition, an analytical criterion for chaotic motion is obtained. In physical systems with small forcing and damping, the criterion usually takes the form of a functional relationship between a critical forcing amplitude and the damping and other system parameters.

REFERENCES

1. Hagedorn, P. (1981). *Nonlinear Oscillations*. Clarendon Press, Oxford.
2. Pippard, A.B. (1987). *The Physics of Vibration*. Cambridge University Press, New York.
3. Stoker, J.J. (1950). *Nonlinear Vibrations*. Interscience, New York.
4. Moon, F.C. (1992). *Chaotic and Fractal Dynamics*. John Wiley & Sons, New York.
5. Asokanthan, S.F., Meehan, P.A. (2000). Non-linear vibration of a torsional system driven by a Hooke's joint. *Journal of Sound and Vibration*, 233(2), 297–310.
6. Meehan, P.A. (2020). Investigation of chaotic instabilities in railway wheel squeal. *Nonlinear Dynamics*, 100, 159–172.
7. Newton, I. (1687). *Philosophiae Naturalis Principia Mathematica*. London: Royal Society of London.
8. Slotine, J.-J. E., Li, W. (1991). *Applied Nonlinear Control*. Prentice Hall, Englewood Cliffs, NJ.
9. Junkins, J.L., Kim, Y. (1993). *Introduction to Dynamics and Control of Flexible Structures*. AIAA, Washington, DC.

10. Mukherjee, R., Chen, D. (1992). An asymptotic stability theorem for autonomous systems. *Journal of Guidance, Control, and Dynamics*, 16(5), 961–963, preprint 1992, to appear, AIAA Journal of Guidance, Control and Dynamics.
11. Thompson, J.M.T., Stewart, H.B. (1986). *Nonlinear Dynamics and Chaos*. Wiley, Chichester, p. 4.
12. Henon, M. (1978). A two dimensional mapping with a strange attractor. *Communications in Mathematical Physics*, 50, 69–77.
13. Nusse, H.E., Yorke, J.A. (1994). *Dynamics: Numerical Explorations*. Springer-Verlag, New York, p. 257.
14. Hilborn, R.C. (1994). *Chaos and Nonlinear Dynamics: An Introduction for Scientists and Engineers*. Oxford University Press, Oxford.
15. Feigenbaum, M.J. (1978). Qualitative universality for a class of nonlinear transformations. *Journal of Statistical Physics*, 19(1), 25–52.
16. Newhouse, S., Ruelle, D., Takens, F. (1978). Occurrence of strange AxiomA attractors near quasi periodic flows on T^m, $m \geqq 3$. *Communications in Mathematical Physics*, 64, 35–40.
17. Manneville, P., Pomeau, Y. (1990). Different ways to turbulence in dissipative dynamical systems. *Physica D*, 1, 219–226.
18. Benettin, G., Galgani, L., Giorgilli, A. et al. (1980). Lyapunov characteristic exponents for smooth dynamical systems and for hamiltonian systems; a method for computing all of them. Part 1: Theory. *Meccanica*, 15, 9–20.
19. Shimada, I., Nagashima, T. (1979). A numerical approach to ergodic problem of dissipative dynamical systems. *Progress of Theoretical Physics*, 61(6), 1605–1616.
20. Wolf, A., Swift, J.B., Swinney, H.L., Vasano, J.A. (1985). Determining Lyapunov exponents from a time series. *Physica*, 16D, 285–317.
21. Mandelbrot, B.B. (1977). *Fractals, Form, Chance and Dimension*. W. H. Freeman, San Francisco, CA.
22. Baker, G.L., Gollub, J.P. (1996). *Chaotic Dynamics: An Introduction*, 2nd Ed. Cambridge University Press, Cambridge.
23. Moon, F.C., Li, G.-X. (1985). The fractal dimension of the two-well potential strange attractor. *Physica D: Nonlinear Phenomena*, 17(1), 99–108.
24. Parker, T.S., Chua, L.O. (1989). *Practical Numerical Algorithms for Chaotic Systems*. Springer-Verlag, Berlin etc.
25. Kaplan, J.L., Yorke, J.A. (1979). Chaotic behavior of multidimensional difference equations. In: Peitgen, H.O., Walther, H.O. (eds) *Functional Differential Equations and Approximation of Fixed Points*. Lecture Notes in Mathematics, vol 730. Springer, Berlin, Heidelberg, pp. 204–227.
26. May, R.M. (1976). Simple mathematical models with very complicated dynamics. *Nature*, 261, 459–467.
27. Szemplinska-Stupnicka, W. (1995). The analytical predictive criteria for chaos and escape in nonlinear oscillators: A survey. *Nonlinear Dynamics*, 7, 129–147.
28. Kolmogorov, A.N. (1954). On the preservation of quasiperiodic motions under a small variation of Hamilton's function. *Doklady Akademii Nauk. SSSR*, 98, 525.
29. Arnold, V.I. (1963). Small denominators and the problem of stability of motion in classical and celestial mechanics. *Russian Mathematical Surveys*, 18, 85–191.
30. Moser, J. (1962). On invariant curves of area-preserving mappings on an annulus. *Nachrichten der Akademie der Wissenschaften in Göttingen. II. Mathematisch-Physikalische*, K1, 1.
31. Wiggins, S. (1988). *Global Bifurcations and Chaos*. Springer-Verlag, New York.
32. Wiggins, S. (1990). *Introduction to Applied Nonlinear Dynamical Systems and Chaos*. Springer-Verlag, New York.
33. Guckenheimer, J., Holmes, P. (1983). *Nonlinear Oscillations, Dynamical Systems and Bifurcations of Vector Fields*. Springer Verlag, New York.

3 Predicting Classical Instabilities in Engineering Systems

An important unstable process which affects most of our lives is investment growth. In a typical investment, the interest grows unstably (approximately exponentially) as it is accrued based on a % of the growing capital that is mathematically modelled and predicted. Hence, we base important life decisions on the mathematical prediction of this unstable process. In this chapter, we will survey classical instability modelling for prediction of the occurrence and growth of instabilities in a range of engineering systems. The phenomena will range from the high-pitched squealing noise that sometimes occurs when a train negotiates a curve/corner, or from car brakes, to the cochlear amplification in our ears that is used to understand a whisper. Subsequently, in Chapter 4, we will look at the prediction of chaotic phenomena in the same range of systems which can occur following the onset of classic instability. By systematically examining instabilities across a spectrum of engineering scenarios, this chapter provides insights into predictive methodologies crucial for predicting when and how classical engineering instabilities occur and grow.

Mathematical modelling enables any instability, in any field of learning, to be transcribed into a universal language that we can analyse, predict and control using the common tools developed in Chapters 2 and 5. In such a way, we will codify, in detail, a range of engineering systems into dynamical equations of motion. From this, stability criteria can be developed to predict under what conditions, classical (non-chaotic) instabilities occur.

3.1 RAILWAY WHEEL SQUEAL

Railway wheel squeal research has identified two main mechanisms driving the instability: falling friction and mode coupling [1], which in turn has two types. These have been identified over a long history of fruitful research primarily after identification of the "falling friction" mechanism [2], as reviewed in useful literature reviews such as [3–5], amongst others. It was probably first confirmed that wheel squeal grows from a local instability to a stable limit cycle by [6] and [7] with important behaviour identified in much research that followed ie [8–11]. There appears to be evidence first of mode coupling in wheel squeal in [12], where rail dynamics was included, but it was not investigated deeply. Plenty of further research followed such as [13–15].

The following efficient modelling of squeal is primarily based on [16–18], and we will use it to explain the analytical methodology for predicting the occurrence and nonlinear amplitude of wheel squeal. Then, conditions and vibration amplitudes for

DOI: 10.1201/9781003496533-3

falling friction and the two mode coupling mechanisms (determined by the coupled structural complex stiffness and damping) are then detailed. We will discuss and compare each mechanism.

Model for Wheel Squeal Instability

Squeal instability vibration may be modelled according to the fundamental interacting mechanics shown in Figure 3.1.

Figure 3.1 illustrates that squeal arises from an instability leading to a limit cycle oscillation due to nonlinearities. It is driven by a complex interaction among three key factors: the vehicle curving mechanics (governing the speed, angle of attack and contact position angle), the rolling contact mechanics (determining the traction-creep parameters) and the vibrational modal dynamics, including coupled wheel/rail interactions. The coupled modal dynamics determine the system's modal damping, c_{si}, which depends on both vehicle curving dynamics (affecting the contact angle, θ_c) and contact mechanics (impacting contact stiffness and friction). Falling friction originates from the contact mechanics and contributes to negative system damping. The squeal noise we hear originates mainly from the wheel vibrations that send sound pressure fluctuations through the air to our ear. The vehicle curving mechanics, contact mechanics and coupled modal dynamics of Figure 3.1 can be described by the reduced model of Figure 3.2. The top view of a single wheel negotiating a curve is shown in Figure 3.2a causing an angle of attack, θ_A, between the wheel and the rail, implying there is a lateral or crabbing velocity, V_c, perpendicular to the rolling velocity. This causes lateral sliding and a friction force F_f in addition to the normal contact force F_N. Therefore, variations in these variables excite two dominant vibration modes that are spragged at an angle θ to each other, shown in Figure 3.2b. If θ is non-zero (or not 90°), then mode coupling is possible. Wheel and rail modes are shown in this case, but other forms of friction-induced mode coupling may be represented, such as brake squeal. The contact plane is oriented at an angle, θ_c from the lateral (or horizontal) direction.

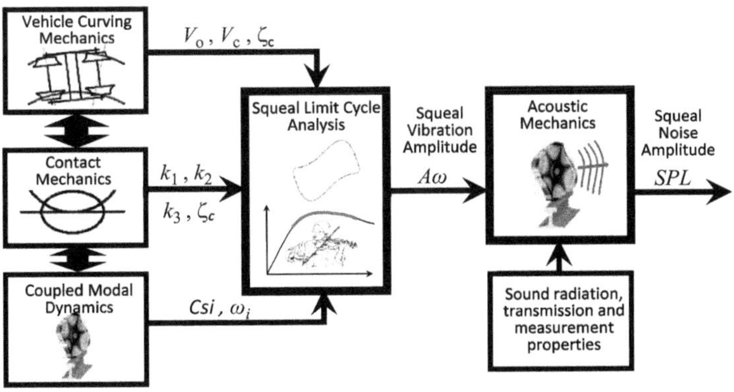

FIGURE 3.1 Wheel squeal instability and noise amplitude mechanics including coupled wheel/rail dynamics [70].

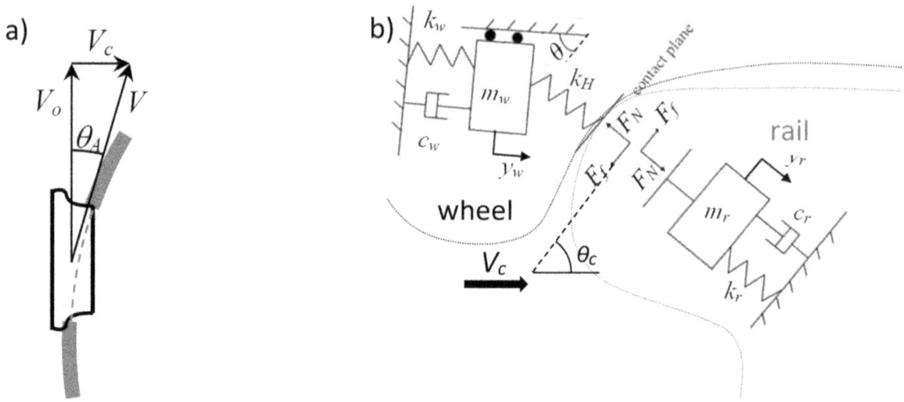

FIGURE 3.2 (a) Model of wheel rolling on a curved rail with an angle of attack; (b) two-mode coupled dynamic model of wheel rail contact [70]. Note $\theta = \theta_c$ for an axial wheel mode.

According to Figure 3.2a, the quasistatic angle of attack, θ_A, gives rise to a lateral crabbing velocity, V_c, and nominal lateral creep, ς_o, that may be modelled as

$$V_c = \theta_A V_0 \quad \text{and} \quad \zeta_o = \theta_A / \cos(\theta_c), \tag{3.1}$$

where the creep is the ratio of the wheel velocity relative to the rail in the contact plane and the rolling speed, V_0. Hence, the corresponding dynamic lateral creepage, ζ, includes variations due to the wheel vibration and may be described by

$$\zeta = \left((V_c / \cos(\theta_c)) + (\delta_w \dot{y}_w(t) / \cos(\theta)) \right) / V_0. \tag{3.2}$$

The lateral creep causes dynamic frictional forces that excite the wheel squeal 2-degree-of-freedom dynamics, shown in Figure 3.2b, that can be expressed as

$$\begin{aligned} m_w \ddot{y}_w(t) + c_w \dot{y}_w(t) + k_w y_w(t) &= -F_N \sin(\theta) - F_f \cos(\theta), \\ m_r \ddot{y}_r(t) + c_r \dot{y}_r(t) + k_r y_r(t) &= F_N, \end{aligned} \tag{3.3}$$

$$\text{where} \quad F_N = k_H (y_w \sin(\theta) - y_r), \quad F_f = \delta_w \ \mu(\zeta)(N + F_N). \tag{3.4}$$

In equations (3.3), y, m, k and c and are the modal displacements, masses, stiffnesses and damping coefficients where the subscripts r and w represent the rail and wheel, respectively. The dominant (uncoupled) mode parameters can be experimentally [19] or numerically [16] determined and may vary with the contact angle as investigated in [20]. N, F_N and F_f are the static and dynamic normal and frictional forces, and δ_w is the unit sliding direction (shown as +1 in Figure 3.2b). The lateral traction creep is nonlinear for large oscillations and is presently modelled as

$$\mu(\zeta) = \mu_s \tanh(0.9\zeta') \left\{ 1 - 0.5 e^{-0.138/|\zeta V_0|} - 0.5 e^{-6.9/|\zeta V_0|} \right\}, \zeta' = \zeta k_{\zeta 3} / N. \tag{3.5}$$

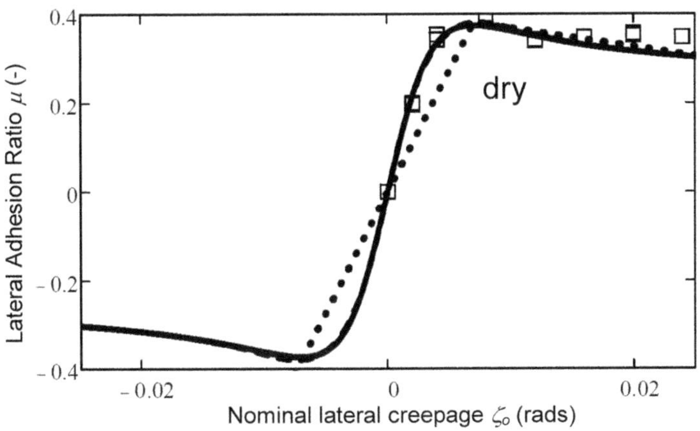

FIGURE 3.3 The creep curve at a rolling speed of 17.8 m/s; simulated using equation (3.5) (-) and overlapping [12,18] (--), analytical (..) and experimentally measured (□) [70].

This wheel squeal creep model is quantitatively very similar to the two-part model of de Beer et al. [12] but is mathematically more concise and continuous. A plot of the creep curves of (3.5) and [12] is provided in Figure 3.3 for direct comparison with a simple bilinear approximation and experimental measurements [18].

The curves of Figure 3.3 show a negatively sloped region for large creepage causing negative system damping (i.e., instability). The full nonlinear curves and simplified analytical approximation are verified by the experimental measurements in the nonpositively sloped region, validating their usage. The simulated creep contact parameter $k_{\varsigma 3}$, encapsulates the contact mechanics, as determined by Hertz [21] and detailed in [18,20] and [22]. The sound pressure level, *SPL*, of the radiated wheel vibration amplitude, $A_w\omega$, through the air may be simply modelled as [23,24]

$$\mathrm{SPL} = 20\log(A_w\omega) + \mathrm{SPL}_{\mathrm{const}}, \tag{3.6}$$

where $\mathrm{SPL}_{\mathrm{const}}$ depends on the transmission and radiation components which should be constant for a nominal site sound recording (i.e., microphone setup).

The full nonlinear equations of motion, (3.1)–(3.6), can be used to predict the occurrence of railway wheel squeal instability. For instance, they may be analytically solved to predict the stability and limit cycle analysis of wheel squeal in what follows.

Prediction of Wheel Squeal

The main mechanisms driving wheel squeal instability may be captured, combined and compared using negative damping stability analysis. In particular, the local stability of the coupled equations of motion (3.1)–(3.6) may be determined by complex eigenvalue analysis which quantifies the system negative damping levels of the modes of vibration. An approximate analytical solution was obtained in [20] using the undamped eigenmodes. In particular, the coupled equations (3.3) and (3.4) can be decoupled in matrix form as

$$\mathbf{p_L}^{\mathrm{T}}\mathbf{Mp}\ddot{\mathbf{Y}} + \mathbf{p_L}^{\mathrm{T}}\mathbf{Cp}\dot{\mathbf{Y}} + \mathbf{p_L}^{\mathrm{T}}\mathbf{KpY} = \mathbf{p_L}^{\mathrm{T}}\mathbf{F}, \quad \mathbf{Y} = \mathbf{p}^{-1}\begin{bmatrix} y_w & y_r \end{bmatrix}^{\mathrm{T}},$$
$$\mathbf{F} = \begin{bmatrix} -\delta_w \mu(\zeta) N\cos(\theta) & 0 \end{bmatrix}^{\mathrm{T}} \tag{3.7}$$

where **M**, **C**, **K** and **F** are the mass, damping, stiffness matrices and forcing vector of equations (3.3) and (3.4), respectively. **Y** is the modal displacement vector from the decoupling transformation of the wheel and rail displacements, y_w and y_r, respectively, using the left and right undamped eigenvectors $\mathbf{p_L}$ and $\mathbf{p}$, defined as

$$\mathbf{p_L} = \begin{bmatrix} 1 & p_{L2} \\ p_{L1} & 1 \end{bmatrix} \text{ where } \mathbf{p_L}^{\mathrm{T}}\mathbf{K} = \lambda^2 \mathbf{p_L}^{\mathrm{T}}\mathbf{M},$$
$$\mathbf{p} = \begin{bmatrix} 1 & p_2 \\ p_1 & 1 \end{bmatrix} \text{ where } \mathbf{Kp} = \lambda^2 \mathbf{Mp}. \tag{3.8}$$

The p variables determine the modal damping, which in turn determines the stability of each of the system modes in **Y**. They can be solved in closed form as a function of the system parameters as [20]

$$p_1 = \left[\omega_w^2 - \omega_r^2 + \sqrt{\Delta}\right] \Big/ \left[2f(\mu,\theta)\,k_H / (m_w \sin(\theta))\right],$$
$$p_2 = 1/\bar{p}_1,\; p_{L1} = -m_w/\bar{p}_1 m_r,\; p_{L2} = -m_r/\bar{p}_2 m_w$$
$$\text{therefore } p_L p_i = 4f(\mu,\theta)\left(k_H/m_j\right)^2 \Big/ \left[\omega_w^2 - \omega_r^2 - \sqrt{\Delta}\right]^2, \text{ where} \tag{3.9}$$
$$f(\mu,\theta) = \sin(\theta)\left(\sin(\theta) + \delta_w \mu \cos(\theta)\right), \quad \omega_w^2 = \left(k_w + k_H f(\mu,\theta)\right)/m_w,$$
$$\omega_r^2 = (k_r + k_H)/m_r, \quad \Delta = \left(\omega_w^2 - \omega_r^2\right)^2 + 4f(\mu,\theta)\,k_H^2/(m_w m_r),$$

where $^{-}$ is the conjugate, i is the mode number and $j = r,w$. Note that $f(\mu,\theta)$ encapsulates Spurr's [25] criteria for spragging. The discriminant Δ determines the complexity of the eigenvectors and hence the type of mode coupled squeal, i.e., viscous type if positive or stiffness type if negative. The analytical solution assumes the undamped and damped eigenvectors are similar due to small or sufficiently proportional structural damping. The solution to the left and right eigenvectors gives the system modal damping from the structural coupling. To determine the creep curve contribution to the system modal damping, analytically, the nonlinear creep behaviour of (3.5) is approximated within bounds as a bilinear curve with slopes, $k_{\varsigma 1}$ and $k_{\varsigma 2}$, for the slip and full sliding regions, respectively, according to

$$\mu(\zeta) = \begin{cases} k_{\zeta 1}\zeta' & \text{for } \zeta' \le 1 \\ k_{\zeta 1} + k_{\zeta 2}(\zeta' - 1) & \text{for } \zeta' > 1 \end{cases} \quad \text{where } \zeta \ge -\zeta_c. \tag{3.10}$$

Equations (3.3), (3.7) and (3.10) may be simplified for small oscillations about a nominal crabbing velocity, V_c, for each mode $i = 1,2$, as

$$m_i\ddot{y}_i(t) + \left(c_{\text{si}} + k_{\text{Fi}}k_{\varsigma 1,2}k_{\varsigma 3}/V_0\right)\dot{y}_i(t) + k_i y_i(t) = 0 \quad \text{where} \quad k_{F1} = 1, k_{F2} = p_{L2}p_2, \tag{3.11}$$

where the system modal mass, stiffness and damping terms are given by

$$m_i = \mathbf{p_L}^{\mathrm{T}}\mathbf{M}\mathbf{p}_{i,i} \quad k_i = \mathbf{p_L}^{\mathrm{T}}\mathbf{K}\mathbf{p}_{i,i} \quad c_i = \mathbf{p_L}^{\mathrm{T}}\mathbf{C}\mathbf{p}_{i,i} \tag{3.12}$$

and c_{si} is the system modal damping from the structural coupling that depends on the complexity of the eigenvectors as

$$c_{\text{si}} = \left\{ \begin{array}{ll} c_i & \text{if } p_i, p_{Li} = \text{Re}(p_i), \text{Re}(p_{Li}) \\ \text{Re}\left(c_i/(2m_i) \pm \sqrt{c_i^2/(2m_i)^2 - k_i/m_i}\right)2m_i & \text{if } p_i, p_{Li} \neq \text{Re}(p_i), \text{Re}(p_{Li}) \end{array} \right\}, \tag{3.13}$$

and $k_{\varsigma 1,2}$ is the local slope of the creep curve. The system modal equation (3.11) is in the form of a single-degree-of-freedom oscillator (i.e., spring mass damping system), so its stability is determined by the sign of the effective system damping term $c_{\text{si}} + k_{\text{Fi}}k_{\varsigma 1,2}k_{\varsigma 3}/V_0$. Equation (3.13) highlights that the system modal damping, c_{si}, depends on the complexity of the eigenmodes and therefore the type of mode coupling (i.e., viscous or stiffness). Since the structural modal damping factors are usually well below critical, it may be deduced from (3.13) that stiffness mode coupling (second line) provides a much larger contribution to the system modal damping (and local eigenvalue magnitude) than viscous mode coupling (first line). The stability and hence occurrence of wheel squeal can then be simply expressed according to when the effective system damping becomes negative and analytically determined using equations (3.11) and (3.13) as

$$\text{Instability/squeal when:} \quad c_{\mathit{eff}} = \left(\underline{c_{si}} + \underline{k_{Fi}k_{\mu 1,2}N/(\zeta_c V)}\right) < 0 \tag{3.14}$$

Structural damping (-ve if Spragging and Mode Coupling) → c_{si}

Friction slope damping (-ve if Falling Friction) → $k_{Fi}k_{\mu 1,2}N/(\zeta_c V)$

The wheel squeal criterion (3.14) concisely combines and superposes the effects of mode coupling and falling friction via the effective system damping, c_{eff}. Squeal instability will initially grow exponentially (like a savings account!) when the effective system damping is negative, but this growth will change with amplitude. In particular, when the sliding oscillation amplitude encroaches on the friction stick region below critical friction, positive damping will be added. At a particular oscillation amplitude, $A_w\omega_i$, where the power from the negative damping exactly balances the positive damping, a steady-state limit cycle will occur according to [17]

$$A_w \omega_i = \left((V_c / \cos(\theta_c)) - \zeta_c V_0\right)\cos(\theta) \big/ \cos(\omega_i t_c)$$
$$\text{where} \quad \sin(2\omega_i t_c) - 2\omega_i t_c = 2\pi\left(V_0 c_{si} + k_{Fi} k_{\zeta 2} k_{\zeta 3}\right) \big/ \left(k_{Fi} k_{\zeta 3}\left(k_{\zeta 1} - k_{\zeta 2}\right)\right), \tag{3.15}$$

where ς_c and t_c are the critical (peak) creep and time in the squeal cycle, respectively. Equation (3.15) is a closed-form analytical solution for wheel squeal vibration (or limit cycle) amplitude. It includes both mode coupling and falling friction mechanisms and assumes bounded positive creep oscillations. Hence, the stability and limit cycle solutions of the simplified two-mode squeal system have been provided in closed form as equations (3.14) and (3.15) assuming creep oscillations are small enough to be approximated by a bilinear friction law (3.10). However, equation (3.13) indicates there are also two types of wheel squeal instability due to mode coupling. More details of these types are now provided in the following section [20].

Spragging and Mode Coupling

One key aspect of the reduced model in Figure 3.2 is the explicit definition of the spragging angle θ. This allows for a clear analytical relationship between spragging and mode coupling to be established through the closed-form solution. By examining equations (3.9), (3.13) and (3.14) and recognising that the parameter $f(\mu,\theta)$ encapsulates the spragging criterion, it becomes evident that spragging and mode coupling mechanisms determine whether the structural modal damping, c_{si}, is positive or negative based on the necessary conditions [20].

$$\text{Mode coupling amplification } c_{si} < 0 \text{ when } \delta_w = -1 \quad \text{and} \quad 0 < \tan(\theta) < \mu$$
$$\text{Mode coupling decay } c_{si} \geq 0 \text{ when } \delta_w = 1 \quad \text{or} \quad \tan(\theta) > \mu. \tag{3.16}$$

The necessary conditions for squeal, given in the first line of equation (3.16), directly correspond to the conditions for spragging [25]. Specifically, the direction, δ_w, determines whether the mode displacement, y_w, is "digging in" (−1) or sliding with (+1) the lateral sliding velocity, $V_{c,o}$. Additionally, for mode coupling amplification (and the potential for squeal) to occur, the well-known spragging angle, θ, must be small, specifically less than atan(μ). Therefore, spragging is a necessary condition for mode coupling squeal. The type of mode coupling—whether viscous or stiffness-based—is then determined by whether the eigenvectors are real or complex [20], as discussed in the following section.

Viscous Mode Coupling

If the discriminant Δ in (3.9) is positive, then viscous mode coupling occurs, and the modal damping is given by $c_{si} = c_i$ (3.13) (i.e., the eigenvectors are real). By analysing these equations, we can assess the impact of mode coupling on squeal amplification or decay based on whether the modal damping ratio, c_i/k_{Fi}, exceeds or falls below the wheel damping, c_w. This leads to the following criteria:

$$\text{Mode coupling amplification } c_i / k_{Fi} < c_w \text{ when } \delta_w = -1 \quad \text{and}$$
$$0 < \tan(\theta) < \mu, \quad \Delta \geq 0$$
$$\text{Mode coupling decay } c_i / k_{Fi} > c_w \text{ when } \delta_w = 1 \quad \text{or} \quad \tan(\theta) > \mu, \quad \Delta \geq 0. \tag{3.17}$$

Equation (3.17), physically indicates that mode coupling will amplify wheel squeal only for the leading outer wheel (i.e., when $\delta_w = -1$) and when the angle θ is small, falling in the range, $0 < \theta < \tan^{-1}(\mu)$. Under all other conditions (excluding cases where there is no effect), mode coupling contributes to positive system damping, leading to the decay of wheel squeal. The maximum viscous mode coupling amplification and decay values have also been determined in [20] as

$$\text{Max mode coupling amplification } \delta_w = -1 \text{ at } \theta_{\max a} = \frac{1}{2}\tan^{-1}(\mu)$$

$$\text{when } f(\mu, \theta_{\max a}) = \frac{1}{2}\left(1 + \sqrt{1-\mu^2}\right) \quad \Delta \geq 0,$$

$$\text{Max mode coupling decay } \delta_w = -1 \text{ at } \theta_{\max d} = \frac{\pi}{2} \text{ when } f(\mu, \theta_{\max d}) = 1 \quad \Delta \geq 0, \tag{3.18}$$

$$\delta_w = 1 \text{ at } \theta_{\max d} = \frac{1}{2}\left(\pi - \tan^{-1}(\mu)\right) \text{ when } f(\mu, \theta_{\max d}) = \frac{1}{2}\left(1 + \sqrt{1+\mu^2}\right).$$

Equation (3.18) physically indicates the maximum mode coupling amplification occurs at a small angle dependent only upon the friction coefficient. Also, the maximum decay occurs at high contact angles indicative of flanging in wheel-rail contact. This agrees with field observations that wheel squeal is predominantly a top-of-rail phenomenon.

Stiffness Mode Coupling

If the discriminant Δ in (3.9) is negative, then stiffness mode coupling occurs. This happens when spragging occurs and the uncoupled modes are close and the solution mathematically becomes complex, causing negative damping via complex stiffness. Physically, this means the coupled modal frequencies converge and one pumps energy from the frictional force into the other mode via a phase difference between the (complex) modal motions. Here, negative system modal damping from the structural coupling, c_{si}, occurs under [20]:

$$\text{Mode coupling amplification when } \delta_w = -1, 0 < \tan(\theta) < \mu, \text{ and } \Delta < 0. \tag{3.19}$$

Equation (3.19) means that stiffness mode-coupled squeal physically occurs when the leading outer wheel (i.e., when $\delta_w = -1$) has a small angle θ in between 0 and $\tan^{-1}(\mu)$ and close uncoupled modes. As previously mentioned, whether the amplification or decay caused by the stiffness mode coupling causes the existence or non-existence of squeal is determined by the effective total system modal damping (3.14). This also takes into account the negative damping from the falling friction slope of the creep curve.

It is of interest to determine the conditions under which maximum stiffness mode coupling amplification occurs. This may be determined [20] as

$$\text{Max mode coupling amplification } \delta_w = -1 \text{ at } \theta_{\max a} = \frac{1}{2}\tan^{-1}(\mu),$$
$$\Delta = 2\left(1-\sqrt{1+\mu^2}\right)k_H^2/(m_w m_r)$$
$$\text{when } \mathrm{Re}\left(\sqrt{-k_i/m_i}\right)2m_i \tag{3.20}$$
$$= \pm 2m_w\sqrt{2\min(\omega_{w,r})}\sqrt{\sqrt{1-2\left(1-\sqrt{1+\mu^2}\right)k_H^2/\left(m_w m_r\left(2\min(\omega_{w,r})\right)^2\right)}-1},$$

where $\min(\omega_{w,r})$ is the minimum undamped natural frequency of the wheel or the rail (with Hertzian stiffness included). Physically, this means that the maximum mode-coupled negative damping component increases with higher Hertzian contact stiffness, wheel mass and friction coefficient, while it decreases with greater matched undamped frequencies and rail mass. By inspection of (3.19), it may be deduced that instability due to stiffness mode coupling is caused by the closeness of the uncoupled modes on the leading outer wheel in a small range of low angles θ defined by the friction coefficient.

Results of Wheel Squeal Predictions

The predictive models developed can be tuned to real field conditions using the literature ie [27] and [28]. As an example, these results were first compared to field measurements of squeal noise as well as Rudd's [2] simplified prediction for validation. In particular, the analytical predictions of squeal amplitude were plotted with field recordings of freight trains [17] and Rudd's predictions, as shown in Figure 3.4. For larger angle of attacks, the crabbing velocity was also plotted.

Figure 3.4 indicates the trend of squeal noise level with angle of attack is well predicted by the analytical results. For lower angle of attacks below 10 mrad, the field results' trend diverges from the analytical prediction due to the presence of other

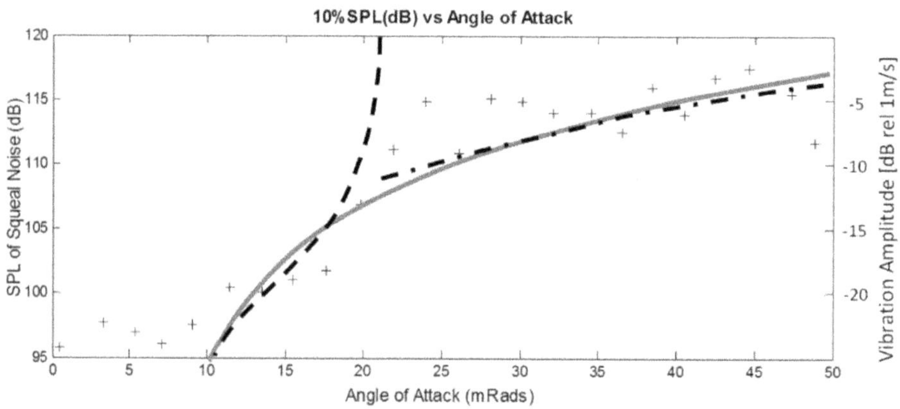

FIGURE 3.4 Sound pressure level of squeal noise versus angle of attack; field measurements (+) analytical prediction (-) modified Rudd [3] prediction (- -) and crabbing velocity (- . -) [17].

sources of noise including flanging and rolling noise. By inspection of the analytical solution (3.15), the trend in Figure 3.4 is due to the squeal vibration amplitude being directly proportional to the difference between the angle of attack and the critical creep. Specifically, the squeal noise amplitude is directly proportional to how much the angle of attack exceeds its critical creep value. The simplified modified prediction from Rudd [2] provides a reasonable prediction for low angles of attack. However, it greatly diverges and over approximates the squeal amplitude for angles of attack approaching three times the critical angle of attack. For higher angles of attack, the squeal amplitude approaches and then exceeds the crabbing velocity. The effect of speed on squeal noise was investigated directly, as shown in Figure 3.5.

The results highlight that the squeal noise increases monotonically with speed, as predicted well by the analytical solution. In particular, the creepage relation (3.2) predicts that the squeal vibration velocity is approximately proportional to the speed when the modal damping is small compared to the negative damping from the creep curve slope. Variations in the field results are most likely mainly due to the non-uniform number of samples over the speed range and field parameter variations. By comparison of the y-axes of Figures 3.4 and 3.5, it may be inferred that the squeal noise level is more strongly affected by the angle of attack than speed. Specifically, higher angle of attacks more consistently lead to higher noise levels than higher speeds. The strong correlations between the predictions and field results provide a sound basis for use of the analytical prediction for efficiently investigating the means of mitigating squeal noise.

The mathematical modelling and stability analysis can be used to make instantaneous predictions of wheel squeal instability behaviour compared to numerical simulations. For example, Figure 3.6a shows the numerical simulations (of equations (3.1)–(3.6)) of wheel squeal at a nominal angle of attack of 16 mrad under falling friction ($\theta_c = 0$) and maximum viscous mode coupling ($\theta_c = 0.16\,\text{rad}$) mechanisms [20]. The analytical predictions of the squeal amplitude and its growth are also shown for comparison in Figure 3.6b. Figure 3.6a shows a numerically verified analytical solution predicting squeal growing faster and settling on a slightly higher steady-state amplitude under viscous mode coupling amplification compared to just the falling friction mechanism.

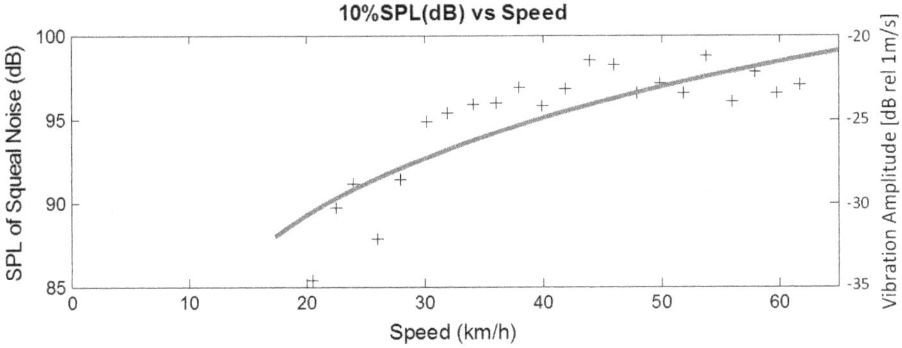

FIGURE 3.5 Sound pressure level of squeal noise versus; field measurements (+) and analytical prediction (–) [17].

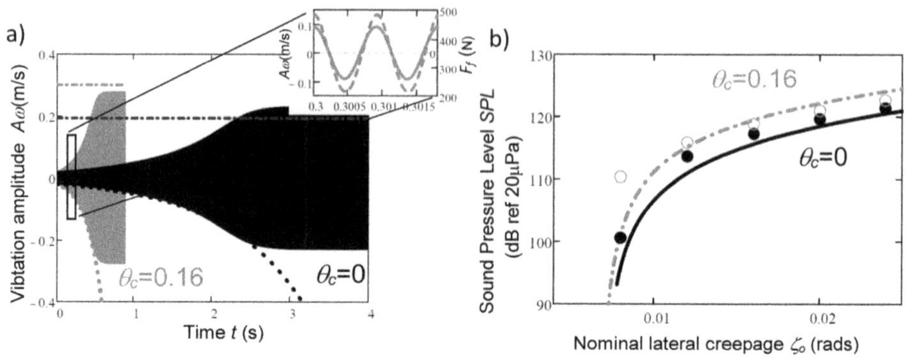

FIGURE 3.6 (a) Simulated time histories of squeal vibration growth under no mode coupling, $\theta_c = 0$ (-) and maximum viscous mode coupling, $\theta_c = 0.16$ rad (-) conditions with zoomup at 0.3 second including frictional force (--) and corresponding analytical predictions of squeal amplitudes under no mode coupling (.-.) and maximum viscous mode coupling (.-.) conditions, and for initial amplitude growth under no mode coupling (..) and maximum viscous mode coupling (..) conditions; (b) sound pressure level vs nominal lateral creepage at a rolling speed of 17.8 m/s; under no mode coupling, $\theta_c = 0$, analytical (-) and simulated (•) result and under maximum viscous mode coupling, $\theta_c = 0.16$ rad, analytical (.-.), simulated (O) result. Leading outer wheel [96].

The increase in squeal growth is due to viscous mode coupling, which reduces the system's modal damping from $c_{s1} = 103$ Ns/m (in the uncoupled case) to $c_{s1} = -490.3$ Ns/m in the coupled case. A zoomed-in view at 0.3 second shows that the initially sinusoidal wheel vibration velocity is in phase with the frictional force, creating a 90° phase difference between the frictional force and tangential deflections, thereby supplying the system with the necessary power for sustained squeal. As the system reaches its limit cycle, the frictional force becomes increasingly nonsinusoidal due to nonlinear creep oscillations. The squeal frequency (1102 Hz, as seen in Figure 3.4a) is determined by the coupling of the wheel and rail mode frequencies ($\omega_w = 1097$ Hz and $\omega_r = 1309$ Hz), along with contact stiffness and system damping. Further analytical predictions and simulations were conducted over a range of nominal lateral creepage angles, as shown in Figure 3.6b. The increase in squeal amplitude from no coupling conditions is relatively minimal (<5 dB). This is because the additional creep oscillations must predominantly balance the energy dissipation from the highly positively sloped region of the creep curve. The differences between the analytical predictions of the growth amplitudes shown in Figure 3.6a and the numerically simulated growth are due to the linearised representation of the slopes of the creep curve (see equation (3.20)) and nonlinear changes in the friction coefficient. However, the analytical model provides a reasonably accurate prediction of squeal occurrence and growth that is extremely efficient (runtime <20 ms for Figure 3.6b compared to 1 hour for only five numerical simulation case solutions).

This validation via numerical solution motivates further wide-ranging parameter investigations of the effects of contact angle and rail damping on wheel squeal, as shown in Figure 3.7.

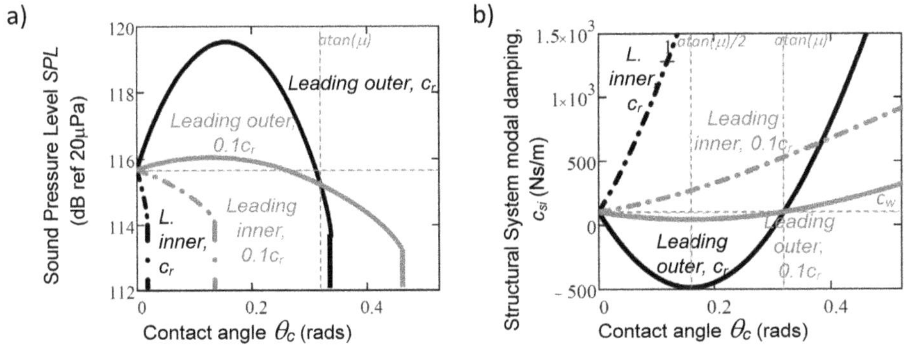

FIGURE 3.7 (a) Change in squeal sound pressure level and (b) the system modal damping contributed from the structural coupling due to contact angle for different wheels and rail damping levels, at a nominal lateral creep of 16 mrads and rolling speed of 17.8 m/s. Leading outer (-) and inner (.-.) wheel at nominal c_r=2.21E4 Ns/m. Leading outer (-) and inner (.-.) wheel at 10%c_r. No mode coupling occurs at $\theta_c=0$ and atan(μ) [20].

Figure 3.7a and b illustrate how the squeal sound pressure level (SPL) and system modal damping vary with contact angle. They show that viscous mode coupling can either amplify or dampen conventional falling-friction squeal, depending on the contact angle and wheel position. For the leading outer wheel, the SPL reaches a maximum at a contact angle of approximately atan(μ)/2, as predicted by equation (3.18). This peak amplification of 4.2 dB (simulated in Figure 3.6) decreases with lower rail damping, which is around 10% of the critical uncoupled rail damping. Figure 3.7b confirms that the maximum squeal SPL amplifications occur midway between points where mode coupling has no effect ($\theta_c=0$ and atan(μ)), in line with equation (3.18). It also highlights a significant negative damping contribution from viscous mode coupling, which can independently trigger wheel squeal, particularly at low nominal creepages, as seen in Figure 3.6b. For contact angles exceeding atan(μ), viscous mode coupling strongly suppresses squeal for both inner and outer wheels. At lower contact angles, it amplifies (and can even be the primary cause of) squeal for the leading outer wheel while damping or eliminating squeal for the inner wheel. These effects intensify as rail damping increases, aligning with analytical predictions.

The effect of stiffness mode coupling over a range of nominal lateral creepages (≈ angles of attack) for the leading outer wheel is also shown in Figure 3.8. Stiffness mode coupling cannot occur for the leading inner wheel in accordance with (3.19).

Figure 3.8a shows the numerical simulations of squeal at a nominal angle of attack of 16 mrad under falling friction only ($\theta_c=0$) and stiffness mode coupling at ($\theta_c=0.16$ rad) as well as analytical predictions of the growth and steady amplitudes. The squeal amplitude and growth rate under mode coupling greatly exceeds those of the no coupling case (reshown from Figure 3.6a) but decrease on the smaller time scale for comparison of the initial growth rates. In particular, the mode coupling case reaches an offset amplitude approaching 0.4 m/s in approximately 0.4 second, whereas the no mode coupling case takes over 2.5 second to reach a smaller amplitude of 0.23 m/s. The zoomup during the squeal growth at 0.016 second again confirms the squeal is powered by the frictional force that is approximately in phase with the wheel vibration velocity.

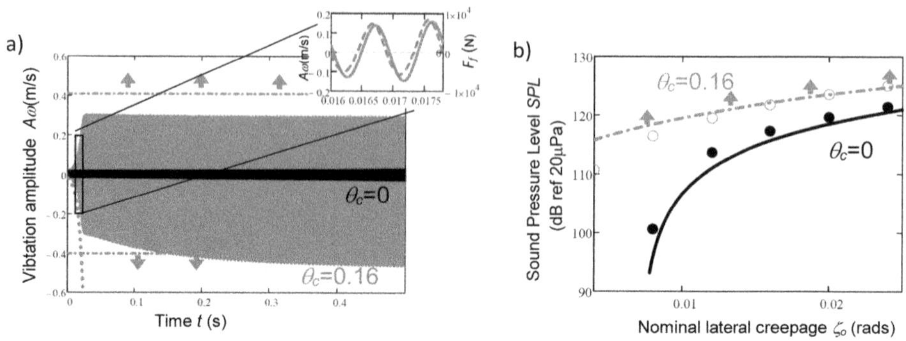

FIGURE 3.8 (a) Simulated time histories of squeal vibration growth under no mode coupling, $\theta_c = 0$ (-) (identical result as Figure 3.6a but on a smaller time scale) and stiffness mode coupling, $\theta_c = 0.16$ rad (-) conditions with zoomup at 0.016 second including frictional force (--). Corresponding analytical predictions of squeal amplitudes under stiffness mode coupling (.-.) conditions, and for initial amplitude growth under stiffness mode coupling (..) conditions; (b) sound pressure level vs nominal lateral creepage at a rolling speed of 17.8 m/s; under no mode coupling, $\theta_c = 0$, analytical (-) and simulated (•) result and under stiffness mode coupling, $\theta_c = 0.16$ rad, analytical (.-.), and simulated (O) result which exceeded 130 dB. Leading outer wheel. Arrows indicate the analytical prediction exceeded the maximum limit of the simplified creep model scope of applicability (3.10) [20].

In this case, the wheel and rail vibrations are not in phase due to the complex nature of the eigenvalue solution. The predicted analytical amplitude growth from equation (3.11) closely matches numerical simulations until large nonlinear oscillations emerge around 0.02 second. Beyond this point, the analytical steady-state squeal amplitude prediction exceeds the creep limit boundary, entering full negative sliding (represented by the dashed–dot lines in Figure 3.8a), which is beyond the scope of the bilinear creep approximation in equation (3.10). The simulations and analytical results confirm that stiffness mode coupling amplifies the squeal amplitude by at least 1.5 times in this case. Figure 3.8b further demonstrates that the simulated squeal reaches a higher steady-state amplitude, at least 5 dB greater, across a wide range of angles of attack. This occurs even for lower angles below the critical creep threshold, indicating that the negative damping from stiffness mode coupling overrides the positive damping effect of the high-slope slip region of the creep curve. Stiffness mode coupling significantly reduces the system's modal damping, lowering it from the uncoupled case ($c_{s1} = c_w = 103$ Ns/m) to $c_{s1} = -3.05 \times 10^4$ Ns/m in the coupled case. This is due to the complex phase relationship in the motion, which results in a large negative damping contribution from the complex modal stiffness in equation (3.13). The prediction error increases at low creepage levels due to the linear approximation of the bilinear creep law (3.10) not fully capturing the nonlinear behaviour shown in Figure 3.1.

To further examine the magnitude of amplification caused by stiffness mode coupling, the impact of contact angle and the proximity of undamped modes was analysed, as presented in Figure 3.9. Figure 3.9a shows how the squeal system modal damping changes with contact angle under stiffness mode coupling conditions. It confirms that stiffness mode coupling causes amplification of falling friction squeal for the leading outer wheel within a small range of contact angles defined by $\Delta < 0$

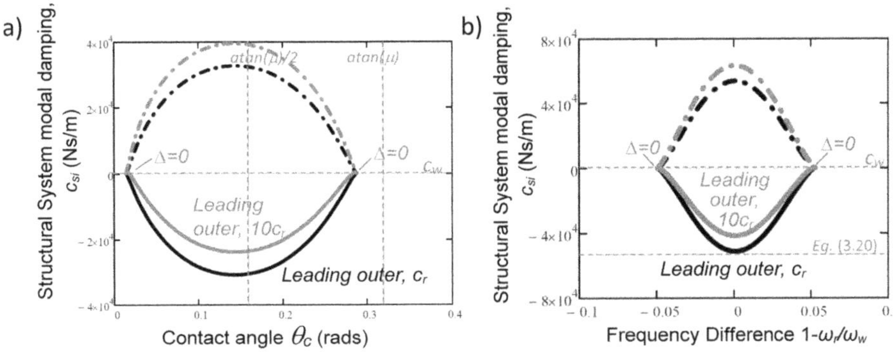

FIGURE 3.9 (a) Change in the system modal damping contributed from the structural coupling due to (a) contact angle and (b) closeness of the uncoupled modes for different wheels and rail damping levels, at nominal lateral creep of 16mrads and rolling speed of 17.8 m/s. Negatively damped mode (-)(-) and positively damped mode (.-.)(.-.) at nominal $c_r = 1097$ Ns/m and $10c_r$ respectively. No stiffness mode coupling occurs at $\Delta \geq 0$ [20].

in accordance with the analytical predictions (3.20). The minimum system modal damping occurs near the middle of this range at a contact angle of approximately atan(μ)/2 in accordance with (3.20). Note that a highly damped mode is also present in Figure 3.9a at a very similar natural frequency in accordance with the stiffness mode coupling phenomena under damping described in [29]. The optimum modal damping levels are approximately 200 times greater in amplitude than the uncoupled structural damping from the wheel alone and are shown to increase moderately (~20%) with a large increase (10x) in rail damping. Hence, in contrast with viscous mode coupling, rail damping has a relatively small but stabilising effect on squeal due to stiffness mode coupling in accordance with [29]. The effect of closeness of the uncoupled modes on the optimum modal dampings is shown in Figure 3.9b.

The optimal conditions for stiffness mode coupling occur when the uncoupled frequencies are matched ($\omega_r = \omega_w$), as predicted by equation (3.20). Under these conditions, the system reaches a maximum negative modal damping of −51.2 kNs/m, closely aligning with the analytical prediction of −52.4 kNs/m when damping is not considered. This value is more than 500 times greater in magnitude than the uncoupled wheel's structural damping, explaining why stiffness mode coupling leads to such large nonlinear squeal amplitudes, exceeding the bounds of the simplified creep model. Equation (3.20) further suggests that stiffness mode coupling amplifies squeal to much greater levels than other types of squeal observed in this study. This occurs because the negative damping is driven by the complex system stiffness ($\propto k_H/\omega_w$), which is significantly larger than the negative damping from either the creep curve slope or structural damping. Since both stable and unstable modes coexist under stiffness mode coupling, the unstable mode ultimately dominates due to its higher growth rate. Figure 3.9b highlights that increased rail damping has a stabilising effect on stiffness mode coupling, similar to the effect seen in Figure 3.9a, but to a lesser degree. This reduced impact is due to the fact that the magnitude of system modal damping in this case is even larger relative to rail damping.

Hence, a very efficient model for wheel squeal prediction has been developed and validated using mathematical modelling of the mechanism and then stability analysis to develop predictive criteria. Much new insights have been gained from this predictive analysis. A similar approach for other instabilities in engineering is demonstrated subsequently.

3.2 BRAKE SQUEAL

Brake squeal research has identified several mechanisms driving the instability, but it mainly shares two with wheel squeal: falling friction and mode coupling. This is not so surprising as it arises from the same dynamic interactions between friction and structural vibrations. Whereas, practically, falling friction seems to dominate field occurrences of wheel squeal, mode coupling appears to be the most prevalent mechanism in brake squeal. In 1963, Spurr thought that through "careful choice of dimensions in the design … I hope brake squeal will accordingly soon be a thing of the past" [30].

Unfortunately, even after more than half a century, brake squeal remains a major challenge in the automotive industry [31]. Numerous theories have been proposed to explain this phenomenon, as extensively reviewed in [32]. Key mechanisms include falling friction with increasing sliding speed, sprag-slip [33] and modal coupling [34,35] (or "binary flutter"). Spurr [33] introduced the "sprag-slip" theory for brake squeal based on studies of squeal in railway drum disc brakes. To illustrate this mechanism, he used a semi-rigid strut inclined against a moving surface, where θ represents the angle between the beam and the vertical plane. When the beam angle was adjusted, it would "sprag" or dig into the surface when $\tan(\theta)=\mu$, i.e., at the "sprag angle." Dynamic slipping would then occur once the applied forces exceeded static friction [36]. The theory was supported using "pin on disc" testing [37]. Hoffmann [38] found sprag–slip oscillations in a beam-on-belt model including elasticity and finite beam thickness. Fieldhouse [39,40] then related the caliper mounting and geometry of the brake to the sprag angle. The results showed squeal is more likely with a leading centre of pressure ("digging-in" of the brake pad). North [34] identified mode coupling in automotive brake squeal and termed it "binary flutter" due to analogies with the aerodynamic instability. Further numerical research showed nonhomogenous damping may increase squeal propensity [29,41]. Much research has followed ie [42–62]. Recently, Meehan [63] obtained an analytical solution for squeal occurrence and its noise amplitude that included and combined the effects of "falling friction," spragging and mode-coupled conditions. The following efficient modelling of squeal and its prediction is primarily based on this work. For clarity and insight, the similarities with wheel squeal are used to develop the closed-form solutions. We will discuss and compare each mechanism.

Model for Brake Squeal Instability

Brake squeal is an instability arising from an interaction between the mechanics of the brake pad, frictional contact and brake system dynamics, as shown in Figure 3.10. Note the similarities with the conceptual model for wheel squeal (see Figure 3.1). When the brake pad contacts the spinning rotor, its mechanics influence key factors

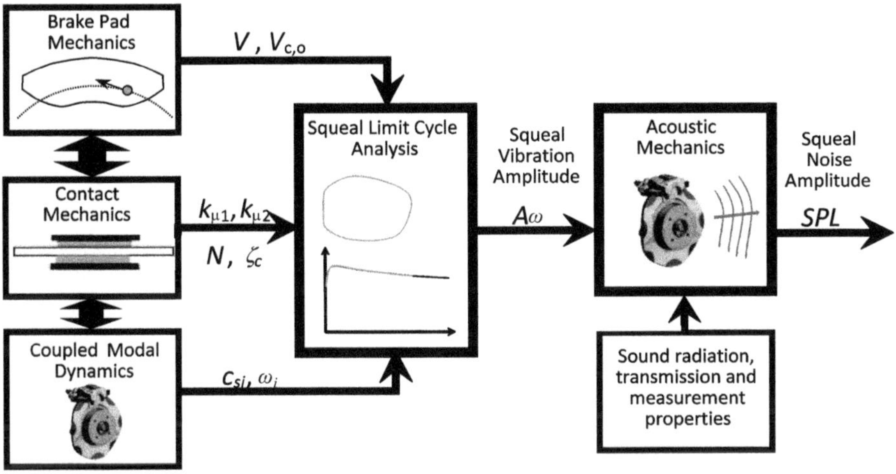

FIGURE 3.10 Conceptual model of brake squeal noise [63].

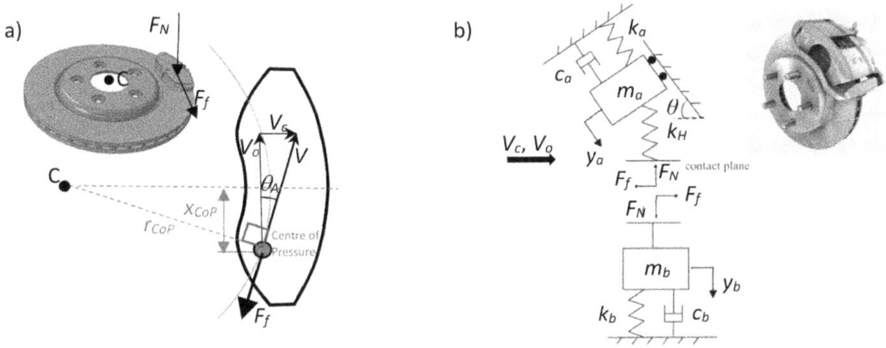

FIGURE 3.11 Two-degree-of-freedom model for brake squeal consisting of (a) brake pad mechanics with sliding and friction forces acting through the centre of pressure and (b) coupled modal dynamics. Insets: Abaqus & Free3D Car Brake System models [63].

such as the centre of pressure, sliding velocity and the velocity component in the direction of the squealing mode (or crabbing velocity) between the pad and rotor. This is analogous to the vehicle curving mechanics in wheel squeal.

The brake pad mechanics dictate the friction forces, which are governed by the friction curve. These forces excite the coupled modal dynamics of the brake system, leading to vibrations that can develop into a squeal limit cycle, generating sound pressure fluctuations in the air [64]. The reduced model in Figure 3.11 illustrates this behaviour. As shown in Figure 3.11a, the sliding velocity components at the off-centred [39,61] brake pad centre of pressure create a brake pad angle of attack, θ_A, between the local tangential and pad longitudinal components due to a nonzero pad lateral crabbing velocity, V_c. This local sliding – whether lateral or longitudinal – induces a friction force, F_f, which alongside the normal contact force, F_N, dynamically excite two dominant modes at a sprag angle, θ, relative to each other, as shown in Figure 3.11b.

Mode coupling can occur when θ is nonzero, with the sprag angle explicitly appearing in the model. Additionally, the model is generalisable to represent any two brake system modes in any direction, including tangential, normal and radial components.

According to Figure 3.11a, the quasistatic brake pad angle of attack, θ_A, is determined geometrically as

$$\sin(\theta_A) = V_c/V = x_{\text{CoP}}/r_{\text{CoP}}, \tag{3.21}$$

where x_{CoP} and r_{CoP}, define the position of the Centre of Pressure. Note that tangential or radial pad motion, which have both been associated with brake squeal [31], can be investigated using this model. The dynamic sliding ratio which includes the vibration component in the contact plane is defined as

$$\zeta = \left(V_{c,o} + \left(\delta_a \dot{y}_a(t)\cos(\theta)\right)\right)/V, \tag{3.22}$$

where δ_a is the sliding direction defined as +1 in Figure 3.11b. Note this direction determines whether there is spragging or not, like in wheel squeal. The contact mechanics couples two dominant modes at a sprag angle according to Figure 3.11b, modelled as

$$\begin{aligned} m_a\ddot{y}_a(t) + c_a\dot{y}_a(t) + k_a y_a(t) &= -F_N\sin(\theta) - F_f\cos(\theta), \\ m_b\ddot{y}_b(t) + c_b\dot{y}_b(t) + k_b y_b(t) &= F_N, \end{aligned} \tag{3.23}$$

$$\text{where, } F_N = k_H\left(y_a\sin(\theta) - y_b\right), F_f = \delta_a\ \mu(\zeta)(N + F_N), \tag{3.24}$$

where, y, m, k and c are the modal displacements, masses, stiffnesses and damping coefficients, respectively. The recently published research of the author showed that one of the key mechanisms governing the linear instability occurrence, under different braking pressure regimes, is the pressure-dependent contact stiffness, k_H, between the brake pad and disc due to contact area and multi-scale effects. This was fully investigated and validated with full brake system measurements in [65]. A smooth nonlinear friction law is used from [66] to represent both large and small vibrations as a function of the dynamic sliding ratio,

$$\mu(\zeta) = \mu_s\tanh(\alpha_\mu\zeta)\left[\mu_k + (1-\mu_k)\text{sech}(\beta_\mu\zeta)\right]. \tag{3.25}$$

where μ_s and μ_k represent the static and kinetic friction coefficients, respectively, and α_μ and β_μ determine the shape of the friction curve. The friction curves of (3.25) are shown in the results as Figure 3.12, for a fundamental and real brake system that has been tuned to measurements in [67].

The curves of Figure 3.12 show the experimental friction curve including falling friction can be well represented by the full nonlinear curve (3.25) and the analytical bilinear approximation with less than 2% and 5% error, respectively. The nominal higher static friction level of the real brake system model is also reasonably represented, with the falling friction notably having a higher slope according to (3.25).

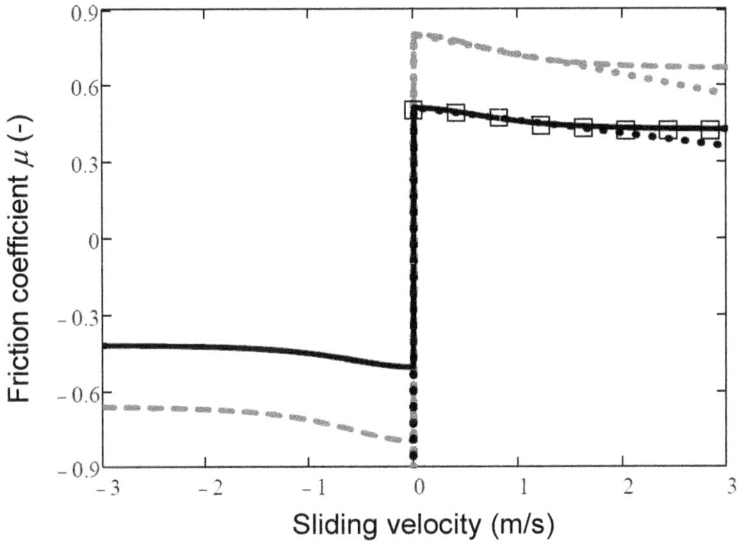

FIGURE 3.12 The friction curve for $\mu_s = 0.5$; simulated using equation (3.25); fundamental and real brake system models (-),(- -), analytical bilinear approximation at 1 m/s (..),(..) and experimentally measured (averaged) for Volvo 850/S70 X8 pad(□) [67], ch.4 [34].

The sound pressure level of brake squeal has already been predicted in [63,68] including comparison of predictions with experimental measurements. Also, [17] confirmed that the sound pressure level may be predicted assuming the sound radiation is proportional to the squared vibration velocity normal to the squealing surfaces. The equations (3.1)–(3.5) represent the full nonlinear equations of motion of the system that can be solved numerically. The model has already been validated to existing brake squeal theory and full brake squeal measurements in [63]. Analytical solutions for predicting the occurrence, growth and amplitude of squeal limit cycles are described subsequently here for convenience.

Prediction of Brake Squeal

The local stability and growth of brake squeal can be determined using a complex eigenvalue analysis. Note this is more complex than linear modal analysis ie [69] where the modes are not coupled through the friction. This is typically performed numerically due to mathematical difficulties; however, an analytical solution was recently obtained in [64] under the assumptions of small nonproportional damping. In such a way, the coupled equations (3.3) and (3.4) can be decoupled to the form

$$\mathbf{p_L}^{\mathrm{T}}\mathbf{M}\mathbf{p}\ddot{\mathbf{Y}} + \mathbf{p_L}^{\mathrm{T}}\mathbf{C}\mathbf{p}\dot{\mathbf{Y}} + \mathbf{p_L}^{\mathrm{T}}\mathbf{K}\mathbf{p}\mathbf{Y} = \mathbf{p_L}^{\mathrm{T}}\mathbf{F},$$
$$\boldsymbol{Y} = \mathbf{p}^{-1}\begin{bmatrix} y_a & y_b \end{bmatrix}^T, \quad \mathbf{F} = \begin{bmatrix} -\delta_a\,\mu(\zeta)N\cos(\theta) & 0 \end{bmatrix}^T, \tag{3.26}$$

where **Y** is the decoupled modal displacement vector and **M**, **K**, **C** and **F** are the, mass, stiffness, damping matrices and forcing vector of equations (3.23) and (3.24), respectively. The left and right undamped eigenvectors $\mathbf{p}_\mathrm{L}$ and **p,** are defined as

$$\mathbf{p}_\mathrm{L} = \begin{bmatrix} 1 & p_{L2} \\ p_{L1} & 1 \end{bmatrix} \quad \text{where} \quad \mathbf{p}_\mathrm{L}{}^\mathrm{T}\mathbf{K} = \lambda^2 \mathbf{p}_\mathrm{L}{}^\mathrm{T}\mathbf{M},$$

$$\mathbf{p} = \begin{bmatrix} 1 & p_2 \\ p_1 & 1 \end{bmatrix} \quad \text{where} \quad \mathbf{Kp} = \lambda^2 \mathbf{Mp}, \tag{3.27}$$

where p_i and p_{Li} are the eigenvector elements that determine the eigenvalues and therefore the system modal damping as a function of system parameters, as [64]

$$p_1 = \left[\omega_a^2 - \omega_b^2 + \sqrt{\Delta}\right] / \left[2f(\mu,\theta)\, k_H / (m_a \sin(\theta))\right],$$

$$p_2 = 1/\bar{p}_1, \quad p_{L1} = -m_a/\bar{p}_1 m_b, \quad p_{L2} = -m_b/\bar{p}_2 m_a$$

$$\text{therefore } p_L p_i = 4f(\mu,\theta)\left(k_H/m_j\right)^2 / \left[\omega_a^2 - \omega_b^2 - \sqrt{\Delta}\right]^2, \tag{3.28}$$

$$\text{where } f(\mu,\theta) = \sin^2(\theta)(1 + \delta_a \mu / \tan(\theta)),$$

$$\omega_a^2 = \left(k_a + k_H f(\mu,\theta)\right)/m_a, \quad \omega_b^2 = \left(k_b + k_H\right)/m_b,$$

$$= \left(\omega_a^2 - \omega_b^2\right)^2 + 4f(\mu,\theta)\, k_H^2 / (m_a m_b).$$

Here, i is the mode number, $j = b,a$, $^{-}$ is the conjugate, $f(\mu,\theta)$ represents Spurr's [25,39,40] criteria for spragging and Δ is the discriminant that determines whether the eigenmodes are real or complex and therefore the type of mode coupling (see [63] for more details). In particular, research [29,20,63] has more clearly identified and characterised two types of mode coupling. The first is stiffness mode coupling where the closeness of the modal frequencies causes instability coupling via the stiffness matrix. The second is viscous mode coupling where the modal damping causes instability coupling via the damping matrix. Stiffness mode coupling is the well-recognised form of mode coupling instability where the natural frequencies of the uncoupled system are close. Viscous mode coupling is less well recognised as it can occur when the uncoupled natural frequencies are not close. More details are provided in [63] and [64]. Mathematically, if Δ is positive, the eigenvalues are real and viscous mode coupling occurs (coupling determined by the damping matrix). Conversely, if Δ is negative, the eigenvalues are complex and stiffness mode coupling occurs (coupling determined by the stiffness matrix). The additional effect of falling friction on the stability is determined by the friction curve. For localised behaviour, the nonlinear friction law of (3.25) can be approximated as a curve with bilinear slopes, $k_{\mu 1}$ and $k_{\mu 2}$, as

$$\mu(\zeta) = \begin{cases} k_{\mu 1}\zeta' & \text{for } \zeta' \le 1 \\ k_{\mu 1} + k_{\mu 2}(\zeta' - 1) & \text{for } \zeta' > 1 \end{cases} \quad \text{where } \zeta \ge -\zeta_c \quad \text{and} \quad \zeta' = \zeta/\zeta_c, \tag{3.29}$$

where, ζ_c, is the critical sliding ratio at the peak friction that is approximated to within 5% error as [63]

$$\zeta_c \approx W_0\left(8(\alpha_\mu/\beta_\mu)^2/(1-\mu_k)\right)/(2\alpha_\mu) \text{ if } \beta_\mu/\alpha_\mu \le 0.25, \tag{3.30}$$

where $W_0()$ is the positive solution to the Lambert W function. The approximation (3.29) has been tuned to measured brake friction curves, as shown in Figure 3.12. The approximations (3.23), (3.26) and (3.29) can be combined to provide analytical equations of motion for the decoupled modes $i = 1,2$, as

$$\begin{aligned} &m_i\ddot{y}_i(t) + c_{\text{eff}}\dot{y}_i(t) + k_i y_i(t) = 0, \\ &\text{where} \quad c_{\text{eff}} = c_{\text{si}} + k_{\text{Fi}}k_{\mu 1,2}N/(\zeta_c V), \quad k_{F1} = 1, \quad k_{F2} = p_{L2}p_2, \end{aligned} \tag{3.31}$$

and c_{eff} is the effective system damping (including falling friction), $k_{\mu 1,2}$ is the local friction curve slope at sliding velocity $V_{o'c}$ and the coupled system modal parameters are

$$m_i = \mathbf{p_L}^{\mathrm{T}}\mathbf{M}\mathbf{p}_{i,i}, \quad k_i = \mathbf{p_L}^{\mathrm{T}}\mathbf{K}\mathbf{p}_{i,i}, \quad c_i = \mathbf{p_L}^{\mathrm{T}}\mathbf{C}\mathbf{p}_{i,i}, \tag{3.32}$$

and c_{si} represents the system modal damping of the coupled structure (without falling friction)

$$c_{\text{si}} = \left\{ \begin{array}{ll} c_i & \text{if } p_i, p_{Li} = \text{Re}(p_i), \text{Re}(p_{Li}) \\ \text{Re}\left(c_i/(2m_i) \pm \sqrt{c_i^2/(2m_i)^2 - k_i/m_i}\right)2m_i & \text{if } p_i, p_{Li} \neq \text{Re}(p_i), \text{Re}(p_{Li}) \end{array} \right\}. \tag{3.33}$$

Equation (3.33) highlights that the system modal damping, c_{si}, depends on the complexity of the eigenmodes and therefore the type of mode coupling (i.e., viscous or stiffness). Since the structural modal damping factors are usually well below the critical threshold, it may be deduced from (3.13) that stiffness mode coupling (second line) provides a much larger contribution to the system modal damping (and local eigenvalue magnitude) than viscous mode coupling (first line). The stability and hence occurrence of brake squeal can then be simply expressed according to when the effective system damping becomes negative and analytically determined using equations (3.31) and (3.33) as

Instability/squeal when: $\quad c_{eff} = \left(\underline{c_{si}} + \underline{k_{Fi}k_{\mu 1,2}N/(\zeta_c V)}\right) < 0 \quad$ (3.34)

Structural damping (-ve if Spragging and Mode Coupling) ↗ c_{si}

↖ Friction slope damping (-ve if Falling Friction)

The brake squeal criterion (3.14) concisely combines and superposes the effects of mode coupling and falling friction via effective system damping, c_{eff}. Squeal instability will initially grow exponentially when the effective system damping is negative, but this growth will change with amplitude. Hence, brake squeal can generally be categorised into two main mechanisms: squeal caused by falling friction and squeal

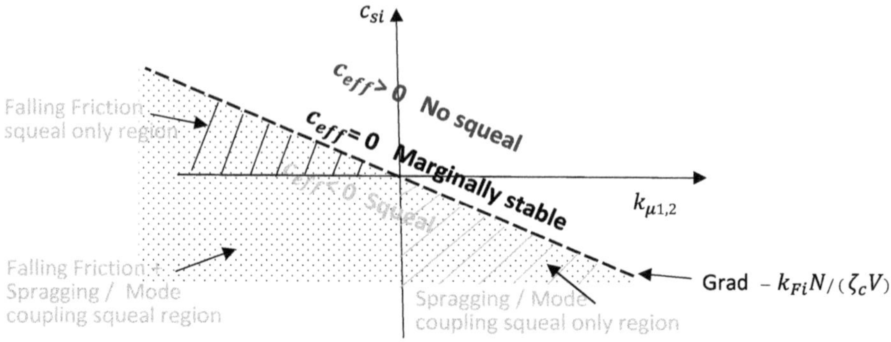

FIGURE 3.13 Plot of the squeal criterion (3.34) revealing regions of different types of squeal [63].

resulting from negative structural damping due to spragging and mode coupling. A graphical representation of how these mechanisms combine and dominate within the parameter space is illustrated in Figure 3.13.

Details of the fundamental mechanisms of Figure 3.13 are also described and clarified in the previous section for wheel squeal via direct analogy [64]. In addition, the minimum friction level required for stiffness mode coupled squeal to occur (in the absence of damping) can be determined analytically by solving the sprag conditions (3.20) along with $\Delta = 0$ to obtain

$$\mu_{\min} = \tan(2\theta_{\min})$$

$$\theta_{\min} = \frac{1}{2}\cos^{-1}\left(\left(2\left(\sqrt{(k_a/k_H)-(k_b/k_H)(m_a/m_b)} - \sqrt{(m_a/m_b)}\right)^2 + 1\right)^{-1}\right). \tag{3.35}$$

Equation (3.35) is a useful result that enables stiffness mode coupling to be predicted and avoided by knowing and controlling the uncoupled modal masses, undamped natural frequencies and the friction coefficient. Also, as for wheel squeal, when the sliding oscillation amplitude encroaches on the friction stick region below critical friction, positive damping will be added. At a particular oscillation amplitude, $A_a\omega_i$, where the power from the negative damping exactly balances the positive damping, a steady-state limit cycle will occur according to [63]. But in this brake squeal case, the static friction slope, $k_{\mu 1}$, approaches infinity as the maximum friction approaches the static friction, so the limit cycle amplitude simplifies to

$$A_a\omega_i = V_{c,o}/\cos(\theta) \quad \text{if} \quad \zeta_c \cong 0, \tag{3.36}$$

where t_c is the time in the squeal cycle that the sliding ratio reaches the critical peak of the friction curve. The closed-form solution of (3.36) assumes a linear sinusoidal limit cycle. This assumption may be broken under larger amplitude nonlinear phenomena such as chaotic motion [31,66,70]. Therefore, equations (3.28), (3.31)–(3.36) provide closed-form analytical solutions to brake squeal limit cycle occurrence, initial growth and amplitude under the instability mechanisms of falling friction, mode coupling and spragging.

Results of Brake Squeal Predictions

The mathematical modelling and stability analysis can be used to make instantaneous predictions of brake squeal instability behaviour that can be validated with numerical and experimental measurements. In doing so, it was important [47] to model the contact stiffness as depending upon brake pressure. This is due to the contact mechanics of rough elastic surfaces and non-elastic behaviour of the brake system components, [63,71], modelled as

$$k_H(p) = k_{\mathrm{Ho}} p^{pf}, \tag{3.37}$$

where k_{Ho} is a coefficient tuned to the contact stiffness at a nominal brake pressure, p represents the brake pressure and pf is an exponential coefficient associated with the contact roughness fractal distribution. In this case, $pf = 1/2$ is selected, representing a typical roughness distribution as per [72]. This implies that both the contact stiffness and the out-of-plane modal parameters, b, vary as functions of pressure. The stability of the equations of motion was analysed across a range of friction coefficients using the analytical solutions from equations (3.33) and (3.28). These results were then compared with numerical complex eigenvalue solutions and experimental measurements [47], as illustrated in Figure 3.14.

Figure 3.14 demonstrates that the proposed model and analytical solution closely match the numerical complex eigenvalue solution and align well with experimental squeal measurements from [47]. In particular, Figure 3.14a and b show that the two modes have distinct frequencies and exhibit equal damping up to a brake pressure of 4.8 psi. Stiffness mode coupling, where the discriminant Δ becomes negative

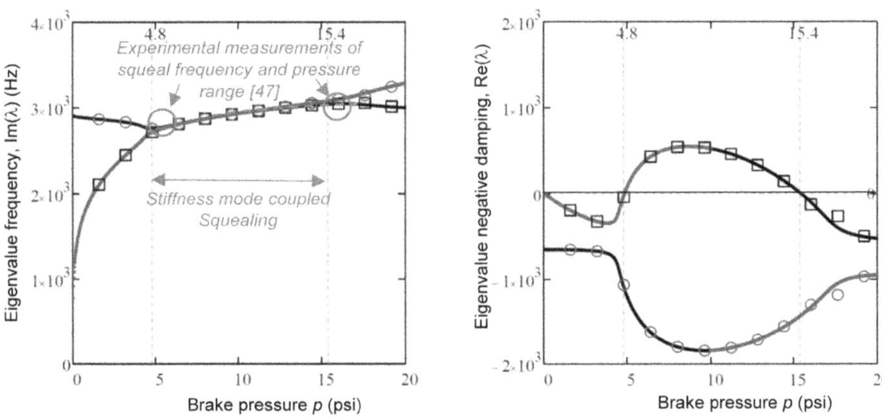

FIGURE 3.14 Complex eigenvalue, numerical (-)(-) and analytical (3.27), (3.28) (o) (□), solutions as a function of brake pressure, p, for brake squeal; (a) frequency Im(λ), (b) negative damping Re(λ). The vertical markings (--) show the two distinct frequencies converge when stiffness mode coupling squeal occurs in the brake pressure analytical range $4.8 < p < 15.4$ psi and experimentally measured (O) range $5.5 < p < 16$ psi. Constant friction at $\mu = 0.5$ according to the experimentally tuned model. Note the analytical solution is only plotted at discrete points for clarity [63].

(as per equation (3.19)), occurs just before this pressure level due to the presence of damping; without damping, squeal would occur exactly when Δ is first negative. At higher brake pressures, the modes converge at approximately 3 kHz with system negative damping, leading to squeal. However, beyond 15.4 psi, the pressure-dependent change in stiffness causes the modes to separate again, becoming distinct and positively damped. Experimental measurements of squeal frequency and pressure ranging from quarter-vehicle dynamometer and static assembly modal testing [47] are included in Figure 3.14a for comparison, showing that the analytical predictions fall within 5% error. In this analysis, friction was assumed to be constant. However, incorporating the falling friction curve from Figure 3.12 would introduce additional negative damping, lowering the friction level at which squeal first occurs. These critical constant friction levels were further examined using the analytical solutions from equations (3.33) and (3.28) across different braking pressures and sprag angles θ, as illustrated in Figure 3.15.

The stability maps in Figure 3.15 show the necessary conditions for brake squeal and why and when it occured. In particular, Figure 3.15a confirms that the experimental squeal observed in [47] results from stiffness mode coupling under the required spragging conditions. Squeal occurs when the mode-coupled structural damping becomes negative, as defined by equations (3.33) and (3.19). This corresponds to the vertical axis in the spragging/mode coupling squeal regions shown in Figure 3.13. Figure 3.15a also demonstrates that as friction levels decrease from 0.5, the brake pressure range at which squeal occurs is reduced and can be eliminated. The stability map in Figure 3.15b indicates that squeal could still occur at lower friction levels – for example, at a friction coefficient of 0.29 – if the sprag angle is reduced to 8°, as predicted by equation (3.20). Additionally, if damping is removed, the lowest friction level at which squeal can occur is predicted by equation (3.35) to be 0.147, at a sprag angle of 4.19°. These findings provide valuable insights into the sprag angles and friction levels necessary for the occurrence and prevention of brake squeal. For instance, by inspection of Figure 3.15a, if the pressure was increased to $p = 25$ psi and sprag mode damping, c_a, is dropped by 80%, viscous mode coupling squeal would

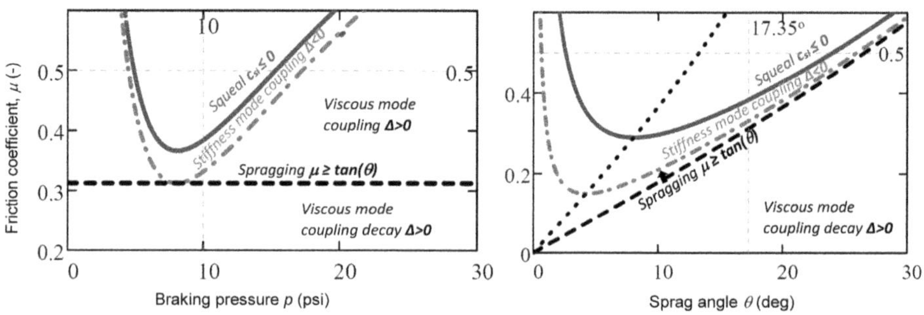

FIGURE 3.15 Analytical squeal stability maps of critical constant friction vs. (a) braking pressure p; (b) sprag angle θ. (-) boundary for squeal under constant friction, i.e., $c_{si} \le 0$ (.-.) boundary for stiffness mode coupling, i.e., $\Delta < 0$, (--) boundary for spragging, i.e., $\mu \ge \tan(\theta)$, (..) line of max mode coupling amplification, i.e., $\mu = \tan(2\theta)$. The markings (--) show the nominal parameter values based on the experimental measurements [47] and determined modal parameters of Figure 3.9. Note the numerical solutions match the analytical solutions shown [63].

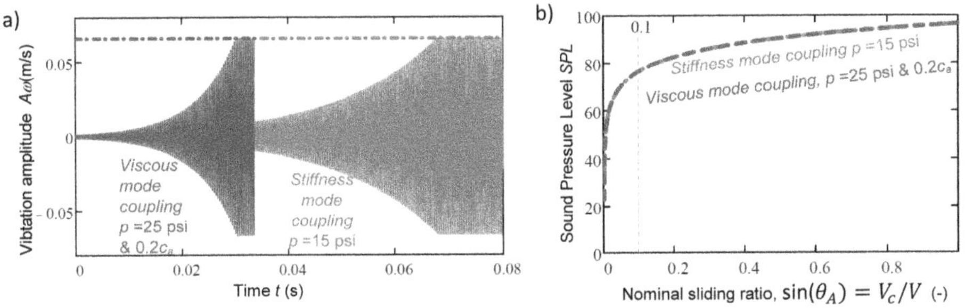

FIGURE 3.16 (a) Simulated time histories of squeal vibration growth under stiffness mode coupling, $p = 15$ psi (-) and viscous mode coupling, $p = 25$ psi & $0.2c_a$ (-) and corresponding analytical predictions of squeal amplitudes under stiffness mode coupling (.-.) and viscous mode coupling, (.-.) (b) Corresponding sound pressure level vs sliding ratio (or brake pad angle of attack); stiffness mode coupling, (.-.) and viscous mode coupling, (- -) and analytical and simulated results overlapped. Nominal speed of $V = 0.628$ m/s, friction coefficient $\mu_s = 0.5$ and $SPL_{const} = 100$ dB to match [47,63].

occur. This is illustrated and compared to nominal stiffness mode coupling conditions in Figure 3.16, including sound pressure levels for a range of sliding ratios (or brake pad angles of attack).

Figure 3.16a highlights that squeal can occur under both stiffness and viscous mode coupling conditions depending on brake pressure and mode damping. Note that falling friction squeal could also occur if the damping of both modes was reduced even further to 0.5% of their nominal values (although practically unlikely). The time history growth behaviour of the squeal under nominal conditions is similar to that measured in [47] with a similar time to reach maximum amplitude (although dependent upon initial conditions). The different growth rates under different mode coupling mechanisms are found to settle at the same analytically predicted limit cycle amplitude in accordance with (3.36) since the sprag angle is unchanged. This is because this squeal limit cycle amplitude is determined primarily by the nonlinearity of the friction curve and hence the sliding ratio of the brake pad angle of attack (as confirmed in Figure 3.16b). Note that in this case, the brake pad angle of attack is determined by the position of the pad centre of pressure so that it is predicted as follows: as the centre of pressure moves away from the centre line, the squeal amplitude increases.

In such a way, mathematical modelling and stability analysis has been used to develop an efficient validated model for brake squeal prediction. Much new insights have been gained from this analysis, which has been shown to be analogous to many aspects of the mechanism of wheel squeal. A similar approach for other instabilities in engineering is demonstrated subsequently. Finally, we recall that the mechanism of mode coupling in brake squeal was first termed “binary flutter” by North [34]. In the next section, we explain why.

3.3 AERODYNAMIC FLUTTER

The analysis, modelling and stability of aerodynamic flutter historically precede much of the previous research on other instabilities like squeal. This is most likely because the mode coupling mechanism was first identified as the main mechanism in flutter.

Theodorsen first provided a general theory for flutter of flat plate-modelled airfoils with closed-form solutions for unsteady aerodynamic terms [105] and experimental validation [106]. Pines further contributed by deriving a closed-form solution for the critical flutter speed, assuming steady flow and neglecting structural damping [73]. Numerous other theories have been developed to enhance the understanding of flutter, as extensively reviewed in [108]. In its simplest form, flutter typically arises from an unstable interaction between the pitching and plunging motions of a cross-section due to aerodynamic forces [105,108]. However, aerodynamic instability can also occur without such coupling, often resulting from negative stiffness or damping effects in a single-degree-of-freedom system [74,108]. Given the nonlinearities present in both aerodynamic and structural dynamics, extensive research continues, across many applications with ongoing studies improving numerical prediction accuracy and deepening the understanding of flutter's complex and variable nature. Hence, presently, we only highlight examples of some recent research pertaining to the fundamental prediction of its onset, limit cycle oscillation and nonlinear behaviour.

Since flutter is an instability, it initially grows exponentially until it is bounded by nonlinearity in the structure dynamics and/or aerodynamics and typically forms a limit cycle oscillation. Intermittent limit cycle oscillations may also be a precursor to flutter [78]. The typical aerodynamic nonlinear mechanisms that cause a flutter limit-cycle oscillation include shock waves, boundary layer separation and transitions. These typically require numerical computational fluid dynamics (CFD), for modelling, to avoid overly conservative design and envelope restrictions for aircraft [76] and structures. For instance, CFD is typically used to model stall flutter limit cycles due to a nonlinear aerodynamic reaction to the motion of the airfoil/structure [77,108]. Wind-induced nonlinear oscillations of twin-box girder bridges have been modelled using CFD with integrated finite-element analysis (FEA) and have shown significant impacts of shape on the limit cycle oscillations and failure modes [98]. Recently, computational fluid and structural dynamics coupling was used to consider both structural free-play and aerodynamic nonlinearity for three-dimensional aeroelastic analyses. The occurrence of limit cycles was shown to be below the linear flutter boundary and was determined from subsonic to transonic Mach numbers including the transonic boundary dip [79]. Much other recent research studies focused on nonlinearities in flutter have also been performed ie [80–95].

Quadratic and cubic pitching structural nonlinearities and accurately modelled unsteady aerodynamic loads were used to predict flutter speed, instability type and limit cycle dynamics in [91]. In a flexible cantilevered wing, limit cycle oscillations were observed after flutter onset due to geometric nonlinear coupling between rotation and bending modes [92]. In biological systems, cubic nonlinearity has been used to model limit cycle oscillations in dragonfly wings, where an anatomical feature called the stigma may help suppress these oscillations [93]. Similarly, structural cubic nonlinearity has been leveraged in piezo-aeroelastic energy harvesters to convert aerelastic vibrations from limit cycle oscillations into usable energy [94,95]. Recent studies have also identified flutter in an airfoil without requiring structural or thermal nonlinearity [89]. This phenomenon was linked to similarities between bilinear lift aerodynamics and friction-induced mode coupling observed in railway wheel squeal and brake squeal [96,97].

This section first introduces a binary mode mathematical model for flutter, incorporating a simplified nonlinear lift law. Closed-form solutions are derived for the onset and amplitude of limit cycle flutter vibrations. The model is then validated using classical flutter theory and Theodorsen's experimental results [105,106], before extending the analysis to viscous mode-coupled flutter and the effects of an angle-of-attack offset.

Model for Flutter Instability

In this section, a two-degree-of-freedom (binary) flutter model is introduced, incorporating mode-coupled dynamics and generalised aerodynamics. This model is inspired by analyses of mode-coupled railway wheel and brake squeal. It features a mass-decoupled framework with generalised flutter derivatives in the aerodynamic formulation. Flutter vibrations emerge from dynamic instability caused by the interaction between the structure's pitching and plunging motions and aerodynamic forces, as illustrated in Figure 3.17.

Referring to Figure 3.17, the aerodynamics cause lift and pitching moments, expressed as a summation of flutter derivative components, H_{1-4} and A_{1-4}, that are dependent on the structure's shape. These aerodynamic forces interact with the coupled modal dynamics of the structural system, which can be characterised by effective modal damping, c_{effi}, and natural frequency, ω_i, and an initial steady-state offset angle of attack, α_o. These interactions can lead to unstable structural vibrations that may evolve into a flutter-induced limit cycle or chaotic oscillations.

Figure 3.18, describes the degrees of freedom; pitching (with angle of attack), α, and plunging, h, of an airfoil. These are excited by a moment, M and lift, L, (defined in [99]) due to local aerodynamic flow speed, W. The airfoil has effective structural vertical and torsional stiffnesses, k_h and k_α, determined by the natural frequencies and sectional mass, m, and total chord length, $B=2b$. Its pitching and plunging motions occur with reference to the elastic support centre O, located at a distance, ba, from the midchord. They are coupled by, S_x, which depends on the displacement, S_x/m, between the centre of gravity CG and the support centre O. When $S_x=0$, there is no coupling (i.e., no mode-coupled flutter), meaning the system is mass-balanced.

The equations of motion of the airfoil described in Figure 3.18 may be determined as [74,108]

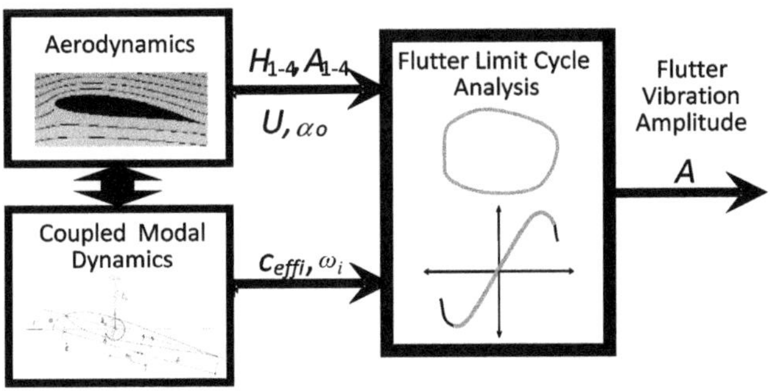

FIGURE 3.17 Conceptual model of flutter vibrations [89].

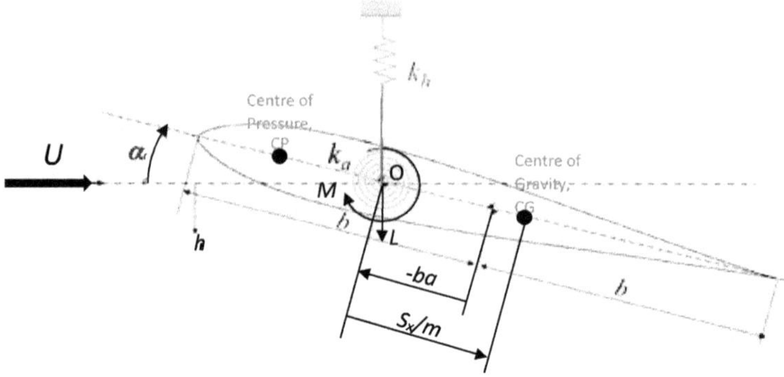

FIGURE 3.18 Binary (2 degree of freedom) coupled wind turbine blade section under local aerodynamic flow [90].

$$m\ddot{h}(t)+c_h\dot{h}(t)+S_x\ddot{\alpha}(t)+k_h h(t)=L,$$
$$J_\alpha\ddot{\alpha}(t)+c_\alpha\dot{\alpha}(t)+S_x\ddot{h}(t)+k_\alpha\left(\alpha(t)-\alpha_o\right)=M, \tag{3.38}$$

where, m, J, c *and* k, are the mass, mass moment of inertia, damping and stiffness constants referred to O, and the subscripts h and α are the degrees of freedom. These parameters can be approximated analytically like for a wind turbine section in [90], experimentally [106] and/or via finite-element and modal analysis. The modal plunging force and pitching moment are determined by the airfoil shape flutter derivatives [99] as

$$L=\frac{1}{2}\rho U^2 B\left(KH_1^*\frac{\dot{h}}{U}+KH_2^*\frac{B\dot{\alpha}}{U}+K^2\left(H_{3\alpha}^*(\alpha)-H_{3\alpha}^*(\alpha_o)\right)+K^2H_4^*\frac{h}{B}\right),$$
$$M=\frac{1}{2}\rho U^2 B^2\left(KA_1^*\frac{\dot{h}}{U}+KA_2^*\frac{B\dot{\alpha}}{U}+K^2A_3^*(\alpha-\alpha_o)+K^2A_4^*\frac{h}{B}\right), \tag{3.39}$$

where ρ is the air density and $K=B\omega/W$ is a reduced frequency, varying with, ω, the frequency of vibration. The appendices of [89,90] and [99] detail the flutter derivatives for a thin airfoil (flat plate flow). They may be numerically or experimentally obtained for any shape, e.g., [100–103]. The lift coefficient function, $H_{3\alpha}^*(\alpha)$, dominates the flutter limit cycle and can be expressed continuously as [89]

$$H_{3\alpha}^*(\alpha)=H_3^*\left(\begin{array}{l}\alpha\left(\begin{array}{l}-k_{\text{stall}}+\left(1+k_{\text{stall}}-\alpha_c\left(1+k_{\text{stall}}\right)/\mid\alpha\mid\right)\\ \left(1-\tanh\left(\beta\left(|\alpha|-\alpha_c\right)\right)\right)/2\end{array}\right)\\ +\operatorname{sgn}(\alpha)\alpha_c\left(1+k_{\text{stall}}\right)\end{array}\right). \tag{3.40}$$

In Equation (3.40), k_{stall}, represents the absolute stall slope factor, meaning that when the critical angle of attack, α_c, is exceeded, the lift coefficient, H_3^*, transitions to a

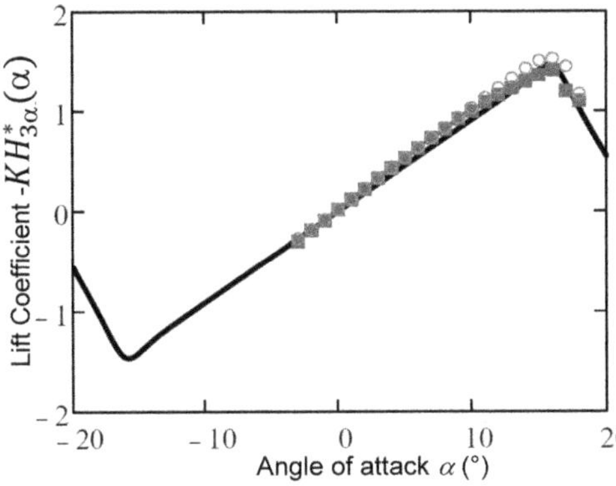

FIGURE 3.19 The lift coefficient curve for NACA 0012 symmetric airfoil; simulated using equation (3.40) (-) with smoothing factor $\beta = 50$, and experimentally measured in wind tunnel for $Re = 1.3 \times 10^6$ (■) and $Re = 2.4 \times 10^6$ (O) [89,104].

negatively sloped value of $-k_{\text{stall}} H_3^*$. Additionally, a smoothing factor β is used to adjust between pre- and post-stall conditions. Figure 3.19 illustrates the continuous bilinear lift coefficient curve derived from equation (3.40), based on experimental data. Specifically, this nonlinear lift model was calibrated using wind tunnel measurements of a NACA 0012 symmetric airfoil [104] ($\alpha_c = 16°$ and $k_{\text{stall}} = 2.5$).

The trends in Figure 3.19 depict an initial increase in the lift coefficient, corresponding to a region of negative stiffness, followed by a decline with positive stiffness beyond the critical angle of attack. The nonlinear model closely aligns with wind tunnel measurements, with deviations within 5% under flutter flow conditions across the two measured Reynolds numbers [104].

To verify flutter predictions, the full nonlinear equations of motion for an airfoil (3.38)–(3.40) can be solved numerically through integration. However, before numerical methods are applied, stability analysis provides an efficient way to analytically determine flutter onset.

Prediction of Flutter Instability

Determining the critical flutter speed, which marks the onset of flutter instability, typically requires numerical methods such as complex eigenvalue analysis or other stability techniques applied to the coupled equations of motion (3.38)–(3.40) i.e. [105,106]. However, [89] introduced an approximate analytical solution that incorporates damping, assuming small nonproportional damping [20,89]. The first step in this approach is to express the equations of motion (3.38)–(3.40) in a mass-decoupled form

$$\mathbf{M}\ddot{\mathbf{X}} + \mathbf{C}\dot{\mathbf{X}} + \mathbf{K}\mathbf{X} = 0, \quad X = \left[x_h = h \;\; x_\alpha = \alpha S_x / m\right]^T, \tag{3.41}$$

where x_α, is the displacement at CG due to the angle of attack, α, as part of the degrees-of-freedom vector, $\mathbf{X}$. The mass, damping and stiffness matrices $\mathbf{M}$, $\mathbf{C}$ and $\mathbf{K}$ of the equations of motion, respectively, may be determined as

$$\mathbf{M} = \begin{bmatrix} m_h & 0 \\ 0 & m_\alpha \end{bmatrix} = \begin{bmatrix} m + m_{ae} - S_x^2/J_\alpha & 0 \\ 0 & (J_\alpha + J_{ae})(m/S_x)^2 - m \end{bmatrix},$$

$$\mathbf{C} = \begin{bmatrix} c_{11} & c_{12} \\ c_{21} & c_{22} \end{bmatrix}$$

$$= \begin{bmatrix} c_h - \frac{1}{2}\rho UKB\left(H_1^* - A_1^* BS_x/J_\alpha\right) & -c_\alpha m/J_\alpha - \frac{1}{2}\rho UKB^2\left(H_2^* - A_2^* BS_x/J_\alpha\right)m/S_x \\ -c_h + \frac{1}{2}\rho UKB\left(H_1^* - A_1^* Bm/S_x\right) & c_\alpha (m/S_x)^2 + \frac{1}{2}\rho UKB^2\left(H_2^* - A_2^* Bm/S_x\right)m/S_x \end{bmatrix},$$

$$\mathbf{K} = \begin{bmatrix} k_{11} & k_{12} \\ k_{21} & k_{22} \end{bmatrix}$$

$$= \begin{bmatrix} k_h - \frac{1}{2}\rho U^2K^2\left(H_4^* - A_4^* BS_x/J_\alpha\right) & -k_\alpha m/J_\alpha - \frac{1}{2}\rho U^2K^2B\left(H_3^* - A_3^* BS_x/J_\alpha\right)m/S_x \\ -k_h + \frac{1}{2}\rho U^2K^2\left(H_4^* - A_4^* Bm/S_x\right) & k_\alpha (m/S_x)^2 + \frac{1}{2}\rho U^2K^2B\left(H_3^* - A_3^* Bm/S_x\right)m/S_x \end{bmatrix},$$

(3.42)

Here, m_h and m_α are decoupled masses at the centre of mass and can include aerodynamic inertia and mass terms J_{ae} and m_{ae}. The fluid structure coupling effects may be identified directly in the components of the mass decoupled mass, damping and stiffness matrices **M**, **C** and **K,** respectively. Flutter onset may be inferred more easily, due to the mass decoupling, by inspection of the sign of the components in matrices, **C** and **K** in (3.42). In particular, the mode coupling flutter mechanism occurs when the aerodynamic lift overcomes the reaction force due to the coupled structural torsional stiffness, causing k_{12} to change sign compared to k_{21}. This mechanism driving binary flutter was identified as dynamic divergence in [89] but also requires the effective system modal damping to be negative, as determined by fully decoupling the equations of motion (3.36) and (3.42) into modal form as

$$\mathbf{p_L}^{\mathrm{T}}\mathbf{Mp}\ddot{\mathbf{Y}} + \mathbf{p_L}^{\mathrm{T}}\mathbf{Cp}\dot{\mathbf{Y}} + \mathbf{p_L}^{\mathrm{T}}\mathbf{KpY} = 0, \quad Y = \mathbf{p}^{-1}\begin{bmatrix} x_h & x_\alpha \end{bmatrix}^T, \tag{3.43}$$

where, x_h and x_α, define the modal displacement vector, **Y**, based on the undamped eigenvectors $\mathbf{p_L}$ (left) and **p** (right)

$$\mathbf{p_L} = \begin{bmatrix} 1 & p_{L2} \\ p_{L1} & 1 \end{bmatrix} \quad \text{where} \quad \mathbf{p_L}^{\mathrm{T}}\mathbf{K} = \lambda^2 \mathbf{p_L}^{\mathrm{T}}\mathbf{M},$$

$$\mathbf{p} = \begin{bmatrix} 1 & p_2 \\ p_1 & 1 \end{bmatrix} \quad \text{where} \quad \mathbf{Kp} = \lambda^2 \mathbf{Mp}. \tag{3.44}$$

A closed-form solution for these eigenvectors has been obtained assuming nonproportional damping is small, as [96]

$$p_1 = \left[\omega_h^2 - \omega_\alpha^2 + \sqrt{\Delta}\right] / \left[-2k_{12}/m_h\right],\, p_2 = 1/\overline{p}_1,\, p_{L1} = -m_h/\overline{p}_1 m_\alpha,\, p_{L2} = -m_\alpha/\overline{p}_2 m_h$$

$$\text{therefore} \quad p_L p_i = 4\left(k_{12}k_{21}/m_j^{\,2}\right) / \left[\omega_h^2 - \omega_\alpha^2 - \sqrt{\Delta}\right]^2,$$

$$\text{where } \omega_h^2 = k_{11}/m_h,\ \omega_\alpha^2 = k_{22}/m_\alpha,\ \Delta = \left(\omega_h^2 - \omega_\alpha^2\right)^2 + 4k_{12}k_{21}/(m_h m_\alpha), \tag{3.45}$$

where $\bar{}$ is the conjugate, i is the mode number and $j = \alpha, h$. The type of mode coupling is determined by the discriminant Δ, which is a measure of how close the uncoupled modes, ω_h and ω_α are. In particular, if it is positive, then real eigenmodes occur associated with viscous mode coupling. Alternatively, if it is negative, complex eigenmodes occur associated with stiffness mode coupling, see [20,89] and Section 3.1.2. Equation (3.43) allows closed-form solutions for the effective system damping for each mode, c_{effi}, expressed as

$$m_i \ddot{y}_i(t) + c_{\text{effi}} \dot{y}_i(t) + k_i y_i(t) = 0, \quad \text{where} \quad c_{\text{effi}}\left(H_3^* \mid_{\alpha_o, > \alpha_c}^{\alpha}\right), \tag{3.46}$$

where c_{effi} can be evaluated under pre stall or post stall conditions according to the bilinear flutter derivative H_3^* at $\alpha = \alpha_o$ or at $\alpha > \alpha_c$. The modal parameters can be calculated in closed form

$$m_i = \mathbf{p_L}^{\mathrm{T}} \mathbf{M} \mathbf{p}_{i,i}, \quad k_i = \mathbf{p_L}^{\mathrm{T}} \mathbf{K} \mathbf{p}_{i,i}, \quad c_i = \mathbf{p_L}^{\mathrm{T}} \mathbf{C} \mathbf{p}_{i,i}. \tag{3.47}$$

The effective system damping is dependent on the complexity of the stability solution as

$$c_{\text{effi}} = \left\{ \begin{array}{ll} c_i & \text{if } p_i, p_{Li} = \mathrm{Re}(p_i), \mathrm{Re}(p_{Li}) \\ \mathrm{Re}\left(\begin{array}{c} c_i/(2m_i) \\ \pm\sqrt{c_i^{\,2}/(2m_i)^2 - k_i/m_i} \end{array}\right) 2m_i & \text{if } p_i, p_{Li} \neq \mathrm{Re}(p_i), \mathrm{Re}(p_{Li}) \end{array} \right\}. \tag{3.48}$$

Equations (3.41)–(3.48) are closed-form solutions for localised flutter vibrations and its type (viscous or stiffness) for an airfoil section. Note flutter will initially grow exponentially until the angle of attack oscillations encroach on the negative stall slope (stabilising effective positive stiffness) above α_c. This periodic limit cycle amplitude can be approximated using an energy balance over one flutter oscillation, as described in [89]. In particular, the limit cycle energy during stall is balanced during non-stall when

$$\begin{gathered} A_\alpha = (\alpha_c - \alpha_o)/\sin(\omega_i t_{c1}) \ \text{ where } \ \sin(2\omega_i t_{c2})/\sin(2\omega_i t_{c1}) = (\alpha_c + \alpha_o)/(\alpha_c - \alpha_o) \\ \text{and } \ \sin(2\omega_i t_{c1}) + \sin(2\omega_i t_{c2}) + 2\omega_i(t_{c1} + t_{c2}) \\ = 2\pi\left(c_{\text{effi}}\left(H_3^* \mid_{>\alpha_c}^{\alpha}\right) / (c_{\text{effi}}\left(H_3^* \mid_{\alpha_o}^{\alpha}\right) - c_{\text{effi}}\left(H_3^* \mid_{>\alpha_c}^{\alpha}\right)\right). \end{gathered} \tag{3.49}$$

Here, $t_{c1,2}$ represent the critical moments within the flutter sinusoidal cycle when the angle of attack reaches its critical values, $\pm\alpha_c$, starting from the nominal steady-state angle of attack, α_o.

Equations (3.41)–(3.49) provide analytical predictions for the flutter onset, initial growth and limit cycle amplitude for a general airfoil structure under unsteady aerodynamics. From them, the following summary of the conditions for flutter onset are provided [89] based on a supercritical Hopf bifurcation to flutter, i.e., transition to a stable limit cycle from a stable equilibrium point. In particular, the onset of flutter is governed by the effective system modal damping c_{effi} and stiffness matrix according to the necessary conditions [89]:

$$\text{Flutter if } c_{\text{effi}} < 0 \quad \text{when} \quad k_{12}k_{21} \leq 0, \tag{3.50}$$

where the second criterion of (3.50) represents dynamic divergence conditions more explicitly as

$$\begin{aligned} k_{12}k_{21} \leq 0 \Rightarrow k_\alpha + \frac{1}{2}\rho U^2 K^2 B\left(H_3^* J_\alpha / S_x - A_3^* B\right) \leq 0 \\ \text{x or} \quad k_h - \frac{1}{2}\rho U^2 K^2\left(H_4^* - A_4^* Bm / S_x\right) \leq 0. \end{aligned} \tag{3.51}$$

The first criterion of k_{12} in (3.51) typically causes dynamic divergence for an airfoil because the aerodynamic component increases with the speed squared and is relatively large and negative. Hence, aerodynamic energy input, via this mechanism, drives flutter (if it causes negative system damping, i.e., $c_{\text{effi}} < 0$). The second criterion in equation (3.51) applies similarly to vertical stiffness and moment. These necessary conditions are comparable to spragging [107], a phenomenon in friction instabilities where the normal force on a sliding structure amplifies the tip load instead of counteracting it. As discussed in Sections 3.1 and 3.2, this mechanism is essential for mode-coupled brake squeal [63] and wheel squeal [96]. The complexity of the eigenvectors (real or complex) [96] determines the mode coupling type, as described subsequently.

Viscous Mode-Coupled Flutter

When the discriminant is positive, $\Delta > 0$, the eigenvector solutions are real, and therefore the effective system modal damping reduces to $c_{\text{effi}} = c_{\text{i}}$ (first line of (3.48)). This is known as viscous mode coupling and will cause flutter amplification (or decay) when the modal damping ratio, c_i, is less (or more) than the diagonal damping components, c_{ii} which is dependent upon the dynamic divergence criterion, as

$$\begin{array}{lll} \text{Viscous mode coupling amplification} & c_i < c_{ii} \text{ when } k_{12}k_{21} \leq 0, & \Delta \geq 0 \\ \text{Viscous mode coupling decay} & c_i \geq c_{ii} \text{ when } k_{12}k_{21} > 0, & \Delta \geq 0^{\cdot} \end{array} \tag{3.52}$$

Hence, (3.52) predicts viscous-mode flutter amplification due to energy input via dynamic divergence conditions and sufficiently separated uncoupled modal frequencies, ω_h and ω_α, such that $\Delta \geq 0$. Under viscous mode coupling, the phasing of the aerodynamic forces compared to velocity causes system power input due to dynamic divergence, meaning increasing damping may surprisingly cause flutter under some conditions. In previous research [89], it was found that viscous mode coupling can occur at lower speeds than classical stiffness mode-coupled flutter, described subsequently.

Stiffness Mode-Coupled Flutter

When the discriminant is negative, $\Delta < 0$, under dynamic divergent conditions, then stiffness mode coupling occurs, causing complex undamped eigenmodes and complex stiffness, leading to negative system damping. This is the classical binary flutter mechanism [105,108] that occurs due to mode coupling when the coupled modal frequencies become close, according to

$$\text{Stiffness mode coupling amplification when} \quad k_{12}k_{21} \leq 0 \quad \Delta < 0. \tag{3.53}$$

Hence, (3.53) predicts stiffness mode flutter amplification due to energy input via the combination of dynamic divergence conditions and close uncoupled modal frequencies, ω_h and ω_α, such that $\Delta < 0$. In this case, the system negative damping is determined primarily by the modal stiffness and will likely cause larger phase space expansion than viscous mode coupling. This assumes the modes are not over critically damped. Also, flutter growth tends to increase with speed as there is an W^2 effect. It is possible to solve for the critical speed for the onset of classical (stiffness mode coupled) flutter onset, assuming no damping and dominant angle of attack flutter derivatives as [89]

$$U_f = \frac{m_\alpha}{\frac{1}{2}\rho K^2 B\left(-H_3^* + A_3^* Bm/S_x\right)m/S_x}\left[\begin{array}{l}\left(\sqrt{\dfrac{k_h(1-\Delta p)}{m_h}} \mp \sqrt{\dfrac{k_\alpha (m/S_x)^2}{m_\alpha}}\right)^2 \\ \mp 2\sqrt{\dfrac{k_h k_\alpha (m/S_x)^2}{m_h m_\alpha}}\left(\sqrt{1-\dfrac{S_x^2}{mJ_\alpha}-\Delta U} - \sqrt{1-\Delta p}\right)\end{array}\right]$$

$$\text{where } \Delta p = \frac{1-S_x^2/(mJ_\alpha)}{1-\left(H_3^*/A_3^*\right)\left(S_x/Bm\right)}, \quad \Delta U = \frac{\Delta p}{2}\left[1+\left(1-\frac{\Delta p}{2}\right)\frac{k_h m_\alpha}{k_\alpha (m/S_x)^2 m_h}\right], \tag{3.54}$$

and Δp, ΔU are normalised factors, i.e., $0 \leq \Delta \leq 1$ assuming the elastic centre and centre of pressure are aft of the centre of gravity and elastic centre, respectively. The closed-form solution (3.54) predicts that the classical critical flutter speed increases with increases in rotational inertia, the spacing between uncoupled plunging (bending) and pitching (torsional) natural frequencies and finally decreases in air density. Added damping will tend to increase the critical flutter speed as well. Finally, it is noted that stiffness mode coupling instability realistically has a higher instability growth rate than viscous mode coupling. This is in accordance with the system damping predictions of (3.46) and (3.48) as the damping ratio is below unity for many vibrating structures [89]. This is also important when considering criteria for chaos following flutter that requires high local phase space expansion.

Results of Flutter Predictions

To validate analytical solutions for flutter, full numerical solutions and the results from Theodorsen [105] are examined. The parameters from [90] were used to assess the stability of the equations of motion across various flow velocities, employing the closed-form solutions from equations (3.48) and (3.45). These results were then compared with numerical complex eigenvalue solutions obtained using the full system model matrices from equation (3.42), as illustrated in Figure 3.20.

Figure 3.20 demonstrates close alignment between the analytical closed-form stability with the numerical complex eigenvalue flutter frequency and negative damping solutions. In Figure 3.20a, the modal frequencies approach each other as flow speed increases but do not fully converge due to nonproportional damping, an effect further explored in [96]. Figure 3.20a and b illustrate two distinct modal frequencies that remain widely spaced yet exhibit nearly equal damping as the flow speed increases towards the critical flutter speed, where the discriminant, Δ, reaches 0. Beyond this point, $\Delta < 0$, stiffness mode coupling occurs, causing the modal frequencies to converge with opposite damping characteristics influenced by both structural and aerodynamic effects. The numerically computed critical flutter speed (92.4 m/s), identified when the real part of the eigenvalue first reaches 0, closely matches the analytical prediction from equation (3.54) and [73]. However, this alignment is

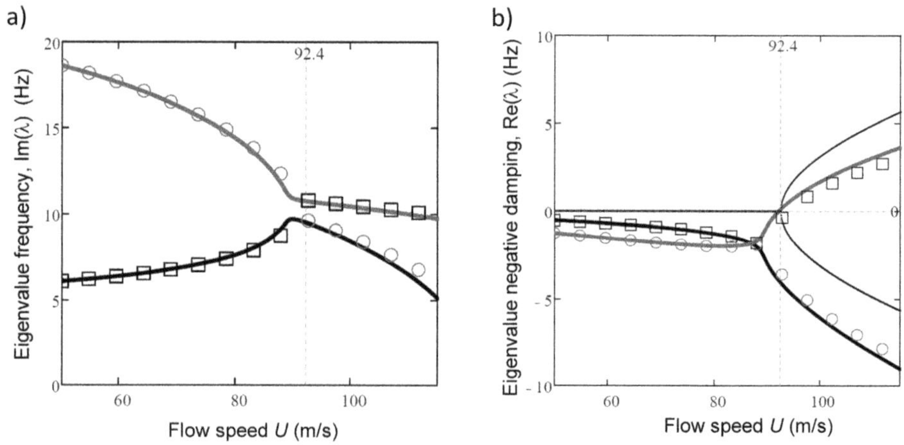

FIGURE 3.20 Complex eigenvalue solutions, numerical (-)(-) and analytical (3.28)–(3.13) (o) (□), as a function of flow speed, U; (a) frequency $\text{Im}(\lambda)$ and (b) negative damping $\text{Re}(\lambda)$ in Hz. The thin black lines (-) show the numerical solution under undamped and angle of attack flutter derivatives only, and the vertical marking (- -) shows the critical flutter speed (92.4 m/s) using (3.54) and [73] under the same conditions. Note the analytical solution closely matches the numerical solution at all points, so only discrete points are plotted for clarity. Assumed constant nominal vibration frequency as flow speed changes [89].

coincidental for the specific parameter values used. A comparison of the thin and thick solid lines reveals that structural damping raises the critical flutter speed by reducing the real part of the eigenvalues, while other flutter derivatives counterbalance this by shifting the bifurcation point to higher flow speeds. These findings emphasise the significant influence of both structural and aerodynamic damping beyond the critical flutter speed. For simplicity, the nominal vibration frequency is assumed to have minimal impact on the results and is held constant at critical flutter speed. This assumption is also applied to the analytical prediction curve for the critical flutter speed in Figure 3.21. The figure compares predicted critical flutter speeds with experimental wind tunnel data from Theodorsen [106], covering three different wing configurations (NACA15840, 15931 and 15842) with varying lengths and tapers (detailed further in [106]).

Figure 3.21 demonstrates a strong correlation between experimental results and both analytical and numerical complex eigenvalue analyses, with errors remaining below 10%, particularly for nominal and higher structural frequency ratio conditions. The simplified analytical prediction for critical flutter speed (equation 3.54), which assumes no damping and a constant nominal vibration frequency, provides an efficient estimate with an error margin of within 7%. In this case, the overall trend is well captured as it is primarily governed by stiffness mode coupling. To further analyse the occurrence of viscous mode coupling flutter, the stability was examined across different flow speeds and structural frequency ratios, as illustrated in Figure 3.22.

Figure 3.22 illustrates the conditions for flutter onset, where the lower flat curve indicates that flutter can only occur when the necessary dynamic divergence condition (as predicted by equation (3.51)) is met. Within this dynamic divergence region, flutter is driven either by viscous-mode coupling at lower speeds or by stiffness

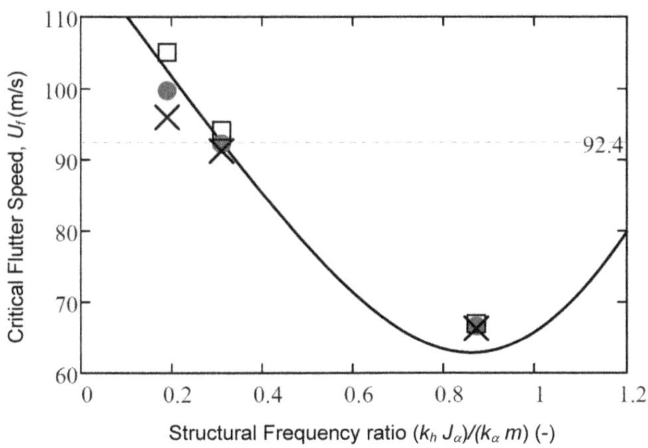

FIGURE 3.21 Critical flutter speed at different structural frequency ratios, experimental wind tunnel tests [106] (x), numerical, direct complex eigenvalue analysis of (3.42) (•), analytical (3.43)–(3.45), (□), simplified under undamped and angle of attack flutter derivatives only using constant nominal vibration frequency 10.867 Hz (3.54) and [73] (-). The horizontal marking (- -) shows the critical flutter speed (92.4 m/s) using (3.54) and [73] under nominal conditions [89].

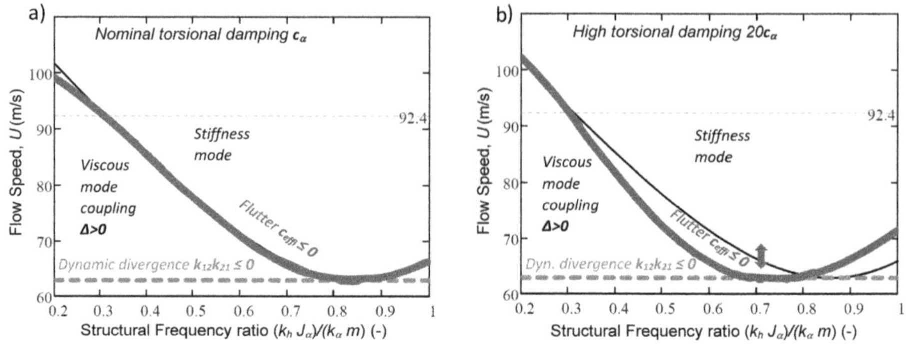

FIGURE 3.22 Flutter stability maps of flow speed vs (a) structural frequency ratio (via k_h variation) (a) nominal conditions; (b) high structural torsional damping $20c_\alpha$, boundary for flutter, i.e., $c_{\text{eff}} \le 0$ (-) boundary for stiffness mode coupling, i.e., $\Delta < 0$, (-) boundary for dynamic divergence, i.e., $k_{12}k_{21} \le 0$, (- -). The markings (--) show the nominal parameter values and where stiffness mode coupling and flutter first occur for the nominal structural frequency ratio. In (b), the double arrows ↕ indicate flutter occurs in regions above (stiffness mode coupled flutter) and below (viscous mode coupled flutter) the stiffness mode coupling boundary (-) that is above the flutter boundary (-) [89].

mode coupling, depending on the sign of the discriminant, Δ (as defined in equations (3.52) and (3.53)). Under nominal conditions (1% critical damping), the transition to flutter closely aligns with the stiffness mode coupling criterion for most structural frequency ratios; however, when torsional damping is significantly increased (20% critical damping), a distinct range of viscous mode coupling occurs (for structural stiffness ratios between approximately 0.32–0.8) at lower flow speeds. This behaviour is not captured by the simplified analytical prediction of the critical flutter speed (3.54) or [73], which assumed flutter occurs only due to stiffness mode coupling. As a result, these simplified models significantly underpredict the critical flutter speed in high-damping cases, particularly for structural stiffness ratios above 0.8. Notably, the dynamic divergence criterion provides a useful lower bound for the critical flutter speed without requiring knowledge of system damping. These findings highlight that, under viscous mode coupling conditions, increased damping can paradoxically destabilise the system, leading to flutter at lower flow speeds.

Figure 3.23 further explores flutter growth into limit cycle oscillations under both stiffness and viscous mode coupling mechanisms across different flow velocities. The time histories in Figure 3.23a confirm that flutter can develop through different mechanisms depending on the structural frequency ratio and damping. As predicted by the stability maps in Figure 3.22, stiffness mode coupling occurs at a flow speed of $U = 95$ m/s under nominal conditions and exhibits the fastest growth rate (with an unstable eigenvalue real part of $\text{Re}(\lambda) = 0.744$ rad/s, consistent with equations (3.41)–(3.48). Again, viscous mode coupling is observed at significantly lower flow speed ($U = 68$ m/s), though still above the dynamic divergence speed, with a significantly slower growth rate ($\text{Re}(\lambda) = 0.029$ rad/s), as predicted in Figure 3.22. Therefore, increased torsional damping, c_α, to 20 times its nominal value and vertical stiffness by a factor of 3.746 led to an unexpected decrease in flutter speed. This counterintuitive effect, where increased damping lowers the flutter speed, is a well-documented characteristic of viscous mode coupling, consistent with prior findings in [96] and [97].

The amplitude of the flutter limit cycle is primarily determined by the critical stall angle of attack, as shown in Figure 3.23a. Once this angle is exceeded, high positive effective stiffness occurs (refer to Figure 3.19), limiting further amplitude growth. Figure 3.23b illustrates that as flow speed increases, the amplitude must also rise to balance the increasing aerodynamic energy input before reaching stall. The rate at which the limit cycle amplitude grows with flow speed depends on the sharpness of the stall slope, k_{stall}. In this case, a relatively high stall slope allows the amplitude to increase gradually to maintain the energy balance. Viscous mode coupling changes the trend of limit cycle amplitude with flow speed but ceases just after a nondimensional flow speed of 1.1, where a transition to stiffness mode coupling occurs.

Figure 3.23 also highlights the impact of an initial steady offset angle of attack, α_o, which is often required to maintain steady lift. In the case of viscous mode coupling, a nonzero, α_o, causes positive torsional flutter oscillations to reach stall before negative oscillations. While this offset does not alter the local dynamic stability or critical flutter speed, it affects the energy balance symmetry over the limit cycle, thereby influencing the steady-state amplitude. A comparison of two viscous mode coupling cases in Figure 3.23b shows that introducing a nonzero steady offset reduces the limit cycle amplitude due to this effect. Additionally, for stiffness mode coupling, a nonzero, α_o, introduces nonsinusoidal fluctuations at lower flow speeds compared to the zero-offset case. The analytical predictions for stiffness mode coupling in Figure 3.23b are within 5% error of full numerical simulations over the given flow speed range. However, errors for viscous mode coupling are higher (within 10%),

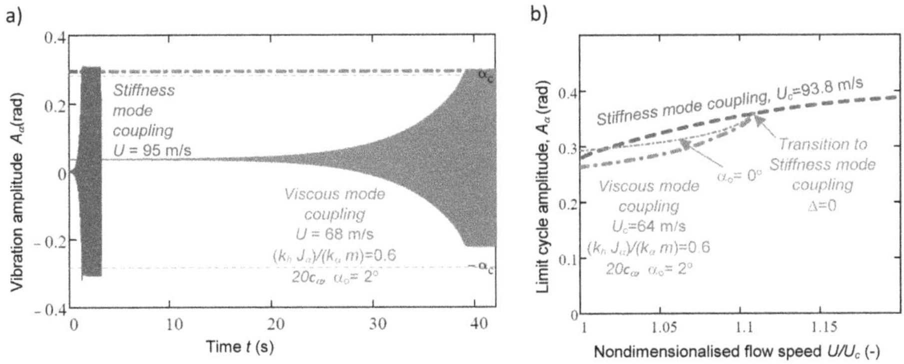

FIGURE 3.23 (a) Simulated time histories of flutter vibration growth under stiffness mode coupling, nominal conditions, $U=95$ m/s, $\alpha_o=0°$ (-) and viscous mode coupling, $U=68$ m/s, $(k_h J_\alpha)/(k_\alpha m)=0.6$, high structural torsional damping $20c_\alpha$, $\alpha_o=2°$ (-) and corresponding analytical predictions of flutter amplitudes under stiffness mode coupling, (.-.)and viscous mode coupling, (.-.) (b) Corresponding limit cycle amplitude vs sliding speed; stiffness mode coupling (- -) and viscous mode coupling, $\alpha_o=2°$ (.-.), $\alpha_o=0°$ (.-.) analytical and simulated results were within 10% error. Vibration frequency (for K) held constant at its flutter speed values for the stiffness and viscous mode coupling cases of (a) [89].

likely due to significant nonproportional damping, which contradicts the assumptions used in the analytical complex eigenvalue solution (equations 3.41–3.49). Despite this, the analytical approximation for flutter amplitude (equation 3.49) provides a highly efficient, near-instantaneous estimate across a range of flow speeds and other parameters. Finally, the range of flow speeds in Figure 3.23b was deliberately limited to approximate sinusoidal limit cycle motion based on the conditions of equation 3.49. In Chapter 4, these assumptions are relaxed to examine more complex nonlinear flutter behaviour.

3.4 ROLLING MILL CHATTER

A typical rolling mill is made up of a series of stands of approximately cylindrical rolls which reduce the thickness of the flat strip.

Rolling mill chatter refers to mechanical vibrations of components, leading to periodic transverse thickness variations or alternating light and dark bands across the strip. Rolling mills typically experience three main types of chatter, as reviewed by Roberts [109]: torsional chatter, third octave chatter and fifth octave chatter. Each type is classified based on the frequency range or musical octave corresponding to the natural frequency of the underlying modal dynamic mechanism. Among these, third octave mode chatter is the most dangerous, characterised by rapidly growing oscillations from an unstable cycle between fluctuations in tension, roll force and exit thickness, see Figure 3.24.

Currently, the most effective way to mitigate chatter in rolling mills is by reducing the operating speed, though this negatively impacts productivity. Figure 3.24 highlights that third octave chatter occurs when vertical oscillations of the roll stack are excited, due to strip deformation interactions that involve interstand dynamics. Practical observations by Paton and Critchley [110] indicate that during chatter, tension oscillations in the strip become significant and correlate with vertical roll stack motion caused by roll force variations. Several researchers, including [111–113] and [114], acknowledge that variations in roll bite geometry can introduce additional damping, when the chatter vibration period is close to the roll gap transit time. However, this is not the case in typical third octave chatter scenarios for thin strip rolling where chatter frequencies are around 150 Hz and mill speeds exceed 800 m/min. This suggests that under these conditions, the roll bite can be considered quasistatic.

In this section, we detail a mathematical model of third octave chatter instability used to derive a closed-form prediction of its occurence. This validated criterion

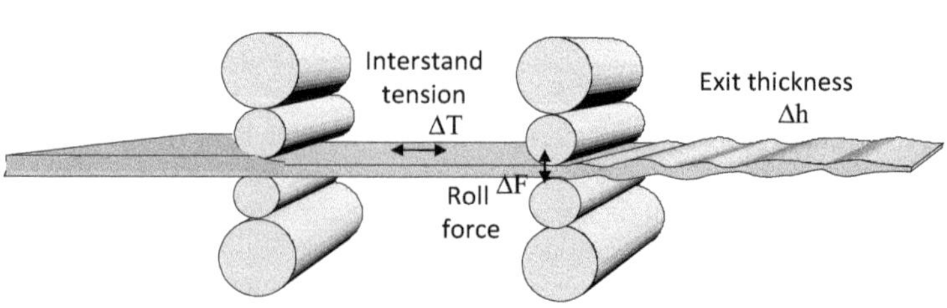

FIGURE 3.24 Simplified diagram of two stands of a tandem rolling mill with Third Octave Chatter.

is then used for predicting the critical rolling speed above which the phenomenon occurs. These results are verified with field measurements of vibrations on a chatter-affected rolling mill.

Model for Rolling Mill Chatter

The vicious cycle driving third octave mode chatter was clearly identified in [115]. This is due to a positive feedback relationship as depicted in the block diagram shown in Figure 3.25. In particular, let's examine the impact of a disturbance in the work roll gap of stand i.

When a perturbation occurs in the work roll gap at stand i, the exit thickness changes. If the exit thickness increases, mass flow continuity requires the strip entry speed to increase by the same proportion. This creates a speed mismatch with the strip exiting the previous stand, leading to a rise in interstand tension. The interstand strip can be modelled as a static elastic spring, where relative movement between its ends affects tension [116]. An increase in entry tension reduces the roll force, which in turn decreases the strip exit thickness as the roll stack compression lessens. This partially counteracts the initial roll gap disturbance. Additionally, the rising tension induces slip changes of opposite signs at the stands on either side of the interstand strip. Over time, the tension response stabilises at a finite value, influenced by both thickness and slip interactions [117]. The response of entry tension to roll gap changes behaves as an approximate first-order linear system with a time constant, τ_i, which is inversely proportional to mill speed. At low speeds, τ_i, is large enough to

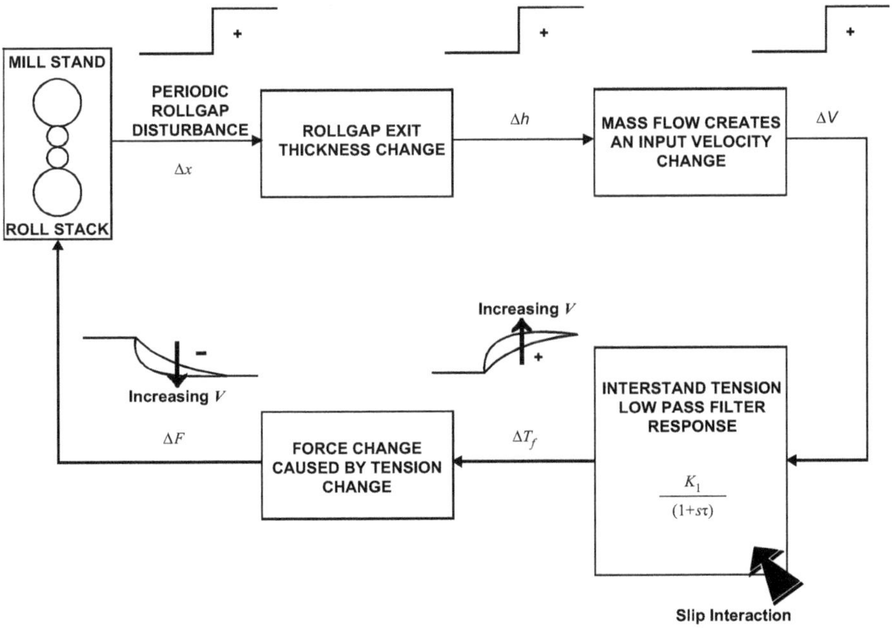

FIGURE 3.25 Simplified block diagram illustrating rollstack vibration and interstand tension interaction for third octave rolling mill chatter [115].

dampen and suppress significant tension fluctuations at the natural frequency of the roll stack. However, as mill speed increases, τ_i, decreases, enabling greater (positive feedback) interaction between roll gap variations and tension fluctuations. This speed-dependent interaction introduces a negative damping effect, which, under certain conditions, can counteract the natural positive damping in the roll stand. When this balance is disrupted, chatter vibrations may develop.

The vertical roll stack dynamics may be modelled by utilising the discrete spring–mass–damper system depicted in Figure 3.26.

The work roll-strip flattening stiffness, k_w, and roll gap damping coefficient are explicitly included in the model, while strip stiffness is incorporated within the roll gap model. Analytical methods for calculating these quasistatic constants are widely available in rolling theory literature, such as [109,116,118]. Additionally, derivations of the roll gap damping coefficient can be found in sources like [114] and [119]. By assuming that the relative movement between the backup chocks and rolls is small and the damping is low, the system shown in Figure 3.26a can be reduced to a 4-degree-of-freedom system, as shown in Figure 3.26b. This figure illustrates each of the modal roll displacements and natural frequencies of the system. It may be seen from Figure 3.26b that modes 2 and 4, being vertically symmetrical about the stand passline, are predominantly excited by roll force variations in the strip and cause significant thickness variations. Only mode 2 has a natural frequency within the range of observed third octave chatter and thus in the present analysis is assumed to be associated with this form of chatter. It may also be noted that modes 3 and 4 are in the frequency range associated with fifth octave chatter.

The fundamental vicious cycle of third octave chatter is represented with mathematical models describing the positive feedback of perturbational responses of vertical roll vibrations, interstand tension dynamics and roll force described previously [115]. The equations governing the interstand tension, mass flow and forward slip are given by

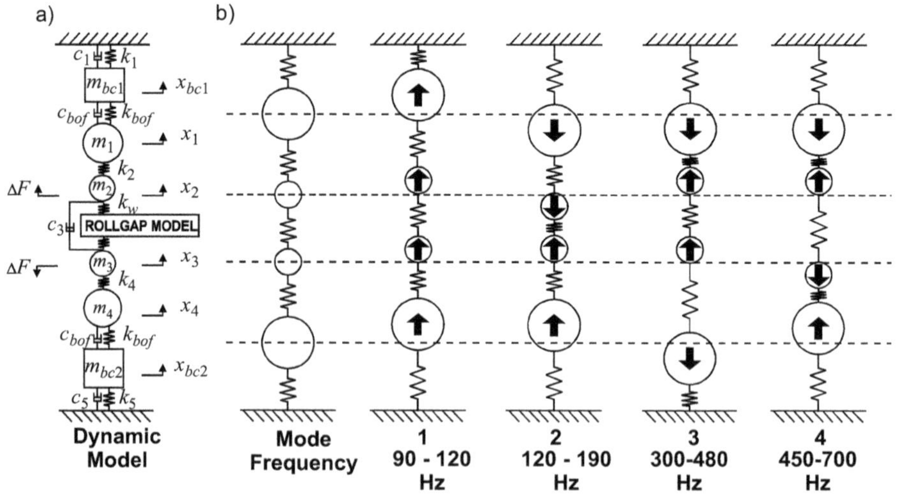

FIGURE 3.26 Four high rollstack of rolling mill; (a) dynamic model; (b) mode shapes [115].

$$\partial T_f/\partial t = \text{WEH}_i\left[V_i - v_{i-1}\right]/L, \tag{3.55}$$

$$\text{VH} = vh, \tag{3.56}$$

$$\Delta f = (\partial f/\partial H)\Delta H + (\partial f/\partial h)\Delta h + (\partial f/\partial T_f)\Delta T_f + (\partial f/\partial t_f)\Delta t_f, \tag{3.57}$$

respectively, where Δ refers to a variation from the nominal value and the forward slip is defined as

$$f = v/V_p. \tag{3.58}$$

V_p is the work roll peripheral velocity. In addition, referring to Figure 3.26, the displacement of the workroll centre due to a change in exit thickness, during chatter vibration described by mode 2, may be expressed as

$$x_2 = \Delta h\left(1+\left(W(\partial P/\partial h)/k_w\right)\right)/2. \tag{3.59}$$

The dynamic equation of motion for the rollstack dominant mode 2 in Figure 3.26 for third octave chatter is

$$\ddot{x}_2 + 2\zeta_o\omega_o\dot{x}_2 + \omega_o^2 x_2 = \Delta F/m, \tag{3.60}$$

where m, ζ_o and ω_o are the modal mass, damping ratio and natural frequency for the third octave vibration mode with displacement vector x_2, respectively. The calculation of modal parameters is described in Thomson [120].

Prediction of Rolling Mill Chatter

The equations of motion may be solved based on the feedback relationship illustrated in the block diagram in Figure 3.25, for the chattering and previous stands. A stability analysis is then conducted to identify the critical mill speed at which this feedback interaction becomes marginally unstable. The stability condition for this system can be determined using various methods. Specifically, equations (3.55)–(3.59) can be solved algebraically for the chattering and penultimate stands (i and $i-1$), yielding a first-order differential equation that describes the interstand tension force.

$$\partial T_f/\partial t = V_i\text{WEH}_i\left[\alpha x_2 - \beta\Delta T_f\right]/L,$$

$$\text{where} \quad \alpha = \frac{2\left(1+h_i\left[(\partial f/\partial h)/f\right]_i\right)}{h_i\left(1+\left(W(\partial P/\partial h)_i/k_w\right)\right)}, \tag{3.61}$$

$$\beta = \left[(\partial f/\partial t_f)/f\right]_{i-1} - \left[(\partial f/\partial T_f)/f\right]_i.$$

This equation describes the response of the entry tension force to a change in roll gap position. By inspection, it may be seen that this response is equivalent to that of a first-order low-pass filter as described in the block diagram of Figure 3.25. The roll force fluctuation, ΔF, arises from the interstand tension variation and is given by the expression $W\left(\partial P/\partial T_f\right)_i \Delta T_f$. Eliminating the tension force from equations (3.60) and (3.61) and expressing the result in Laplace transfer notation give

$$s^2 x_2 + 2\zeta_o \omega_o s x_2 + \omega_o^2 x_2 = \frac{\alpha W\left(\partial P/\partial T_f\right)_i}{m\beta(1+s\tau_i)} x_2, \tag{3.62}$$

where $\tau_i = L/\left(V_i \mathrm{EH}_i W\beta\right)$. Equation (3.62) is the characteristic equation for the chatter feedback model described by Figure 3.25. Applying Routh's stability criteria to (3.62) leads to a criterion for chatter occurrence in terms of the ratio of the critical stand exit speeds, v_c, to the product, $\omega_o L$, as

$$v_c/\left(\omega_o L\right) = a'\left(b \pm \sqrt{b^2 - 1}\right), \tag{3.63}$$

where $a' = 1/\left(h_i E W\beta\right)$. Under the assumptions, $b^4 >> 1$ and $\zeta_o/b << 1$, which are typically valid for the last stands of a tandem mill rolling thin product, the lower critical speed, v_{cl}, may be approximated by series expansion of (3.63) as

$$v_{\mathrm{cl}} = \frac{m\omega_o^3 L\zeta_o\left[1 + W\left(\partial P/\partial h\right)_i/k_w\right]}{EW^2\left(-\partial P/\partial T_f\right)_i\left[1 + h_i\left(\partial f/\partial h\right)_i/f\right]}. \tag{3.64}$$

Similarly, under the same assumptions, it may be determined that the ratio of the upper critical speed to lower critical speed is given by, $4b-1$, and will always be greater than 3. It is of note that Routh's stability criterion provides a necessary and sufficient condition for bounded input–bounded output stability and is therefore valid for all types of bounded inputs such as impulse or harmonic disturbances.

The solution for the critical mill speed is interesting as it condenses multiple independent physical factors into two key dimensionless parameters. The first parameter, a', represents the ratio of the roll stack's natural frequency to the breakpoint frequency of the interstand tension response. It quantifies the degree of dynamic coupling between strip deformation and roll stack vibrations. The second parameter, b, represents the effective negative damping in the system. It is defined as the difference between the positive roll stand damping and the negative damping introduced by interstand tension interactions. When b becomes significantly larger than 1, the lower critical speed approaches 0, leading to instability. Therefore, equation (3.64) suggests that factors reducing the critical speed include lower roll stand damping, roll stack natural frequency, slip sensitivity to tension stress and strip thickness in contrast with higher roll force sensitivity to tension stress and strip thickness.

These predicted trends align with practical observations reported in the literature. The slip sensitivity to tension force is influenced by roll roughness and lubrication, which in turn depend on the friction coefficients in the roll gaps. Higher friction coefficients result in lower values of β and thus higher values of the chatter parameter, b, increasing instability. However, the broader impact of friction changes on rolling parameters must also be considered.

Results of Rolling Mill Chatter Predictions

To further analyse chatter phenomena and validate the stability criterion, a simulator was developed using the equations of motion applied to multiple stands in a tandem mill [121]. The critical speed calculated using both analytical and numerical methods showed close agreement, with discrepancies within 5%. Figure 3.27 illustrates the resulting chatter oscillations in entry tension and exit thickness across each stand of the tandem mill. Chatter can be triggered by various disturbances that naturally occur during normal mill operation. In this case, these oscillations were triggered by an increase in speed, followed shortly by a subsequent decrease.

The figure shows the instability builds up over a period of 5 seconds and gradually decreases in amplitude once the speed is reduced. During the instability, tension and thickness fluctuations are approximately 90° out of phase, consistent with field measurements [110]. This phase relationship is a hallmark of process instability as it signifies energy input into the system. Additionally, the roll motion aligns with the behaviour described by mode 2 in Figure 3.26.

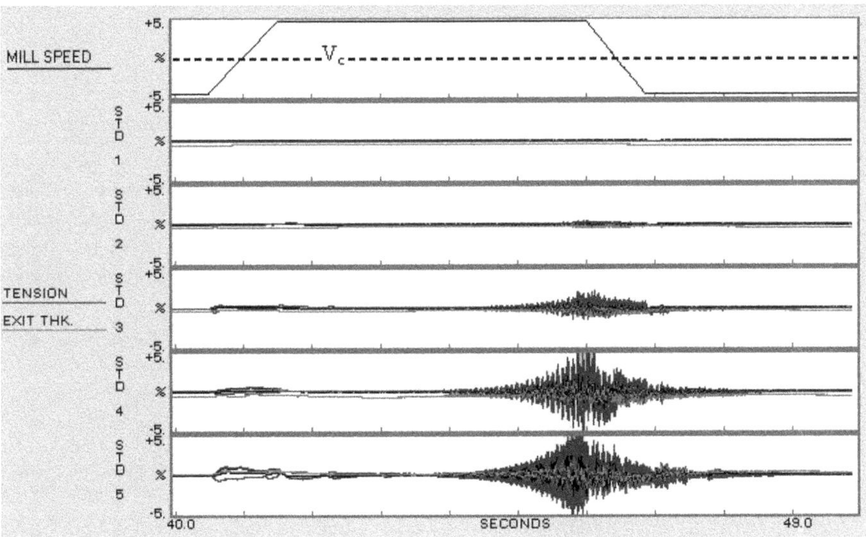

FIGURE 3.27 Simulation of the effect of mill speed on the occurrence of third octave chatter [115].

The analytical predictions of the critical chatter speeds were also compared to measurements from a five-stand thin strip tandem mill. Several products were identified as particularly susceptible to chatter. Table 3.1 presents a rolling schedule of these critical products, highlighting instances where high vibration amplitudes were observed during rolling over the test period. To validate the analytical predictions against field chatter measurements, the critical speeds were calculated based on the rolling schedules. Using the stability criterion from equation (3.63) and the rolling schedule data, the critical speeds for these products were determined and documented under the "Dynamic Characteristics" section in Table 3.1.

Table 3.1 demonstrates a strong correlation between analytical predictions and experimental measurements, with predicted critical speeds matching measured values within 5% error. This margin falls within the expected uncertainty range for friction, damping coefficients and mill speed measurements. For these products, it is evident that mill speed is being intentionally limited due to chatter as the critical mill speed is significantly lower than both the maximum mill speed and the scheduled rolling speed.

Once the chatter parameters a' and b, are established, sensitivity analysis can be applied to predict how changes in operating conditions affect critical speeds (i.e., see [115]). For example, adjusting interstand tensions and lubrication properties for each product could help optimise rolling practices and mitigate chatter more effectively. In addition, it is interesting to note that another type of rolling mill chatter known as fifth octave causes corrugations on the rolls over a much slower growth

TABLE 3.1
Comparison of Analytical and Experimental Results [115]

Rolling Schedule							Dynamic Characteristics			
Grade	Stand	Exit Thick (mm)	Exit Tension (kN/m)	Roll Force (kN/mm)	Friction Coef —	Width (mm)	Damping Coef —	Chatter Frequency (Hz)	Measured Chatter Speed at Stand 5 (m/min)	Calculated Chatter Speed at Stand 5 (m/min)
(b)	4	0.321	51.5	10.05	0.025					
	5	0.223	12.5	11.08	0.018	835	0.15	160	850	845
(a)	4	0.33	48.5	10.81	0.025					
	5	0.223	12.5	11.78	0.018	885	0.15	158	715	737
(c)	4	0.288	52.8	9.62	0.025					
	5	0.193	11.0	11.30	0.018	815	0.15	168	834	868
(d)	3	0.457	80.8	9.80	0.03					
	4	0.308	47.1	10.03	0.025	915	0.15	143	903	931

Note: Interstand length 4.1 m.

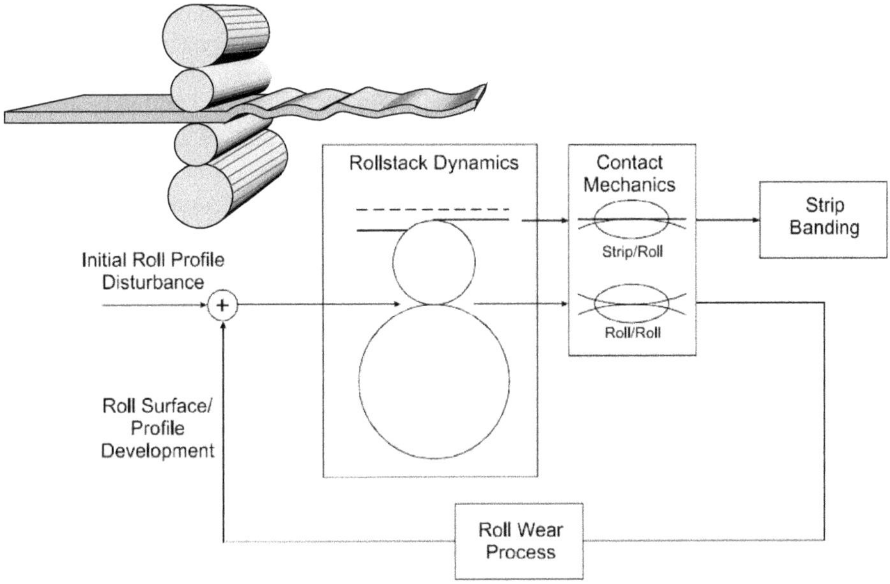

FIGURE 3.28 Fifth octave rolling mill chatter vicious cycle [122].

period than third octave (days rather than seconds). These then transfer to the strip, i.e., see Figure 3.28. It has similarities with rail corrugations, which will be detailed subsequently.

3.5 RAIL CORRUGATION

Rail corrugations induce vibrations as vehicles pass over them, causing excessive noise, restricting running speeds and in some cases causing serious track defects. They are similar to road [123] and rolling mill [124] corrugations and have remained persistent and have grown in prevalence, worldwide, in multiple forms over many decades [125]. Wear-type rail corrugations include those classified as "rutting" and "roaring rails" [125], characterised by both long (100–400 mm) and short pitch wavelengths (25–80 mm). The resultant railway noise due to short pitch wavelengths is in the range of 200–1500 Hz and is particularly undesirable in populated areas. Several techniques [126], such as rail hardfacing and wheel/rail profile control have had some reported success in reducing the growth of corrugations but are not reliable cures for all conditions. At present, the only reliable cure for wear-type rail corrugation is removal by grinding, which costs substantial maintenance expenditure per annum to the railway industry [127]. These costs appear to be increasing in line with the significant increase in usage, development and speed of railways throughout the world.

Much research has been focussed on prediction and prevention of rail corrugation [128]. Research has typically resulted in integrated simulation incorporating complex finite-element models for the dynamics of the track and discrete element models for the rolling contact mechanics, e.g. [129–131]. Wu and Thompson [132] have numerically investigated the effect of multiple wheel/track passes using a frequency domain model. To provide fundamental insights, a number of efforts have also been directed towards obtaining analytical predictions of wear-type rail corrugation [127,133,134]. Frederick [135] was one of the first to propose a feedback loop to model rail corrugation and analyse the response based on linear phase relationships. Muller [127], Nielsen [134] and Bellette [136] have investigated a nonlinear contact mechanics filter to explain reports of the independence of wavelength with speed for shorter pitch corrugations. Bhaskar et al. [133] and Muller [127] investigated the stability of the interaction between the structural dynamics and contact mechanics over one wheelset passage. Much research effort continues to focus on this problem with interesting and notable steps towards developing experimental, simulation and understanding wheel corrugations made by many authors ie [137–139]. Meehan et al. [140,141] extended this research, providing an analytical prediction of the growth of wear-type rail corrugation over multiple wheelset passes independent of initial profile assumptions. The closed-form solution provides an extremely efficient means of estimating corrugation growth and optimum grinding intervals as compared to sophisticated, but computationally expensive, numerical models. They extended this with an investigation of the effect of dynamic interactions between multiple wheel passages in [142]. In the next sections, these models for wear-type rail corrugation will be presented and important analytical and numerical results predicting the growth of the phenomena discussed.

Model for Rail Corrugation

An accurate yet simplified model that predicts the growth of wear-type rail corrugation based on the important track and rolling stock parameters can be developed with understanding of the instability mechanism. Wear-type rail corrugation occurs due to a vicious cycle of vehicle track vibrations causing wear fluctuations that re-excite the same vibrations over the next wheel passage. Figure 3.29 describes the wear-type rail corrugation development feedback mechanism.

The wheelset track vibrational dynamics, I, may be described by a decoupled equation of motion for each mode, in the real analytical form

$$\ddot{y}_i + 2\zeta_i\omega_i\dot{y}_i + \omega_i^2 y_i = k_c z_{in}(p_i - 1)/m_i . \tag{3.65}$$

Note the modal dynamics can be alternatively expressed in the form of complex receptances [143] which are more continuous (frequency domain) representations of the wheel and track dynamics. The receptances may be found from theoretical models or measured experimentally via impact hammer testing. The coordinate transform for the modal displacements, y_i, is given by

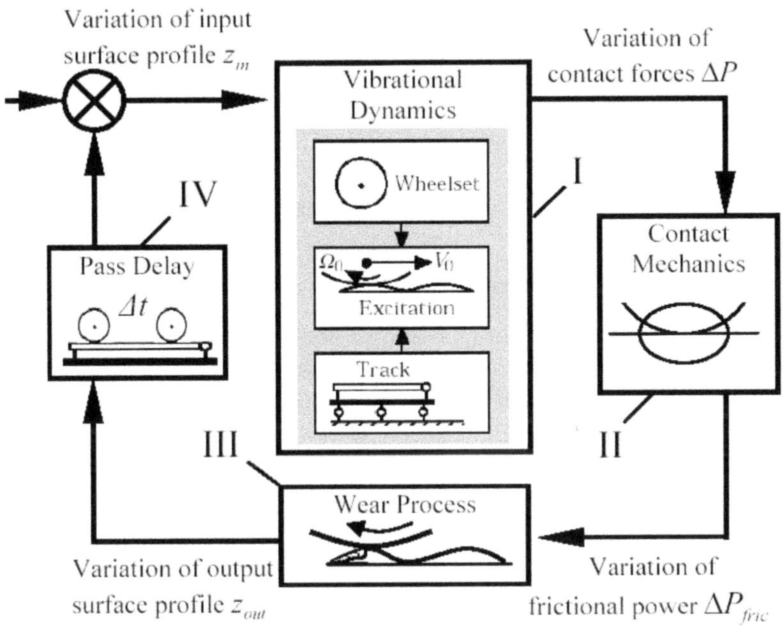

FIGURE 3.29 Feedback model for wear-type rail corrugation [143].

$$y_w = \sum_{i=1}^{n} p_i y_i, \quad y_r = \sum_{i=1}^{n} y_i. \tag{3.66}$$

A modal description of the vibrational dynamics is chosen based on experimental and theoretical evidence of approximately constant corrugation pitch associated with a dominant mode of system vibration. In this case, a vertical vibration mode is considered; however, the modal equations (3.65) and (3.66) may be used to represent other vibrational modes.

The contact mechanics model, II, is based on quasistatic microslip with consideration of small linear variations about nominal nonlinear operating conditions. In this case, variations in longitudinal components of traction, slip and wear are considered the dominant contributions to corrugation growth. However, the linear parameters could be representative of other components if desired, i.e., lateral traction in curves/cornering. The wear process, III, is assumed to be proportional to the frictional power dissipated. The equations governing the contact mechanics, II, and the frictional wear process, III, (see [140,141] for details) can be combined and solved for each mode to give

$$\begin{aligned} &(z_{\text{out}} - z_{\text{in}})/\Delta z_o = \Delta \dot{W}_{\text{frict}}/\dot{W}_{\text{frict}o} = \Delta\xi/\xi_o = C_\xi\left(\Delta P/P_o\right), \\ &\text{therefore} \quad \left(z_{\text{out}_i} - z_{\text{in}_i}\right)/\Delta z_0 = C_\xi k_c\left(y_i\left(1 - p_i\right) + z_{\text{in}_i}\right)/P_0. \end{aligned} \tag{3.67}$$

For multiple wheel passages, the rail profile variation entering the rolling contact region of the wheelset, $z_{in}(x)$, is the same rail profile variation exiting from the previous wheelset pass, $z_{out}(x)$. Therefore, assuming a time interval between wheelset passes, Δt, the profile wear of successive passes of wheelsets is represented by the time delay relationship,

$$z_{in}(t) = z_{out}(t - \Delta t). \tag{3.68}$$

Prediction of Rail Corrugation

The rail corrugation equations of motion over multiple wheel passages can be analysed for stability using many methods. However, the time delay between passes adds a nonlinearity that can be dealt with in the frequency domain. Using the Laplace transform denoted as

$$Ł\{f(\omega_i t)\} = F(S), \tag{3.69}$$

Equations (3.65) and (3.67) may be solved to obtain a characteristic relationship between the input and output wear profile over a given wheel passage as

$$Z_{out_i}/Z_{in_i} = 1 + K_b\left[1 - K_{c_i}/\left(S^2 + 2\zeta_i S + 1\right)\right], \tag{3.70}$$

where

$$K_b = C_\xi k_c \Delta z_0 / P_0, \quad K_{c_i} = k_c(1-p_i)^2/m_i\omega_i^2. \tag{3.71}$$

Equation (3.70) represents the dynamic behaviour of the system over one wheelset passage. K_b represents the sensitivity of wear variations to wheel/rail contact deflection variations. Similarly, K_{c_i} may be shown to represent the modal sensitivity of the wheel/rail relative displacement to a change in the input longitudinal profile. For realistic railway parameters, K_b, K_{c_i} and ζ_i are always positive. Under these assumptions, it may be easily shown, using renowned stability analysis techniques, that the second-order system, (3.70), is always stable, in line with the results of [133].

To investigate the dynamic behaviour over multiple wheelset passages, the passage time delay equation (3.68) is also transformed into the Laplace form

$$Z_{in_i}/Z_{out_i} = e^{-S\omega_i \Delta t}. \tag{3.72}$$

Equations (3.70) and (3.72) describe a single input–single output feedback system that may also be represented by a block diagram (see [140]). The stability behaviour of the system may be determined analytically from the characteristic equation for the complete system, obtained by solving equations (3.70) and (3.72) as

$$1 - (1 + K_b)\left(1 - \frac{\beta_i}{S^2 + 2\zeta S + 1}\right)e^{-S\omega_i \Delta t} = 0, \tag{3.73}$$

where

$$\beta_i = \frac{K_b K_{ci}}{1+K_b}. \tag{3.74}$$

The stability behaviour of the system is determined by the dominant real part of the system closed loop poles, which are the roots to the characteristic equation (3.73), described by the nondimensional expression

$$S = \sigma + \omega_d j, \tag{3.75}$$

where j denotes the imaginary component and σ and ω_d relate to the corrugation growth rate and frequency. The characteristic equation (3.73) may be considered to define both magnitude and phase conditions due to the complex nature of the roots, S. The nontrivial solutions to these conditions are developed under the realistic assumptions

a. $0 < K_b << \zeta_i << 1$,
b. $0 < \beta_i << \zeta_i << 1$,
c. $|\sigma| << \zeta_i << 1$,

and are summarised in the following.

Phase Condition

The phase condition of (3.73) provides the solution for the imaginary component of the closed-loop poles as

$$\omega_d = \pm \tfrac{2\pi}{\omega_i \Delta t} n, \quad n = 0,1,2\ldots. \tag{3.76}$$

Equation (3.76) defines an infinite number of closed loop poles at equally spaced discrete intervals along the imaginary axis of the root locus. Each solution to (3.76) represents a frequency (or corrugation wavelength) that will be present in the response of the system. The infinite number of roots (or order) of the system arises due to the nonlinear passage time delay term. The effect of the magnitude of the wheel pass time delay Δt on the system behaviour via equation (3.76) is investigated and discussed subsequently.

Magnitude Condition

Solution for the magnitude condition of (3.73) yields an expression for the real part of the system poles

$$\sigma \approx \frac{1}{\omega_i \Delta t} \ln\left[(1+K_b)\left(1+\tfrac{\beta_i(\omega_d^2-1)}{(1-\omega_d^2)^2+(2\zeta_i\omega_d)^2}\right)\right]. \tag{3.77}$$

Equations (3.75)–(3.77) define the analytical solution for all the closed-loop system poles (wavelength spectrum). The stability of the system and hence the growth of corrugations may be determined from equation (3.77). In particular, the growth rate of instability for mode i, defined by parameter G_{r_i}, can be expressed as the magnitude of the transfer function

$$\left|Z_{\text{out}_i}/Z_{\text{in}_i}\right| = 1 + G_{r_i} = e^{\sigma\omega_i\Delta t}. \tag{3.78}$$

The dominant pole (or mode) magnitude is determined by finding the critical value for the imaginary component, ω_d, for which the maximum value for σ occurs. This is given by

$$\omega_d^2 = 1 + 2\zeta_i. \tag{3.79}$$

This maximum value for growth is typically not realised exactly as the system phase condition described by (3.76) must be satisfied. If the parameter $\omega_i\Delta t$ is typically very large, the discretisation of ω_d, defined by (3.76), is small enough such that there will always be a pole that approximately satisfies (3.79) to sufficient accuracy. In this case, each wheelset passage occurs after the dynamic effects of the previous wheelset have settled down to a negligible level. This assumption has been made in previous research, i.e., [129,130,133,134], and may be used to simplify the growth rate to the form

$$G_{r_i} \approx K_b\left(1 + K_{c_i}/4\zeta_i\left(1+\zeta_i\right)\right), \tag{3.80}$$

assuming a small k_0. This closed-form analytical prediction not only can be used to immediately determine the sensitivity of corrugation growth to different railway parameters [141] but is also convenient for estimating optimum grinding intervals for corrugation removal. In particular, according to [126], an optimum grinding interval for corrugations, τ, may be determined based on minimisation of rail mass removal and (hence approximate costs). With reference to [126], it is easily shown that the relationship between the optimum grinding interval and the growth parameter is

$$\tau = \frac{\Delta t}{Gr_i}. \tag{3.81}$$

However, for the consideration of adjacent wheels on a vehicle travelling at considerably high speeds, it is expected that the approximation, (3.80), will not be as accurate as for low speeds. Greater accuracy will be obtained if the value for ω_d that satisfies (3.76) and is closest to satisfying (3.79) is used. This is investigated in more detail in [142] and physically is important when the wheel passing frequency is an integer multiple of the natural frequency, ω_i. This is also why fifth octave rolling mill chatter can be delayed by avoiding certain roll speed and diameters.

We may wonder whether this stability analysis applies to both straight and curved track conditions. This question was answered in [143]. There, it was shown that the same stability and corrugation growth results, focussed on longitudinal wear for tangent track corrugations [140,142], are also pertinent to lateral creep and wear in curving/cornering. In particular, it was shown that the dominant mechanism in curving corrugations is the vertical vibrations causing lateral creep oscillations and differential wear. This is because the lateral damping due to the contact mechanics is typically much higher than the lateral impedance of the rail. Therefore, the only quantitative difference with the curving corrugation result is that K_b is dependent upon the lateral creep behaviour, as opposed to longitudinal creep.

Finally, we note the effects of varying parameters during the study of a corrugation growth rate cycle, i.e. in [142–146], which were as follows:

- The effect of sleeper positioning. e.g. [145]
- The effect of wheel passing frequency. e.g. [142]
- The effect of speed variation/distribution. e.g. [143,145]
- The effect of varying contact patch and steady wear rate. e.g. [144]
- The effect of rainfall and humidity. e.g. [146]
- The effect of lubrication. e.g. [144]

Some of these will be discussed as control measures in a later chapter.

Results of Rail Corrugation Predictions

Rail corrugation predictions have been made using the developed models and compared with field and experimental measurements. For instance, using the equations of motion and stability analysis (modified to include measured receptances and speed variation [143]), corrugation growth can be predicted from an initial profile obtained from field measurements, as seen in Figure 3.30. In particular, if the FFT of the growing profiles is taken and the peak value plotted against pass number, the same corrugation growth as the field measurements can be seen in Figure 3.30b.

The growth rate G_r represents the coefficient of exponential growth, as predicted, where the dominant wavelength spectral peak at approximately 100 mm is plotted versus the number of passes. This exponential growth rate has been observed to be largely stationary over subsequent measurement periods between regrinds at this site, for corrugation amplitudes less than 100 μm. Hence, the predictive models are seen to accurately mimic the field-measured corrugation growth behaviour in this case.

Experimental validation of rail corrugation predictions has also been performed using a specially designed two disc rolling contact test rig. Corrugations were initiated on the test rig by creating a defect in the surface of the rail disc approximately 20 mm long and 100 μm deep to provide an initial disturbance to excite the dynamics and growth process. This may represent a typical field corrugation initiator such as a rail joint or a dipped weld. A typical corrugation wear pattern and growth after running a number of experiments can be seen in Figure 3.31. The results for the high and low traction ratio tests can be seen in Figure 3.31c.

In this figure, the results for the test are shown alongside three theoretical predictions based on steady-state wear. In particular, the first prediction uses a calculated constant steady-state change in profile per pass, Δz_0, using a previously measured wear coefficient of 1.6×10^{-9} kg/Nm, the steady frictional power and the initial contact geometry based on a Hertzian contact. Note that the dry coefficient of friction is approximately 0.45. The other two corrugation growth rate predictions are based on the measured average and varying Δz_0 during the test, using the mean and two piecewise linear approximations.

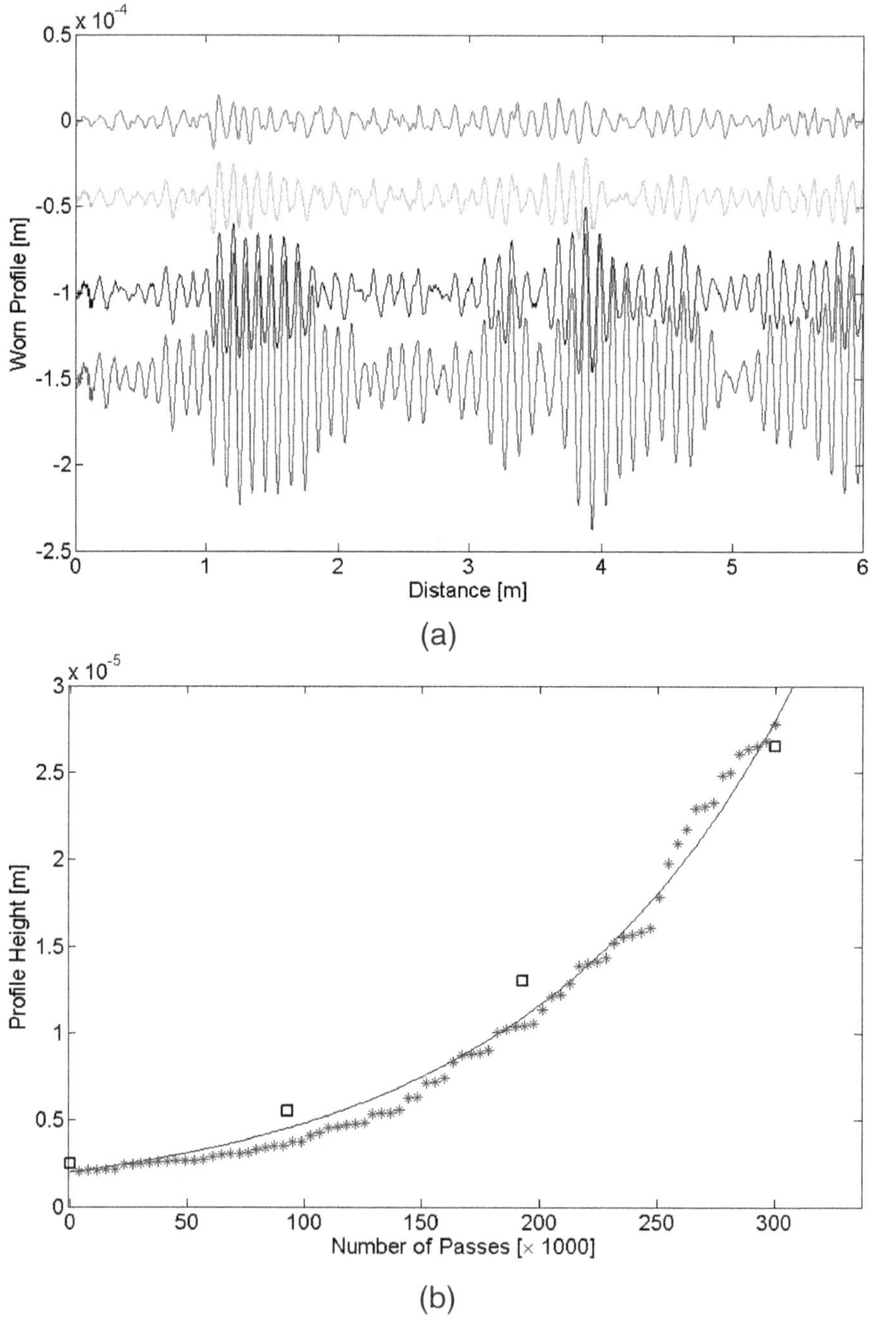

FIGURE 3.30 Measured and simulated profile evolution. (a) Initial measured profile and subsequent simulated profiles at 90,000 wheelset pass intervals; (b) comparison of field measured growth of Corrugation Amplitude FFT Peak. (□) 100 mm Peak Wavelength Value and (*,-) Trendline [143,175].

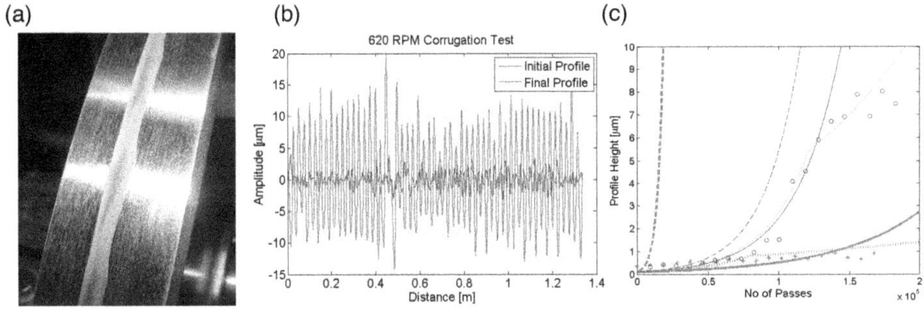

FIGURE 3.31 (a) Typical corrugated wear pattern after corrugation test; (b) for example, dry, constant speed corrugation test result showing initial and final rail disc profiles. (c) Predicted and measured corrugation profile evolution under high (thick lines) and low (thin lines) traction ratio conditions. Profile height refers to the highest peak in the profile amplitude spectrum. - Theoretical prediction based on average measured steady-state change in the profile per pass, Δz_0. . . . Theoretical prediction based on varying measured steady-state change in the profile per pass, Δz_0. - - - - Theoretical prediction based on frictional power, initial contact geometry and previously measured wear coefficient [147]. o Measured corrugation amplitude (low *Q/P*). * Measured corrugation amplitude (high *Q/P*) [144].

It was found that the low traction ratio test yielded a measured growth rate initially in line with the theoretical prediction based on the theoretically calculated Δz_0. However, the high traction ratio case yielded a much lower growth rate than the theoretical estimate. The growth rate estimates based on the actual measured Δz_0 are in much closer alignment to the measured data. The growth rate is in fact lower for the high-traction ratio case, which is opposite to the prediction based on a theoretically predicted Δz_0. This occurs because Δz_0 for the high-traction ratio case is much lower than expected and reaches a small value very quickly. The reason for this counterintuitive result seems to be due to the high frictional power altering the contact geometry on the test rig. The contact quickly increases in conformity, which enlarges the contact patch and lowers the frictional power density. This was confirmed directly by measurements of the rail disc radius change for the duration of the test [144]. Hence, after considering the effect of steady-state wear rate changing during the growth cycle, the predictive models match the experimental results quite well. This predictive insight provided much information on means of mitigating rail corrugations, as discussed in the last chapter.

3.6 TRANSMISSION DRIVELINE INSTABILITY

A commonly used element in drive trains of machinery for the transmission of power is a Hooke's Joint, also known as the universal joint or Cardan joint. A transmission driveline connected by Hooke's joint may exhibit severe shaft twisting vibrations primarily due to the fluctuating angular velocity ratio inherent in such systems. The coupling is used to transmit power through a pair of torsional shafts which are aligned at different angles due to design requirements. Such a torsional system has been used in automobiles and locomotives for many years. A commercial model of this joint used in the automotive field is illustrated in Figure 3.32.

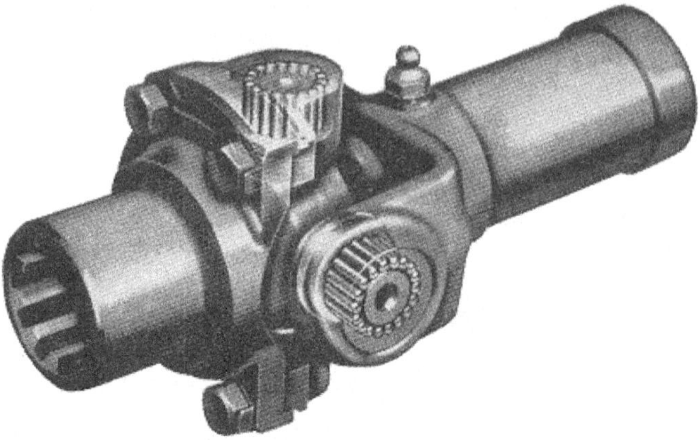

FIGURE 3.32 Hooke-type universal joint [178] p. 45, Fig 2.27.

This transmission driveline instability is highly undesirable, causing noticeable noise, severe mechanical shakings, premature fatigue and catastrophic failures in shafts, dog clutches, gear teeth etc. The occurrence as well as severity of these detrimental torsional vibrations depend on the range of engine or drive shaft speeds. Previous strategies to avoid these vibrations have concentrated on either redesigning the joint to provide a constant velocity ratio, avoiding the critical speed ranges or limiting the angle of the joint. Hence, understanding and predicting the phenomenon are important for insights into these control measures.

Porter [148] investigated this problem considering a linearised 1-degree-of-freedom model and predicted two critical speed ranges at which the linear solution becomes exponentially unstable. It was also shown that the unstable regions are dependent upon system parameters such as the input and output shaft stiffness ratio, damping ratio and Hooke's joint angle. This analysis was later extended by Porter and Gregory [149] to a nonlinear model, for the resulting amplitudes of unstable oscillation to be predicted. Subsequently, Porter [150] considered a 2-degree-of-freedom model and established both the instability conditions as well as the amplitudes of the steady-state motion. However, the occurrence of combination resonances, the corresponding input shaft speed ranges and conditions for instabilities were not predicted in the above studies. Hence, Asokanthan and Hwang [151] employed the method of averaging and established closed-form instability conditions associated with combination resonance. This analysis was extended by Asokanthan and Wang [152] to investigate and confirm the instability behaviour by means of maximal Lyapunov exponent and the instability behaviour confirmed experimentally. From a practical perspective, Birkett [153] and Lu and Sharp [154] have numerically studied undesirable nonlinear effects of Hooke's joint in trucks and locomotives, respectively. However, only the linearised equations were considered. Asokanthan and Meehan [155] addressed this by development and investigation of the full nonlinear equations, as summarised in the following section.

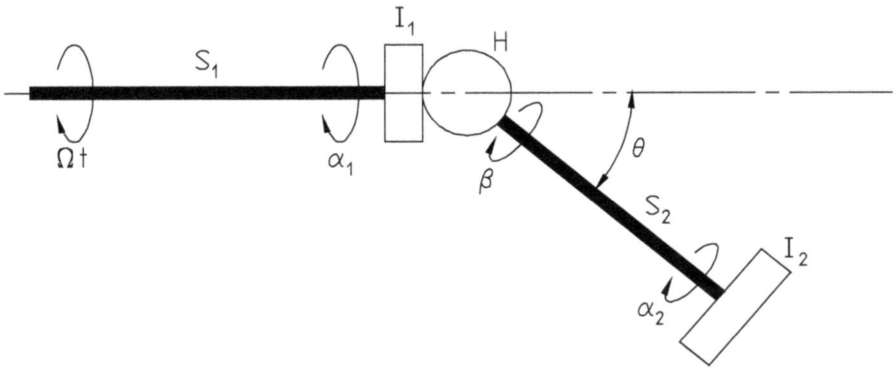

FIGURE 3.33 Driveline incorporating Hooke's Joint [155].

Model for Transmission Driveline Instability

For the purposes of the analysis of the Hooke's joint driveline system, a 2-degree-of-freedom system incorporating Hooke's joint (H) is considered, as illustrated in Figure 3.33. This rotational system having an angular misalignment of θ is driven at a constant angular velocity Ω. The quantities α_1, β represent the input and the output angles of Hooke's joint, respectively, while α_2 symbolises the output angle of the system. The input and the output shafts have been considered flexible and have torsional stiffnesses S_1, S_2, respectively. The moments of inertia associated with the input and output drive systems are denoted by I_1 and I_2, respectively, which also take the driven inertia load into consideration.

It is also assumed that both input as well as output shafts of the system are massless, and their flexural vibrations are considered to be completely prevented by the associated support bearings. The energy dissipation associated with the rotational motions of I_1 and I_2 are taken to be proportional to the viscous damping coefficients C_1 and C_2, respectively.

Equations of motion in terms of the twist variables x and y may be derived using Lagrange's equations (see e.g., Porter [150] and Asokanthan and Hwang [151]). In deriving the equations, the linearised form of the well-known kinematic relationship between the input and output angular velocities across Hooke's joint has been employed.

$$\tan\alpha_1 = \cos\theta\tan\beta. \tag{3.82}$$

The equations of motion for the first and second shaft twisting may be derived as

$$\begin{aligned} &I_1\ddot{\alpha}_1 + C_1\dot{\alpha}_1 + S_1(\alpha_1 - \Omega t) - S_2(\alpha_2 - \beta)\frac{\cos\theta}{1-\sin^2\theta\cos^2\alpha_1} = 0, \\ &I_2\ddot{\alpha}_2 + C_2\dot{\alpha}_2 + S_2(\alpha_2 - \beta) = 0. \end{aligned} \tag{3.83}$$

To make a first-order approximation of the system equations of motion (3.83), it is convenient to introduce the independent shaft twist variables x and y, represented by $\alpha_1 - \Omega t$ and $\alpha_2 - \beta$, respectively. Also, if the following parameters are defined

$$\varepsilon = \sin^2\theta, \quad \omega_{11}^2 = S_1/I_1, \quad \omega_{21}^2 = S_2/I_1, \quad \omega_{22}^2 = S_2/I_2,$$
$$C_1/I_1 = 2\varepsilon\mu_{11}\omega_{11}, \quad C_2/I_2 = 2\varepsilon\mu_{22}\omega_{22}, \tag{3.84}$$

upon substitution and expansion in terms of ε, equation (3.83) becomes

$$\ddot{x} + \omega_{11}^2 x - \omega_{21}^2 y = \varepsilon\left\{\frac{\omega_{21}^2 y\cos(2x+2\Omega t)}{2} - 2\mu_{11}\omega_{11}(\dot{x}+\Omega)\right\} + O(\varepsilon^2)$$
$$\ddot{x} + \ddot{y} + \omega_{22}^2 y = \varepsilon\left\{\begin{matrix}(\dot{x}+\Omega)^2\sin(2x+2\Omega t) \\ -\dfrac{\ddot{x}\cos(2x+2\Omega t)}{2} - 2\mu_{22}\omega_{22}(\dot{x}+\dot{y}+\Omega)\end{matrix}\right\} + O(\varepsilon^2), \tag{3.85}$$

where $O(\varepsilon^2)$ represents the smaller terms of order ε^2 or higher. Equations (3.85) represent the torsional motion of the 2-degree-of-freedom system incorporating Hooke's joint and contain periodically varying coefficients. These coefficients are seen to be nonlinear functions of the state variables. It is known that closed-form solutions to these equations are not available, but insights into the instability behaviour may be obtained by considering the linearised equations.

Prediction of Transmission Driveline Instability

To predict the instability of Hooke's joint Driveline, the equations of motion are examined for stability of their solution. Equations (3.85) may be linearised about the steady-state solution (x_s, y_s) that is of order ε, by expanding the trigonometric functions and neglecting higher-order terms involving powers and products of x, y, x_s, y_s and their derivatives. The linearised homogeneous equations are as follows:

$$\ddot{x} + \omega_{11}^2 x - \omega_{21}^2 y = \varepsilon\left\{\frac{\omega_{21}^2}{2} y\cos(2\Omega\, t) - 2\mu_{11}\omega_{11}\dot{x}\right\},$$
$$\ddot{x} + \ddot{y} + \omega_{22}^2 y = \varepsilon\left\{\begin{matrix}2\Omega^2 x\cos(2\Omega\, t) - 2\mu_{22}\omega_{22}(\dot{x}+\dot{y}) \\ +2\Omega\,\dot{x}\sin(2\Omega t) - \dfrac{\ddot{x}\cos(2\Omega\, t)}{2}\end{matrix}\right\}, \tag{3.86}$$

which contain periodic coefficients of frequency 2Ω.

Before performing a stability analysis, it is first necessary to evaluate the natural frequencies of the torsional modes in terms of other system parameters. The equations corresponding to (3.86) when $\varepsilon = 0$ are considered for this purpose. The natural frequencies can be evaluated to be

$$\omega_{1,2}^2 = \frac{1}{2}\left(\omega_{11}^2 + \omega_{22}^2 + \rho\omega_{11}^2 \mp \sqrt{\left(\omega_{11}^2 + \omega_{22}^2 + \rho\omega_{11}^2\right)^2 - 4\omega_{11}^2\omega_{22}^2}\right). \tag{3.87}$$

In equations (3.87), the quantity ρ represents the stiffness ratio S_2/S_1 , and the upper sign refers to ω_1^2, while the lower sign refers to ω_2^2. The ratio of amplitudes of input and output shaft twist may also be derived using equations (3.87) as

$$\lambda_i = \frac{\omega_{11}^2 - \omega_i^2}{\rho\omega_{11}^2} = \frac{\omega_i^2}{\omega_{22}^2 - \omega_i^2}, \quad i = 1,2. \tag{3.88}$$

It may be seen from solutions (3.87) that the natural frequencies ω_1, ω_2 are always real for any set of system parameter values. Hence, equation (3.86) represents the free motion of a linear non-gyroscopic system. It is known that exact solutions, in a closed form, to the governing equations (3.87) are not available. However, an asymptotic method such as the method of averaging (see e.g., Bogoliubov and Mitropolsky [156]) may be used to obtain an approximate solution. This method has been successfully applied to gyroscopic systems, which represented moving elastic bands and chain drives (see, e.g. Ariaratnam and Asokanthan [157] and Asokanthan and Ariaratnam [158]). The complete description of the method and the analysis have been presented by Asokanthan and Hwang [151]. The main results are summarised here.

It was predicted [151], using the method of averaging, that parametric instabilities characterised by exponential build-up of amplitudes will occur at certain rotation rates. In particular, parametric excitation can occur when the excitation frequency 2Ω is in the neighbourhood of certain multiples and combinations of the natural frequencies ω_1 and ω_2, according to the following cases:

Subharmonic Resonance

In this case, the onset of instability occurs if the excitation frequency 2Ω lies in the neighbourhood of $2\omega_p$, $p = 1,2$. Under this condition, Asokanthan and Hwang [151] obtained the following closed-form expression for the stability condition:

$$\omega_p\left[1 - 2\sqrt{\varepsilon^2\eta_p^2 - (\varepsilon\chi_p)^2}\right] < \Omega < \omega_p\left[1 + 2\sqrt{\varepsilon^2\eta_p^2 - (\varepsilon\chi_p)^2}\right], \tag{3.89}$$

where

$$\eta_p = \frac{1 - 2\rho\lambda_p - \rho\lambda_p\lambda_s}{16(\lambda_p - \lambda_s)(1 - \rho\lambda_p)}$$

$$\varepsilon\chi_p = \frac{(C_1/I_1)(1+\lambda_s) - (C_2/I_2)(1+\lambda_p)}{4\omega_p(\lambda_p - \lambda_s)}, \quad p,s = 1,2, \quad p \neq s.$$

When the damping constants C_1, C_2 are negligible, the instability condition (3.89) reduces to

$$\omega_p[1 - 2\eta_p\varepsilon] < \Omega < \omega_p[1 + 2\eta_p\varepsilon]. \tag{3.90}$$

Combination Resonance

In this case, instabilities are initiated via combination resonance that takes place when the frequency of excitation 2Ω lies in the neighbourhood of the sum or difference of the natural frequencies (i.e., when $\Omega \approx |\omega_p \pm \omega_s|$, $p, s = 1, 2, p \neq s$). The stability condition in this case of additive combination resonance can be shown to take the form:

$$(\omega_1 + \omega_2)\left\{\tfrac{1}{2} - \tfrac{1}{2}\left(r + \frac{1}{r}\right)\varepsilon\sqrt{\eta_3\eta_4 - G_1G_2}\right\} < \Omega$$
$$< (\omega_1 + \omega_2)\left\{\tfrac{1}{2} + \tfrac{1}{2}\left(r + \frac{1}{r}\right)\varepsilon\sqrt{\eta_3\eta_4 - G_1G_2}\right\} \tag{3.91}$$

where

$$G_p = \left[\mu_{11}\frac{1 + \lambda_s}{\sqrt{1 - \rho\lambda_p}} - \mu_{22}\frac{(1 + \lambda_p)^{\frac{3}{2}}}{\sqrt{\lambda_p}}\right]\frac{\omega_p}{\omega_p + \omega_s}, \quad p, s = 1, 2, \quad p \neq s,$$

and $r = \sqrt{G_1/G_2}$ while η_3 and η_4 are given by the following expressions:

$$\eta_3 = \frac{\omega_1}{\omega_1 + \omega_2}\left\{\frac{1}{8(\lambda_1 - \lambda_2)}\left(1 - \frac{(1 + \lambda_2)\lambda_2}{\omega_1^{\,2}}\frac{S_2}{I_1}\right)\right\},$$

$$\eta_4 = \frac{\omega_2}{\omega_1 + \omega_2}\left\{\frac{1}{8(\lambda_1 - \lambda_2)}\left(\frac{(1 + \lambda_1)\lambda_1}{\omega_2^{\,2}}\frac{S_2}{I_1} - 1\right)\right\}.$$

In the absence of damping, condition (3.91) reduces to

$$(\omega_1 + \omega_2)\left\{\tfrac{1}{2} - \varepsilon\sqrt{\eta_3\eta_4}\right\} < \Omega < (\omega_1 + \omega_2)\left\{\tfrac{1}{2} + \tfrac{1}{2}\varepsilon\sqrt{\eta_3\eta_4}\right\}. \tag{3.92}$$

Note that unlike in the sub-harmonic case, in this form of instability, both modes participate simultaneously. It may also be deduced, from the instability conditions, that if the damping constants C_1 and C_2 do not satisfy the condition $r = 1$, then the unstable speed range becomes wider than the corresponding range for the undamped case. In the case of combination resonance of the difference type, i.e., when the excitation frequency 2Ω is in the neighbourhood of $\omega_1 - \omega_2$, it was shown that the system is always stable.

In order to predict all possible critical speed ranges and the associated response amplitudes, the linear analysis is not adequate, and a nonlinear analysis is required. In addition, a nonlinear analysis will identify other unstable behaviours resulting from nonlinearities. It will also serve to confirm the results obtained using the linear analysis, as described in the following section.

Results of Transmission Driveline Instability Predictions

To illustrate the applicability of the analytical results presented in the previous sections, an experimental drive-line that has been developed at the University of Queensland was considered. The parameters chosen for the present study conform to the above experimental system. For these parameters, the natural frequencies ω_1, ω_2,

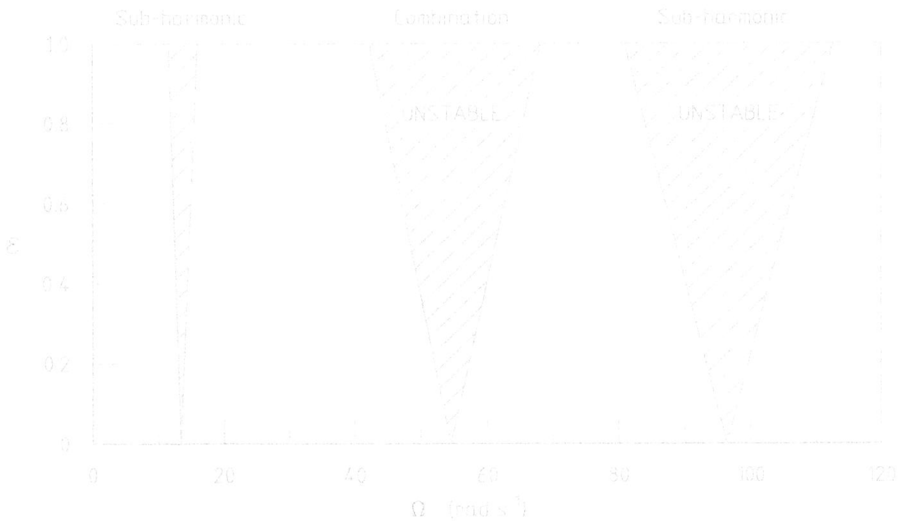

FIGURE 3.34 Instability regions in Hooke's Joint Driveline predicted by linear analysis [155].

associated with the torsional vibration in the first and second modes, take the values 13.78 and 96.61 rad/s, respectively. In this case, the primary combination resonance frequency ω_c takes the value of 55.19 rad/s. Figure 3.34 shows the instability regions for the above system parameter values plotted in the excitation parameter input shaft speed space using equations (3.90) and (3.92).

The straight lines attached to each region represent the stability boundaries associated with the undamped systems. The first and the third regions shown in this figure correspond to the sub-harmonic resonances, while the middle region corresponds to the combination resonance of the additive type. It may be recalled here that the excitation parameter ε is equal to $\sin^2\theta$, θ being the Hooke's joint angle. It may be seen from this figure that the regions become wider with increasing joint angle. It can be shown that these regions become narrower and become elevated in the parameter space if damping is taken into account via equations (3.89) and (3.91). The region for the additive type of combination resonance was identified for the first time in a Hooke's joint drive system in [151].

A summary of the numerical investigation of the driveline using the full nonlinear equations of motion (3.83) of [155] is provided for verification. In particular, instabilities were identified and characterised by varying the amplitude of the input angular velocity using tools developed in [159–161]. Figure 3.35a shows a bifurcation diagram of the time history of the system as the input angular velocity is varied from $\Omega = 5$ to 120 rad/s, for a Hooke's joint angle of 40°.

Figure 3.35 depicts the amplitude of the twist of the output shaft, y, for a range of input angular velocities, Ω after transient conditions have subsided. Instability regions may be identified on the figure as large increases in the amplitude of twist. Note that bounded motion is still present in the instability regions due to the nonlinearities within the system. As illustrated, criteria for the onset of parametric subharmonic resonance can be identified for both natural frequencies in the form

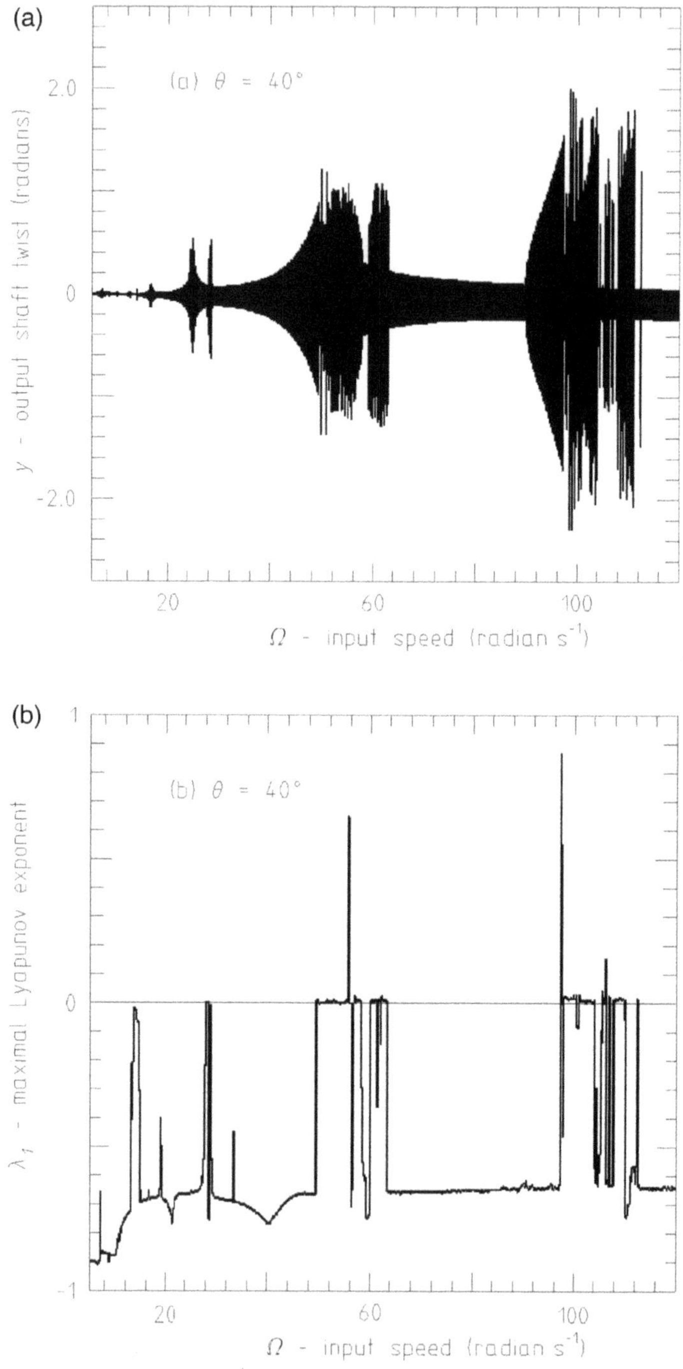

FIGURE 3.35 Bifurcation diagram for $\theta = 40°$ and Ω range of 5–120 rad/s. (a) Time history; (b) maximal Lyapunov exponent [155].

$2\Omega = 2\omega_p/n$; $p = 1,2$ for $n = 1,2,4,6,8$. These critical speeds which correspond to the primary resonances are the same as those predicted by the approximate linear analysis. In addition, a number of combination resonances in the general form of $2\Omega = l\omega_p \pm m\omega_s/n$; $p = 1,2$, $p \neq s,l,m,n$ are positive integers, as may be identified from this figure. Note that the instabilities seen at $\Omega \approx 63$ and 106 rad/s were not predicted by the linear analysis and are the result of nonlinearities within the system. Other forms of this general combination resonance were found to appear as the Hooke's joint angle was increased further. The bifurcation diagram of the maximum Lyapunov exponent for the same set of parameters is seen in Figure 3.35b. From this figure, it is evident that the parametric resonance phenomena described above are indicated by a sudden increase in the maximum Lyapunov exponent of the system. Note that the exponent may not become positive near resonance speeds as the resulting motion is nontransient and bounded due to the nonlinearities in the system. In fact, the present deterministic system has a positive maximum Lyapunov exponent only when the corresponding motion of the system is chaotic. This will be explored further in the next chapter. A zero maximum Lyapunov exponent indicates quasiperiodic motion, which is nonperiodic motion arising from the combination of two or more distinct frequencies. As illustrated in Figure 3.35b, quasiperiodic motion occurs at $\Omega \approx 28$, 52, 63, 106 and 111 rad/s, which are combination resonances. At these input shaft speeds, the motion of the output shaft is a result of incommensurate combinations of the natural frequencies ω_1 and ω_2. Chaotic motion, indicated by a positive maximum Lyapunov exponent, is seen to occur at $\Omega \approx 56,97,106$ rad/s.

Figure 3.36 describes the power spectral density of the system for the same parameters at four different input shaft frequencies $\Omega \approx 13.7$, 40, 55.2, 96.6 rad/s.

Parametric instability is evident when the input frequency is close to each natural frequency $\omega_1 = 13.7$ rad/s and $\omega_1 = 96.6$ rad/s, as indicated in Figures 3.36a and 3.36d. Note the output motion frequency is the same as a natural frequency in both cases. For the stable case depicted in Figure 3.36b, when the input frequency is at 40 rad/s, we notice that the output frequency is two times the input frequency Ω due to the excitation resulting from the Hooke's joint angle, as seen in equations (3.86). This was the case for all stable speeds. The combination resonance condition is indicated on Figure 3.36c for an input frequency of 55.2 rad/s. The frequency spectrum in this case indicates that the resulting motion is made up of a combination of frequencies in the form $\Omega = n\omega_1 \pm m\omega_2$ where n, m are integers starting from 0. The presence of these mixing components in the spectrum indicates that the time-dependent processes are strongly nonlinear. These frequencies are incommensurate as the maximum Lyapunov exponent was found to be 0 for this input frequency, as seen in Figure 3.35b, indicating that the resulting motion is quasiperiodic.

The Hooke's joint angle θ was then decreased to 20°, and a bifurcation diagram of the time history for input frequencies from 5 to 120 rad/s is obtained, as shown in Figure 3.37a. It may be seen that the amplitude of output vibration is lower when compared to the case shown in Figure 3.35a. Also, a number of instability regions have disappeared due to the presence of damping in the system. In fact, for the lower Hooke's joint angle, noticeable parametric instability is only

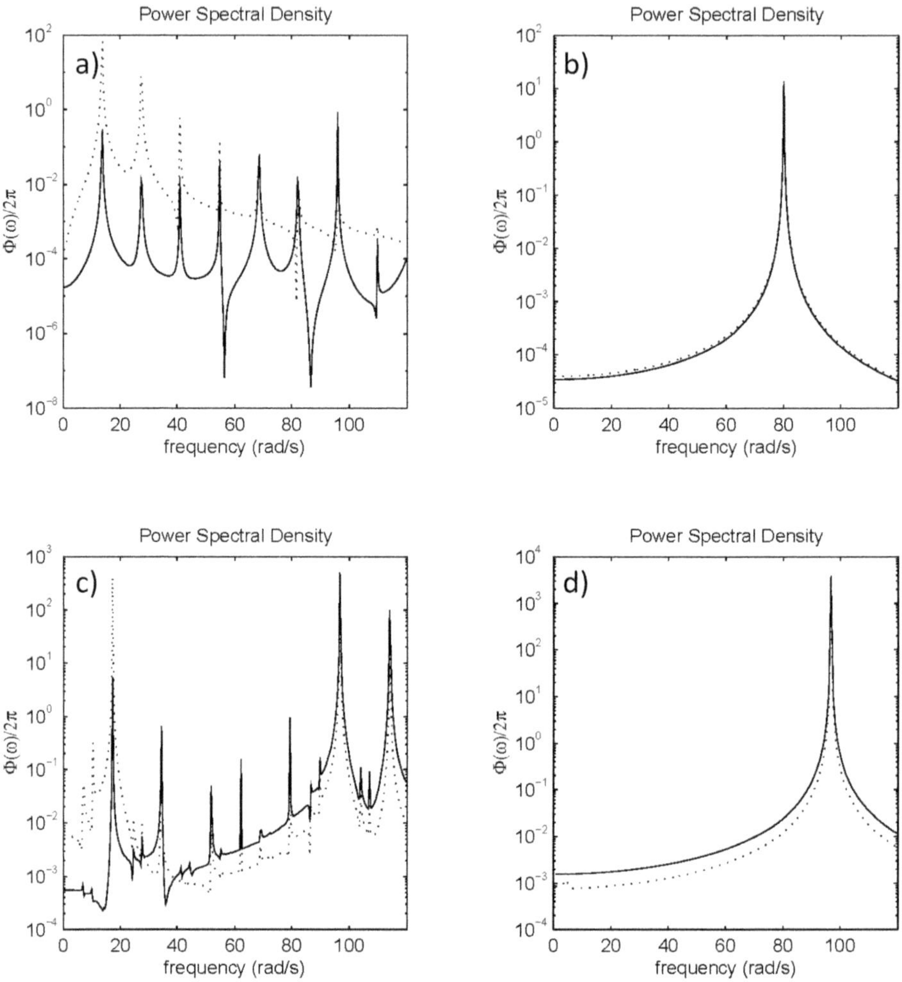

FIGURE 3.36 Power spectral density for $\theta = 40°$. (a) $\Omega = 13.7$ rad/s; (b) $\Omega = 40$ rad/s; (c) $\Omega = 55.2$ rad/s; (d) $\Omega = 96.6$ rad/s [155].

associated with the combination and higher natural frequency parametric resonance cases. From a practical point of view, these results indicate that these cases are of more concern as a vibration problem than the lower natural frequency case, for lower joint angles.

The bifurcation diagram of maximum Lyapunov exponents for the same parameters shown in Figure 3.37b also indicates the same instability regions as sudden increases in the value of the maximum exponent. Comparison between the results depicted in Figures 3.35b and 3.37b indicates the magnitudes of the maximum Lyapunov exponents are also dependent on the Hooke's joint angle. It may also be noted that no chaotic motion is present for a joint angle of 20°, indicating that the onset of chaotic motion is brought about through increased Hooke's joint angle. Further investigation of this nonlinear behaviour following instability is investigated in Chapter 4.

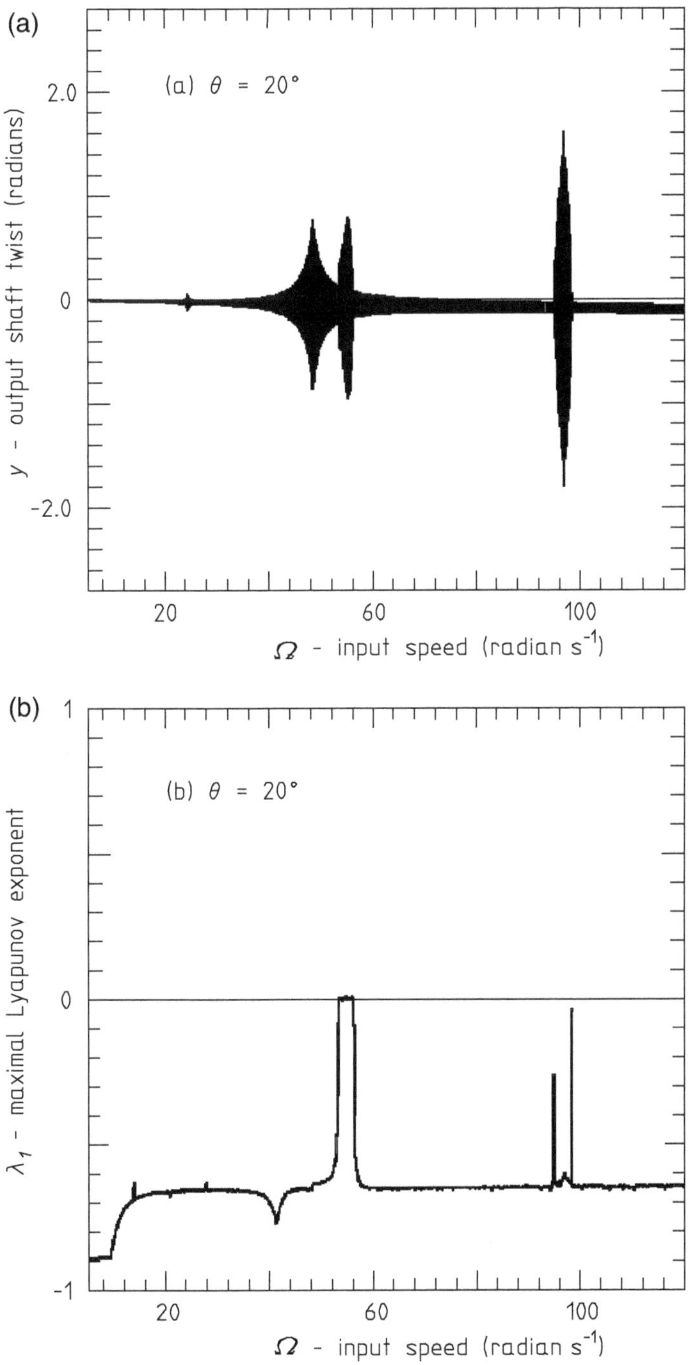

FIGURE 3.37 Bifurcation diagram for $\theta = 20°$ and Ω range of 5–120 rad/s. (a) Time history; (b) maximal Lyapunov exponent [155].

3.7 OTHER INSTABILITY PHENOMENA

There are many other instability phenomena in engineering and other systems that have been mathematically predicted. Here, we focus on a couple of additional processes to those detailed in the previous sections that have been carefully researched and mathematically modelled to provide useful predictions of their phenomena. The reader is directed to the literature referenced if interested in further details.

INNER EAR INSTABILITY AND COCHLEAR AMPLIFICATION

The cochlear amplifier is due to an instability arising from a desirable tuned vicious cycle within the inner ears of mammals. This gives us a distinct advantage in hearing over other animals and allows us to perform efficient complex communication. Specifically, it is essentially a positive feedback loop that amplifies the sound input to our external ear to our inner ear hearing sensors. This happens within the cochlea (organ within the inner ear – see Figure 3.38) that amplifies the travelling wave of vibrations along a tuned basilar membrane. These vibrations are sensed in the inner ear, and then an active force is generated in synchrony to increase the vibrations. This active mechanism within the cochlea was first proposed by Gold [162]. It was

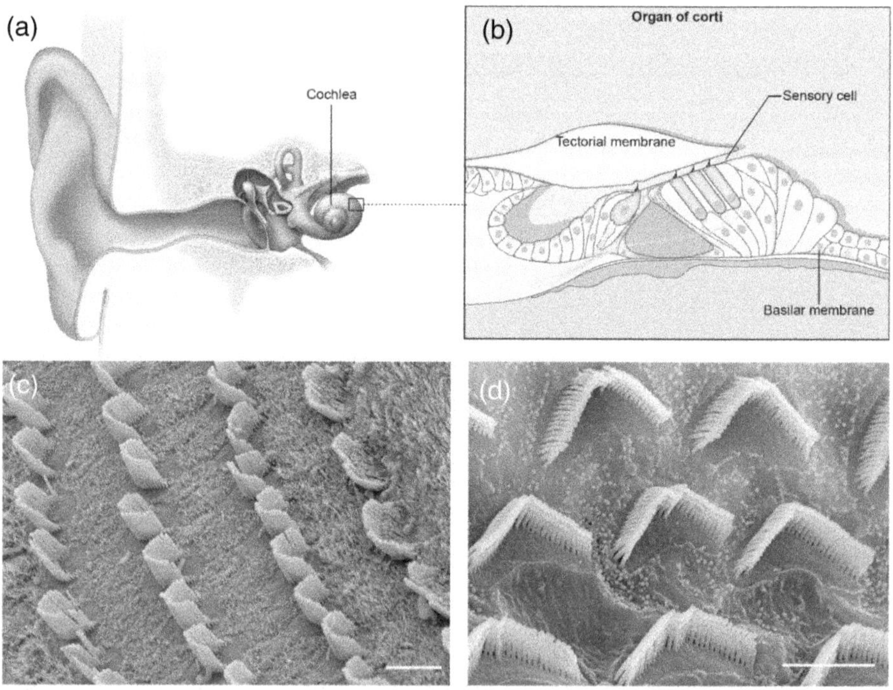

FIGURE 3.38 (a) The ear showing the inner ear's cochlea, (b) Organ of Corti within the cochlea where sound vibrations are sensed by hair cells along the basilar membrane, (c) the three rows of outer hair cells (left) and one row of inner hair cells (right), (d) outer hair cells – which sense and actuate as part of the cochlear amplifier [179].

later validated when it was discovered that sounds can be produced by the inner ear, called otoacoustic emissions [163]. The amplification process in the cochlea arises from electromotility of the outer hair cells, which can act both as sensors and actuators [164–166]. This mechanism is a biologic form of piezoelectricity [167,168].

The precise behaviour of the outer hair cells *in vivo* is largely unknown. However, mathematical models have been developed to capture the cochlear amplifier effects. These include a travelling wave amplification model at the characteristic place along the basilar membrane (where we perceive the pitch of a sound tone). The first modelling effort to define the cochlear amplifier was a simple augmentation of von Bekesy's passive travelling wave [169] with an active component. For more recent mathematical models, readers are referred to the details in Neely and Kim [170], de Boer [171] and Giesler and Sang [172]. One hypothesis of how the cochlear amplifier is tuned is based upon the angle of the OHCs relative to the basilar membrane. The stereocilia are closer to the stapes than is the base of the cell. Thus, the force created by an OHC in response to having its stereocilia deflected acts in a "feed-forward" mechanism to deliver energy to a more apical section of the cochlea, slightly ahead of the travelling wave [172] (see Figure 3.39).

Results for mathematical model predictions of the cochlea amplifier have been verified using measurement of the live cochlea response from animals, as shown in Figure 3.40. Note the cochlear amplifier is an active response, so it only occurs in live mammals. Measurements of basilar membrane vibration in the living cochlea demonstrate dramatically improved sensitivity and sharpness of tuning. In other words, the basilar membrane vibrates more in a living cochlea than in a dead cochlea, particularly in response to quiet sounds, see Figure 3.40.

These precise cochlear tuning characteristics are conveyed to auditory nerve fibres via synaptic transmission from the inner hair cells [173]. With this predictive understanding, scientists and engineers have restored the hearing of deaf people by partially mimicking this important inner ear instability using cochlear implant devices, although we are yet to match humans' hearing performance. The cochlea will remain a focus of attention and a source of wonder and satisfaction at every stage of the solution to better understand how the human body functions and to cure hearing disorders.

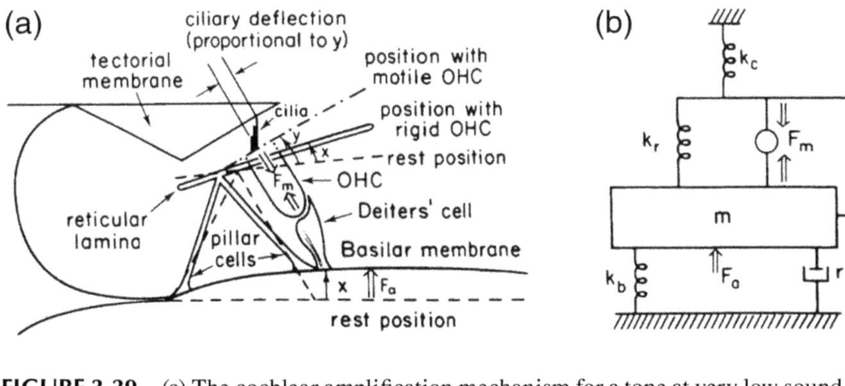

FIGURE 3.39 (a) The cochlear amplification mechanism for a tone at very low sound levels which amplifies the acoustic energy via an active Force input F_m and hence greatly enhances the mammalian hearing range, (b) equivalent mechanical circuit (C. D. Geisler, X. Shan, 1990, Fig 1, p 86).

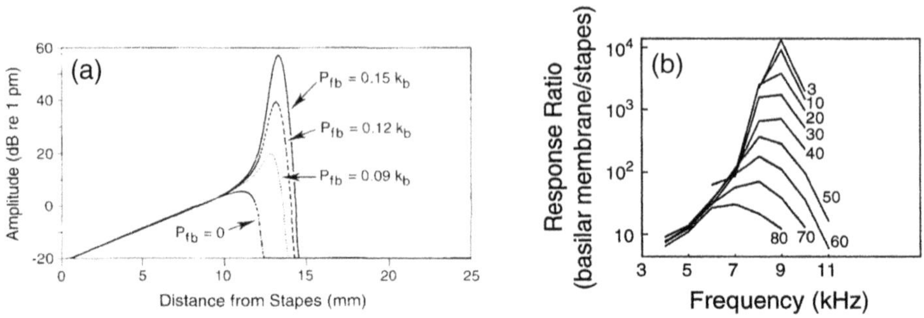

FIGURE 3.40 Response of the basilar membrane due to tones of difference loudness showing cochlear amplification (a) [172]; (b) measured responses from a live chinchilla [180].

Tool Chatter

Tool chatter causes unwanted vibrations and oscillations in cutting tools, resulting in poor surface finish, excessive noise and wear, reduced tool life and/or catastrophic failure [174]. Hence, it is a significant problem in machining processes limiting production and increasing maintenance costs. Understanding tool chatter as an instability process requires predictive mathematical modelling with experimental validation to improve safe machining operating ranges.

The main mechanism of machine tool chatter was perhaps first clearly identified by Tobias and Fishwick in 1958 [175]. They modelled the feedback loop between the metal cutting process and the machine tool structure. The "vicious cycle" comes from the regenerative effect, where the cutting process causes vibrations and ripples on the workpiece (or sometimes the tool itself), which regenerates dynamic forces that synchronise with the tool's motion on the next pass (similar to rail corrugations). This leads to a self-excited vibration phenomenon, resulting in regenerative chatter. There have also been other mechanisms of tool chatter identified, corresponding to "falling friction" and mode coupling [176]. Note that we see these mechanisms in other instability processes such as wheel squeal, flutter and rail corrugations. The mathematical modelling includes parameters such as spindle speed, depth of cut and cutting coefficients, which influence the stability of the machining process, as reviewed in [174,176]. Prediction of tool chatter occurrence is crucial for optimising machining parameters and preventing instability. Analytical methods, such as the Routh–Hurwitz criterion or Nyquist stability analysis, can provide insights into the stability of the system without the need for extensive computation. Resulting stability charts or lobes are commonly employed to visualise the stable and unstable regions in the parameter space. These charts depict combinations of cutting parameters that lead to stable cutting conditions or result in tool chatter. Stability lobes are determined through analytical methods or numerical simulations of the mathematical model. To validate and refine the mathematical models, experimental measurements are essential. Machining experiments, often conducted on dedicated test rigs or CNC machines, help capture the dynamic behaviour of the cutting process. Sensors, such as accelerometers or force transducers, are used to measure vibrations and cutting forces. These experimental data are then compared with the results obtained from

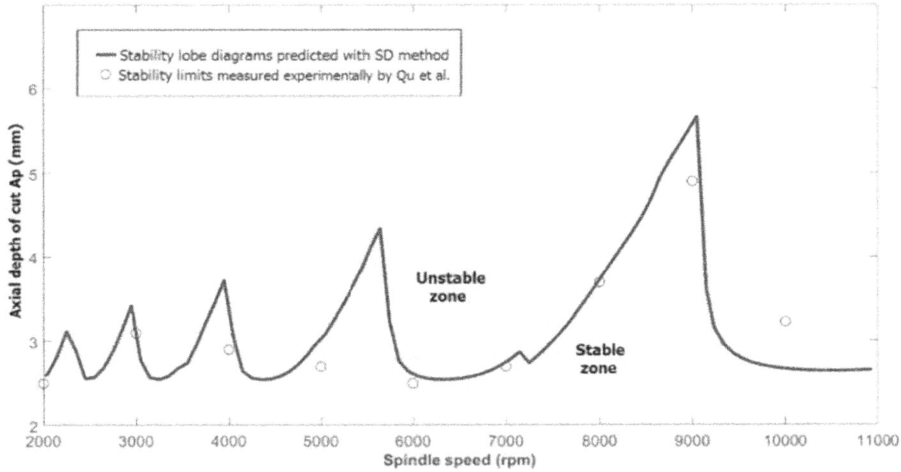

FIGURE 3.41 Tool chatter stability lobes predicted and experimentally measured in high-speed milling of thin-walled parts [181].

the mathematical models to assess the accuracy and reliability of the predictions. An example of comparison between prediction and experimental measurement of chatter stability of thin-walled parts during high-speed milling is shown in Figure 3.41.

These results show a good correlation between the obtained analytical graph of the critical depth of cut vs spindle speed and the experimental results of Qu et al. [177]. This highlights that the stability analysis of the mathematical model with a time delay offers satisfactory predictions of the experimental tests. Advanced experimental or numerical techniques such as modal analysis or frequency response functions can also be employed to improve predictions by identifying the natural frequencies of the machining system and validating the predicted instability regions. The integration of experimental and numerical approaches enhances the understanding of tool chatter as an instability process and contributes to the development of more effective strategies for chatter suppression.

Hence, tool chatter is another instability process that can be predicted and avoided using mathematical modelling, stability methods and experimental validation. The mathematical models describe and provide insights into the dynamic behaviour of the machining system, and stability charts guide the selection of stable cutting conditions. Experimental measurements provide crucial feedback for refining and validating these models, leading to improved predictions and better control of tool chatter in machining processes.

REFERENCES

Wheel Squeal

1. Hoffmann, N., Fischer, M., Allgaier, R., Gaul, L. (2002). A minimal model for studying properties of the mode-coupling type instability in friction induced oscillations. *Mechanics Research Communications*, 29, 197–205.

2. Rudd, M.J. (1976). Wheel/rail noise Part II: Wheel squeal. *Journal of Sound and Vibration*, 46(3), 381–394.
3. Remington, P.J. (1985). Wheel/Rail squeal and impact noise: What do we know? What don't we know? Where do we go from here? *Journal of Sound and Vibration*, 116(2), 339–353.
4. Thompson, D.J., Jones, C.J.C. (2000). A review of the modelling of wheel/rail noise generation. *Journal of Sound and Vibration*, 231(3), 519–536.
5. Thompson, D.J., Squicciarini, G., Ding, B., Baeza, L. (2018). A state-of-the-art review of curve squeal noise: Phenomena, mechanisms, modelling and mitigation. *Notes on Numerical Fluid Mechanics and Multidisciplinary Design*, 139, 3–41.
6. Heckl, M.A., Abrahams, I.D. (2000). Curve squeal of train wheels, Part 1: Mathematical model for its generation. *Journal of Sound and Vibration*, 229(3), 669–693.
7. Chiello, O., Ayasse, J.-B., Vincent, N., Koch, J.-R. (2006). Curve squeal of urban rolling stock—Part 3: Theoretical model. *Journal of Sound and Vibration*, 293(3–5), 710–727.
8. Brunel, J.F., Dufrénoy, P., Naït, M., et al. (2006). Transient models for curve squeal noise. *Journal of Sound and Vibration*, 293(3), 758–765.
9. Heckl, M.A. (2000). Curve squeal of train wheels, Part 2: Which wheel modes are prone to squeal? *Journal of Sound and Vibration*, 229(3), 695–707.
10. Glocker, C., Cataldi-Spinola, E., Leine, R.I. (2009). Curve squealing of trains: Measurement, modelling and simulation. *Journal of Sound and Vibration*, 324(1), 365–386.
11. Meehan, P.A., Liu, X. (2019). Modelling and mitigation of wheel squeal noise under friction modifiers. *Journal of Sound and Vibration*, 440, 147–160.
12. de Beer, F.G., Janssens, M.H.A., Kooijman, P.P. (2003). Squeal noise of rail-bound vehicles influenced by lateral contact position. *Journal of Sound and Vibration*, 267(3), 497–507.
13. Pieringer, A. (2014). A numerical investigation of curve squeal in the case of constant wheel/rail friction. *Journal of sound and Vibration*, 333(18), 4295–4313.
14. Zenzerovic, I., Kropp, W., Pieringer, A. (2018). Influence of spin creepage and contact angle on curve squeal: A numerical approach. *Journal of Sound and Vibration*, 419, 268–280.
15. Cui, X.L., Chen, G.X., Yang, H.G., et al. (2015). Effect of the wheel/rail contact angle and the direction of the saturated creep force on rail corrugation. *Wear*, 330–331, 554–562.
16. Ding, B., Squicciarini, G., Thompson, D. (2019). Effect of rail dynamics on curve squeal under constant friction conditions. *Journal of Sound and Vibration*, 442, 183–199.
17. Meehan, P.A., Liu, X. (2018). Modelling and mitigation of wheel squeal noise amplitude. *Journal of Sound and Vibration*, 413, 144–158.
18. Liu, X., Meehan, P.A. (2016). Investigation of squeal noise under positive friction characteristics condition provided by friction modifiers. *Journal of Sound and Vibration*, 371, 393–405.
19. Liu, X., Meehan, P.A. (2015). Investigation of the effect of crabbing velocity on squeal noise based on a rolling contact two disc test rig. In: *Proceedings of the 10th International Conference of Contact Mechanics and Wear of Rail/wheel Systems CM2015*, Colorado, USA.
20. Meehan, P.A. (2020). Prediction of wheel squeal noise under mode coupling. *Journal of Sound and Vibration*, 465, 115025.
21. Johnson, K.L. (1985). *Contact Mechanics*. Cambridge University Press, Cambridge.
22. Liu, X., Meehan, P.A. (2014). Investigation of the effect of relative humidity on lateral force in rolling contact and curve squeal. *Wear*, 310(1–2), 12–19.
23. Anderson, D., Wheatley, N. (2008). Mitigation of wheel squeal and flanging noise on the Australian Rail Network. In: *Proceedings of the 9th International Workshop on Railway Noise*, Notes on Numerical Fluid Mechanics and Multidisciplinary Design book series (NNFM, volume 99), 2008, 399–405.

24. Thompson, D.J. (2009). *Railway Noise and Vibration: Mechanisms, Modelling and Means of Control*. Elsevier, Oxford.
25. Spurr, R.T. (1961). A theory of brake squeal. *Proceedings of Automotive Division, Institution Mechanical Engineers (AD)*, 1, 33–40.
26. Fourie, D.J., Gräbe, P.J., Heyns, P.S., Fröhling, R.D. (2019). Frequency domain model for railway wheel squeal resulting from unsteady longitudinal creepage. *Journal of Sound and Vibration*, 445, 228–246.
27. Oregui, M., Núñez, A., Dollevoet, R., Li, Z. (2017). Sensitivity analysis of railpad parameters onvertical railway track dynamics. *Journal of Engineering Mechanics*, 143(5), 04017011.
28. Harris, C.M., Piersol, A.G. (2002). *Harris' Shock and Vibration Handbook*, 5th Ed. McGraw Hill, New York.
29. Hoffmann, N., Gaul, L. (2003). Effects of damping on mode-coupling instability in friction induced oscillations. *ZAMM-Zeitschrift fur Angewandte Mathematik und Mechanik*, 83(8), 524–534.

Brake Squeal

30. Jarvis, R.P., Mills, B. (1963). Vibrations induced by friction. *Proceedings of the Institution of Mechanical Engineers*, 178(32), 857–864.
31. Oberst, S., Lai, J.C.S. (2011). Chaos in brake squeal noise. *Journal of Sound and Vibration*, 330(5), 955–975.
32. Kinkaid, N.M., O'Reilly, O.M., Papadopoulos, P. (2003). Automotive disc brake squeal. *Journal of Sound and Vibration*, 267(1), 105–166.
33. Spurr, R.T. (1961). A theory of brake squeal. *Proceedings of Automotive Division, Institution Mechanical Engineers (AD)*, 1, 33–40.
34. North, M.R. (1976). Disc brake squeal. In: *Braking of Road Vehicles, Automobile Division of the Institution of Mechanical Engineers*. Mechanical Engineering Publications Limited, London, England, 169–176.
35. Hoffmann, N., Fischer, M., Allgaier, R., Gaul, L. (2002). A minimal model for studying properties of the mode-coupling type instability in friction induced oscillations. *Mechanics Research Communications*, 29, 197–205.
36. Spurr, R.T. (1971). Brake squeal. Paper No. C95/71, Vibration and Noise in Motor Vehicles, I.Mech.E., 13–16.
37. Earles, S.W.E. (1977). A mechanism of disc-brake squeal. Technical Report 770181, SAE, Warrendale, PA.
38. Hoffmann, N., Gaul, L. (2004). A sufficient criterion for the onset of sprag-slip oscillation. *Archive of Applied Mechanics*, 73, 650–660.
39. Fieldhouse, J., Ashraf, N., Talbot, C. (2008). The measurement and analysis of the disc/pad interface dynamic centre of pressure and its influence on brake noise. SAE Technical Paper 2008-01-0826.
40. Fieldhouse, J., Ashraf, N., Talbot, C., et al. (2006). Measurement of the dynamic center of pressure of a brake pad during a braking operation. SAE Technical Paper 2006-01-3208.
41. Cantone, F., Massi, F. (2011). A numerical investigation into the squeal instability: Effect of damping. *Mechanical Systems and Signal Processing*, 25(5), 1727–1737.
42. Grange, P., Clair, D., Baillet, L., Fogli, M. (2009). Brake squeal analysis by coupling spectral linearization and modal identification methods. *Mechanical Systems and Signal Processing*, 23(8), 2575–2589.
43. Nack, W.V. (2000). Brake squeal analysis by finite elements. *International Journal of Vehicle Design*, 23(3–4), 263–275.

44. Chen, F., Mckillip, D., Luo, J., Wu, S. (2004). Measurement and analysis of rotor in-plane mode induced disc brake squeal and beyond. SAE Technical Paper 2004-01-2798. doi:10.4271/2004-01-2798.
45. Park, C., Han, M., Cho, S.S., et al. (2001). A study on the reduction of disc brake squeal using complex eigenvalue analysis. SAE Technical Paper 2001-01-3141.
46. Papinniemi, A., Zhao, J., Stanef, D., Ding, J. (2005). An investigation of in-plane vibration modes in disc brake squeal noise. SAE Technical Paper 2005-01-3923. doi:10.4271/2005-01-3923.
47. Park, J., Lee, J.C., Cho, S., Yoon, K. (2012). Reduction of brake squeal analyzed in terms of coupling between in-plane and out-of-plane modes. *SAE International Journal of Passenger Cars - Mechanical Systems*, 5(4), 1230–1243.
48. Yokoyama, T., Matsushima, T., Matsui, N., Misumi, R. (2012). A study of reduction for brake squeal in disc in-plane mode. SAE Technical Paper 2012-01-1825.
49. Matsuzaki, M. (1993). Brake noise caused by longitudinal vibration of the disc rotor. SAE Technical Paper 930804.
50. Earles, S., Soar, G. (1974). A vibrational analysis of a pin disc system with particular reference to squeal noise in disc-brakes. In: *1974 Institute of Physics, Stress Analysis Group, Annual Conference*, September, 237–251.
51. Kang, J. (2010). The onset of friction-induced vibration and spragging. *Journal of Sound and Vibration*, 329, 3537–3549.
52. Massi, F., Baillet, L., Giannini, O., Sestieri, A. (2007). Brake squeal: Linear and nonlinear numerical approaches. *Mechanical Systems and Signal Processing*, 21(6), 2374–2393.
53. Stender, M., Tiedemann, M., Hoffmann, L., Hoffmann, N. (2019). Determining growth rates of instabilities from time-series vibration data: Methods and applications for brake squeal. *Mechanical Systems and Signal Processing*, 129, 250–264.
54. Oberst, S., Lai, J.C.S. (2015). Pad-mode-induced instantaneous mode instability for simple models of brake systems. *Mechanical Systems and Signal Processing*, 62–63, 490–505.
55. Nechak, L., Besset, S., Sinou, J.-J. (2018). Robustness of stochastic expansions for the stability of uncertain nonlinear dynamical systems – Application to brake squeal. *Mechanical Systems and Signal Processing*, 111, 194–209.
56. Nobari, A., Ouyang, H., Bannister, P. (2015). Uncertainty quantification of squeal instability via surrogate modelling. *Mechanical Systems and Signal Processing*, 60–61, 887–908.
57. Stender, M., Tiedemann, M., Hoffmann, N., Oberst, S. (2018). Impact of an irregular friction formulation on dynamics of a minimal model for brake squeal. *Mechanical Systems and Signal Processing*, 107, 439–451.
58. Stender, M., Tiedemann, M., Spieler, D., et al. (2021). Deep learning for brake squeal: Brake noise detection, characterization and prediction. *Mechanical Systems and Signal Processing*, 149, 107181.
59. Denimal, E., Sinou, J.-J., Nacivet, S. (2019). Influence of structural modifications of automotive brake systems for squeal events with kriging meta-modelling method. *Journal of Sound and Vibration*, 463, 114938.
60. Murakami, H., Tsunada, N., Kitamura, T. (1984). A study concerned with a mechanism of disc brake squeal. SAE841233.
61. Kim, J., Kim, J., Kim, Y., et al. (2015). An improvement of brake squeal CAE model considering dynamic contact pressure distribution. *SAE International Journal of Passenger Cars - Mechanical Systems*, 8(4), 1188–1195 (SAE Technical Paper 2015-01-2691).
62. Hoffmann, N., Gaul, L. (2003). Effects of damping on mode-coupling instability in friction induced oscillations. *ZAMM-Zeitschrift fur Angewandte Mathematik und Mechanik*, 83(8), 524–534.

63. Meehan, P.A., Leslie, A.C. (2021). On the mechanisms, growth, amplitude and mitigation of brake squeal noise. *Mechanical Systems and Signal Processing*, 152, 107469.
64. Meehan, P.A. (2020). Prediction of wheel squeal noise under mode coupling. *Journal of Sound and Vibration*, 465, 115025.
65. Kinkaid, N.M., O'Reilly, O.M., Papadopoulos, P. (2003). Automotive disc brake squeal. *Journal of Sound and Vibration*, 267(1), 105–166.
66. Feeny, B., Moon, F.C. (1994). Chaos in a forced dry-friction oscillator: Experiments and numerical modelling. *Journal of Sound and Vibration*, 170(3), 303–323.
67. Eriksson, M., Jacobson, S. (2001). Friction behaviour and squeal generation of disc brakes at low speeds. *Proceedings of the Institution of Mechanical Engineers*, 215(D), 1245–1256.
68. Soobbarayen, K., Besset, S., Sinou, J-.J. (2015). A simplified approach for the calculation of acoustic emission in the case of friction-induced noise and vibration. *Mechanical Systems and Signal Processing*, 50–51, 732–756.
69. Blevins, R.D. (1979). *Formulas for Natural Frequency and Mode Shape*. Van Nostrand Reinhold Co., New York.
70. Meehan, P.A. (2020). Investigation of chaotic instabilities in railway wheel squeal. *Nonlinear Dynamics*, 100, 159–172.
71. Johnson, K.L. (1985). *Contact Mechanics*. Cambridge University Press, Cambridge.
72. Pohrt, R., Popov, V. (2013). Contact stiffness of randomly rough surfaces. *Scientific Reports*, 3, 3293.

Flutter

73. Pines, S. (1958). An elementary explanation of the flutter mechanism. In: *Proceedings of Dynamics and Aeroelasticity Meeting*, I. A. S., New York, 1958, 52–59.
74. Blevins, R.D. (1990). *Flow Induced Vibrations*, 2nd Ed. Van Nostrand Renhold, New York, USA
75. Venkatramani, J., Sarkar, S., Gupta, S. (2018). Investigations on precursor measures for aeroelastic flutter. *Journal of Sound and Vibration*, 419, 318–336. doi:10.1016/j.jsv.2018.01.009.
76. de Henshaw, M.J. C., Badcock, K.J., Vio, G.A., et al. (2007). Non-linear aeroelastic prediction for aircraft applications. *Progress in Aerospace Sciences*, 43(4–6), 65–137.
77. Sisto, F., Thangam, S., Abdelrahim, A. (1991). Computational prediction of stall flutter in cascaded airfoils. *AIAA Journals*, 29(7), 1161–1167.
78. Zhou, R., Ge, Y., Yang, Y., et al. (2018). Wind-induced nonlinear behaviors of twin-box girder bridges with various aerodynamic shapes. *Nonlinear Dynamics*, 94(2), 1095–1115. doi:10.1007/s11071-018-4411-y.
79. Huang, C., Huang, J., Song, X., et al. (2021). Three dimensional aeroelastic analyses considering free-play nonlinearity using computational fluid dynamics/computational structural dynamics coupling. *Journal of Sound and Vibration*, 494. doi:10.1016/j.jsv.2020.115896.
80. Zhao, L.C., Yang, Z.C. (1990). Chaotic motions of an airfoil with non-linear stiffness in incompressible flow. *Journal of Sound and Vibration*, 138(2), 245–254.
81. Dai, H., Yue, X., Xie, D., Atluri, S.N. (2014). Chaos and chaotic transients in an aeroelastic system. *Journal of Sound and Vibration*, 333(26), 7267–7285.
82. Ghommem, M., Hajj, M.R., Nayfeh, A.H. (2010). Uncertainty analysis near bifurcation of an aeroelastic system. *Journal of Sound and Vibration*, 329(16), 3335–3347.
83. Xie, D., Xu, M., Dai, H., Dowell, E.H. (2015). Proper orthogonal decomposition method for analysis of nonlinear panel flutter with thermal effects in supersonic flow. *Journal of Sound and Vibration*, 337, 263–283.

84. Brouwer, K.R., Perez, R.A., Beberniss, T.J., et al. (2021). Investigation of aeroelastic instabilities for a thin panel in turbulent flow. *Nonlinear Dynamics*, 104(4), 3323–3346.
85. dos Santos, L.G.P., Marques, F.D., Vasconcellos, R.M.G. (2022). Dynamical characterization of fully nonlinear, nonsmooth, stall fluttering airfoil systems. *Nonlinear Dynamics*, 107(3), 2053–2074.
86. Esbati Lavasani, R., Shams, S. (2020). A new dynamic stall approach for investigating bifurcation and chaos in aeroelastic response of a blade section with flap free-play section. *International Journal of Bifurcation and Chaos*, 30(14), 2050200.
87. Tian, W., Yang, Z., Zhao, T. (2019). Nonlinear aeroelastic characteristics of an all-movable fin with freeplay and aerodynamic nonlinearities in hypersonic flow. *International Journal of Non-Linear Mechanics*, 116, 123–139.
88. Moon, F.C. (1992). *Chaotic and Fractal Dynamics: An Introduction for Applied Scientists and Engineers*. Wiley Interscience, New York.
89. Meehan, P.A. (2022). Flutter prediction of its occurrence, amplitude and nonlinear behaviour. *Journal of Sound and Vibration*, 535, 117117.
90. Meehan, P. A. (2023). Prediction and suppression of chaos following flutter in wind turbines. *Nonlinear Dynamics*, 111(24), 22153–22176.
91. Bouma, A., Yossri, W., Vasconcellos, R., Abdelkefi, A. (2022). Investigations on the interactions between structural and aerodynamic nonlinearities and unsteadiness for aeroelastic systems. *Nonlinear Dynamics*, 107(1), 331–355. doi:10.1007/s11071-021-07011-z.
92. Robinson, B., da Costa, L., Poirel, D., et al. (2020). Aeroelastic oscillations of a pitching flexible wing with structural geometric nonlinearities: Theory and numerical simulation. *Journal of Sound and Vibration*, 484. doi:10.1016/j.jsv.2020.115389.
93. Guo, M., Zheng, G. (2021). Stigma as two degrees of freedom energy sink for flutter suppression. *Journal of Sound and Vibration*, 515. doi:10.1016/j.jsv.2021.116441.
94. Tian, H., Shan, X., Cao, H., et al. (2021). A method for investigating aerodynamic load models of piezoaeroelastic energy harvester. *Journal of Sound and Vibration*, 502. doi:10.1016/j.jsv.2021.116084.
95. Stender, M., Tiedemann, M., Hoffmann, N. (2019). Energy harvesting below the onset of flutter. *Journal of Sound and Vibration*, 458, 17–21.
96. Meehan, P.A. (2020). Prediction of wheel squeal noise under mode coupling. *Journal of Sound and Vibration*, 465, 115025.
97. Meehan, P.A., Leslie, A.C. (2021). On the mechanisms, growth, amplitude and mitigation of brake squeal noise. *Mechanical Systems and Signal Processing*, 152, 107469.
98. Meehan, P.A., Liu, X. (2018). Modelling and mitigation of wheel squeal noise amplitude. *Journal of Sound and Vibration*, 413, 144–158.
99. Scanlan, R.H., Jones, N.P., Singh, L. (1997). Inter-relations among flutter derivatives. *Journal of Wind Engineering and Industrial Aerodynamics*, 69–71, 829–837.
100. Le Maître, O.P., Scanlan, R.H., Knio, O.M. (2003). Estimation of the flutter derivatives of an NACA airfoil by means of Navier–Stokes simulation. *Journal of Fluids and Structures*, 17(1), 1–28.
101. Shi, J., Hitchings, D. (1994). Calculation of flutter derivatives and speed for 2-D incompressible flow by the finite-element method. *Applied Mathematical Modelling*, 18(10), 538–549.
102. Brunton, S.L., Rowley, C.W. (2013). Empirical state-space representations for Theodorsen's Lift Model. *Journal of Fluids and Structures*, 38, 174–186.
103. Perry, B. (2020). Re-computation of numerical results contained in NACA Report No. 685. NASA/TP-2020–220562.
104. Miley, S.J. (1982). Catalog of low-Reynolds-number airfoil data for wind-turbine applications. Texas A and M Univ., College Station (USA). Dept. of Aerospace Engineering, Technical Report RFP-3387 ON: DE82021712, 1982.

105. Theodorsen, T. (1935). General theory of aerodynamic instability and the mechanism of flutter. NACA Report 496.
106. Theodorsen, T., Garrick, I.E. (1940). Mechanism of flutter: A theoretical and experimental investigation of the flutter problem. NACA Report 685.
107. Spurr, R.T. (1961). A theory of brake squeal. *Proceedings of Automotive Division, Institution Mechanical Engineers (AD)*, 1, 33–40.
108. Dowell, E.H., Scanlan, R.H., Sisto, F., et al. (2021). A modern course in aeroelasticity. In: *Solid Mechanics and its Applications*, vol 264, 6th Ed. Springer Science. Dordrecht, Netherlands

Rolling Mill Chatter

109. Roberts, W.L. (1988). *Flat Processing of Steel*. Marcel Dekker, New York.
110. Paton, D.L., Critchley, S. (1985). Tandem mill vibration: Its cause and control. *Iron and Steel Making*, March, 37–43.
111. Tlusty, J., Chandra, G., Critchley, S., Paton, D. (1982). Chatter in cold rolling. *Annals of CIRP*, 31(1), 195–199.
112. Pawelski, O., Rasp, W., Friedewald, K. (1987). Theory of interaction of plastic and elastic deformations. In: *4th International Steel Rolling Conference*, Deauville, France, June 1–3, E.11.1–E.11.5.
113 Tamiya, T., Furui, K., Iida, H. (1980). Analysis of chattering phenomenon in cold rolling. In: *Proceedings of the International Conference on Steel Rolling*, ISIJ, Tokyo, 1191–1202.
114. Yun, I.-S., Wilson, W.R.D., Ehmann, K.F. (1998). Chatter in the strip rolling process. *Transactions of the ASME, Journal of Manufacturing Science and Engineering*, 120(2), 330–348.
115. Meehan, P.A. (2002). Vibration instability in rolling mills: Modelling and experimental results. *The Journal of Vibration and Acoustics*, 124, 221–227.
116. Bryant, G.F. (1973). *Automation of Tandem Mills*. Iron & Steel Inst. Publication No. 160, London.
117. Edwards, W.J. (1975). Design of a cold rolling mill thickness controller incorporating tension variation. *Journal of the Australian Institute of Metals*, 20(1), 59–67.
118. Yarita, I., Furukawa, K., Seino, Y., et al. (1978). An analysis of chattering in cold rolling for ultrathin gauge steel strip. *Transactions of the Iron and Steel Institute, Japan*, 18, 1–10.
119. Guo, R. (1994). Material damping effect in cold rolling process. *Iron and Steel Engineer*, December, 28–35.
120. Thomson, W.T. (1988). *Theory of Vibration with Applications*. Allen & Unwin, London.
121. Cockerell, R.A., Edwards, W.J., Spooner, P.D., Thomas, P.J. (1993). A dynamic rolling mill simulator for all reasons. In: *12th IFAC Congress*, Sydney, 29–36.
122. Boulton, G., Domanti, T., Edwards, J., et al. (2000). Modelling and suppression of rolling mill chatter. Association of Iron and Steel Engineers.

Rail Corrugation

123. Both, J.A., Hong, D.C., Kurtze, D.A. (2001). Corrugation of roads. *Physica* A, 301, 545–559. Yun, I.S., Wilson, W.R.D., Ehmann, K.F. (1998). Review of chatter studies in cold rolling. *International Journal of Machine Tools & Manufacture*, 38, 1499–1530.
124. Shelley, S., Zwart, J., Fournier, A. (1997). New insights into calender barring prevention. In: *Proceedings of the 1997 83rd Annual Meeting of the Technical Section of Canadian Pulp and Paper Association. Part A*. 1997, A251–A258.

125. Grassie, S.L., Kalousek, J., Rail corrugation: characteristics, causes and treatments. *Proceedings of the Institution of Mechanical Engineers, Part F*, 207, 57–68.
126. Nelson, J.T. (1997). *Transit Cooperative Research Program Report 23, "Wheel/Rail Noise Control Manual"*. ch. 10. National Academy Press, Washington, DC.
127. Muller, S. (2011). A linear wheel-rail model to investigate stability and corrugation on straight track. *Wear*, 249, 1117–1127.
128. Sato, Y., Matsumoto, A., Knothe, K. (2002). Review on rail corrugation studies. *Wear*, 253, 130–139.
129. Hempelmann, K., Knothe, K. (1996). An extended linear model for the prediction of short pitch corrugation. *Wear*, 191, 161–169.
130. Igeland, A., Ilias, H. (1997). Rail head corrugation growth predictions based on non-linear high frequency vehicle/track interaction. *Wear*, 213, 90–97.
131. Matsumoto, A., Sato, Y., Tanimoto, M., Qi, K. (1996). Study on the formation mechanism of rail corrugation on curved track. *Vehicle System Dynamics*, 25, 450–465.
132. Wu, T.W., Thompson, D.J. (2003). An investigation into rail corrugation due to microslip under multiple wheel/rail interactions. In: *Proceedings of the 6th International Conference on Contact Mechanics and Wear of Rail/Wheel Systems (CM2003) in Gothenburg*, Sweden, June 10–13, 2003, 59–67.
133. Bhaskar, A., Johnson, K.L., Wood, G.D., Woodhouse, J. (1997). Wheel–rail dynamics with closely conformal contact. Part 1. Dynamic modelling and stability analysis. *Proceedings of the Institution of Mechanical Engineers*, 211(F), 11–26.
134. Nielsen, J.B. (1999). Evolution of rail corrugation predicted with a non-linear wear model. *Journal of Sound and Vibration*, 227, 915–933.
135. Frederick, C.O. (1987). A rail corrugation theory. In: *Proceedings of the 2nd International Conference on Contact Mechanics and Wear of Rail/Wheel Systems in Kingston*, Rhode Island, July, Waterloo Press, 1987, 181–211.
136. Bellette, P.A., Meehan, P.A., Daniel, W.J.T. (2010). Contact induced wear filtering and its influence on corrugation growth. *Wear*, 268(11–12), 1320–1328.
137. Johnson, K.L., Gray, G.G. (1975). Development of corrugations on surfaces in rolling contact. *Proceedings of the Institution of Electrical Engineers*, 189, 567–580.
138. Johansson, A. (2005). Out-of-round railway wheels – causes and consequences. PhD Thesis, Chalmers University of Technology, Goteborg, Sweden, F15–18.
139. Igeland, A. (1996). Railhead corrugation growth explained by dynamic interaction between track and bogie wheelsets. *Proceedings of the Institution of Mechanical Engineers, Part F: Journal of Rail and Rapid Transit*, 210, 11–20.
140. Meehan, P.A., Daniel, W.J.T., Campey, T. (2005). Prediction of growth of wear-type rail corrugation. *Wear*, 258, 1001–1013.
141. Meehan, P.A., Daniel, W.J.T., Campey, T. (2003). Wear-type rail corrugation prediction and prevention. In: *Proceedings of the 6th International Conference on Contact Mechanics and Wear of Rail/Wheel Systems(CM2003) in Gothenburg*, Sweden, June 10–13, 2003, 445–454.
142. Meehan, P.A., Daniel, W.J.T. (2008). Effects of wheel passing frequency on wear-type corrugations. *Wear*, 265(9–10), 1202–1211.
143. Meehan, P.A., Bellette, P.A., Batten, R.D., et al. (2009). A case study of wear-type rail corrugation prediction and control using speed variation. *Journal of Sound and Vibration*, 325, 85–105.
144. Bellette, P.A., Meehan, P.A., Daniel, W.J.T. (2011). Validation of a tangent track corrugation model with a two disk test rig. *Wear*, 271(1–2), 268–277.
145. Batten, R.D., Bellette, P.A., Meehan, P.A., et al. (2011). Field and theoretical investigation of the mechanism of corrugation wavelength fixation under speed variation. *Wear*, 271(1–2), 278–286.

146. Meehan, P.A., Bellette, P.A., Horwood, R.J. (2014). "Does god play dice with corrugations?": Environmental effects on growth. *Wear*, 314(1–2), 254–260.
147. Ding, B., Squicciarini, G., Thompson, D., Corradi, R. (2018). An assessment of mode-coupling and falling-friction mechanisms in railway curve squeal through a simplified approach. *Journal of Sound and Vibration*, 423, 126–140.

Transmission Driveline Instability

148. Porter, B. (1961). A theoretical analysis of the torsional oscillation of a system incorporating a Hooke's joint. *The Journal of Mechanical Engineering Science*, 3, 324–329.
149. Porter, B., Gregory, R.W. (1963). Non-linear torsional oscillation of system incorporating a Hooke's joint. *The Journal of Mechanical Engineering Science*, 5, 191–200.
150. Porter, B. (1964). Non-linear torsional oscillation of a two-degree-of-freedom system incorporating a Hooke's joint. *Proceedings of the Royal Society of London Series A*, 277, 92–106.
151. Asokanthan, S.F., Hwang, M.C. (1996). Torsional instabilities in a system incorporating a Hooke's Joint. *ASME Journal of Vibration and Acoustics*, 118, 368–374.
152. Asokanthan, S.F., Wang, X.H. (1996). Characterisation of torsional instabilities in a Hooke's joint driven system via maximal Lyapunov exponent. *Journal of Sound and Vibration*, 194(1), 83–91.
153. Birkett, C.A. (1989). Computer simulation of driveline vibration due to universal joints in heavy and medium duty trucks. *SAE Transactions on Commercial Vehicles*, 100, 642–646.
154. Lu, G., Sharp, R.S. (1990). A simulation study of some starting behaviour problems of diesel hydraulic locomotives. *Proceedings of the Institution of Mechanical Engineers: Part F: Journal of Rail and Rapid Transit*, 204(F2), 81–90.
155. Asokanthan, S.F., Meehan, P.A. (2000). Non-linear vibration of a torsional system driven by a Hooke's joint. *Journal of Sound and Vibration*, 233(2), 297–310.
156. Bogoliubov, N.N., Itropolsky, Y.A.M. (1961). *Asymptotic Methods in the Theory of Non-linear Oscillations*. Gordon and Breach, New York.
157. Ariaratnam, S.T., Asokanthan, S.F. (1987). Dynamic stability of chain drives. *Journal of Mechanisms, Transmissions and Automation in Design (ASME)*, 109, 412–418.
158. Asokanthan, S.F., Ariaratnam, S.T. (1994). Flexural instabilities in axially moving bands. *ASME Journal of Vibration and Acoustics*, 116, 275–279.
159. Nusse, H.E., Yorke, J.A., Kostelich, E. (1994). *Dynamics: An Interactive Program for IBM-Compatible PCs and for; NIX Computers (Computer File)*. Springer Verlag, New York.
160. Wolf, A., Swift, J.B., Swinney, H.L., Vasano, J.A. (1985). Determining Lyapunov exponents from a time series. *Physica* D, 16, 285–317.
161. Hilborn, R.C. (1994). *Chaos and Nonlinear Dynamics: An Introduction for Scientists and Engineers*. Oxford University Press, New York.

Other Instability Phenomena

162. Gold, T. (1948). Hearing, II. The physiological basis of the action of the cochlea. *Proceedings of the Royal Society of Edinburgh*, 135, 492–498.
163. Kemp, D.T. (1978). Stimulated acoustic emissions from within the human auditory system. *Journal Acoustical Society of America*, 64, 1386–1391.
164. Dallos, P., Evans, B.N., Hallworth, R. (1991). Nature of the motor element in electrokinetic shape changes of cochlear outer hair cells. *Nature*, 350, 155–157.

165. Huang, G., Santos-Sacchi, J. (1994). Motility voltage sensor of the outer hair cell resides within the lateral plasma membrane. *Proceedings of the National Academy of Sciences of the United States of America*, 91, 12268–12272.
166 Zheng, J., Shen, W., He, D.Z., et al. (2000). Prestin is the motor protein of cochlear outer hair cells. *Nature*, 405, 149–155.
167. Dong, X.X., Ospeck, M., Iwasa, K.H. (2002). Piezoelectric reciprocal relationship of the membrane motor in the cochlear outer hair cell. *Biophysical Journal*, 82, 1254–1259.
168. Spector, A.A., Brownell, W.E., Popel A.S. (2003). Effect of outer hair cell piezoelectricity on high-frequency receptor potentials. *Journal of the Acoustical Society of America*, 113, 453–461.
169. von Bekesy G. (1960). *Experiments in Hearing*. McGraw-Hill, New York.
170. Neely, S.T., Kim, D.O. (1986). A model for active elements in cochlear biomechanics. *The Journal of the Acoustical Society of America*, 79(5), 1472–1480.
171. de Boer, E. (1995). On equivalence of locally-active models of the cochlea. *Journal of the Acoustical Society of America,* 98, 1400–1409.
172. Geisler, C.D., Sang, C. (1995). A cochlear model using feed-forward outer-hair-cell forces. *Hearing Research*, 86, 132–146.
173. Geisler, C.D. (1998). *From Sound to Synapse: Physiology of the Mammalian Ear.* Oxford University Press, New York.
174. Basit, A., Khan, N.B., Ali, S., et al. (2024). Chatter detection and suppression in machining processes: a comprehensive analysis. *International Journal of Interactive Design and Manufacturing*, 18, 3751–3771.
175. Tobias, S.A., Fishwick, W. (1958). A theory of regenerative chatter. The Engineer-London.
176. Zhang, X., Wan, L., Ran, X. (2024). Research progress on the chatter stability in machining systems. *International Journal of Advanced Manufacture Technology*, 131, 29–62.
177. Qu, S., Zhao, J., Wang, T. (2016). Three-dimensional stability prediction and chatter analysis in milling of thin-walled plate. *International Journal of Advanced Manufacture Technology*, 86, 2291–2300.
178. Mabie, H.H., Reinholtz, C.F. (1987). *Mechanisms and Dynamics of Machinery.* John Wiley & Sons, New York, 45.
179. Schwander, M., Kachar, B., Müller, U. (2010). The cell biology of hearing. *The Journal of Cell Biology*, 190, 9–20.
180. Ruggero, M.A., Robles, L., Rich, N.C., Recio, A. (1992). Basilar membrane responses to two-tone and broadband stimuli. *Philosophical Transactions of the Royal Society B*, 336, 307–315.
181. Damous, M., Zeroudi, N., Chellil, A., et al. (2023). Prediction of a chatter stability of thin-walled parts during high-speed milling considering a variation of dynamic characteristics based on surface roughness measurement. *International Journal of Advanced Manufacture Technology*, 127, 5729–5744.

4 Predicting Chaotic Instabilities in Engineering Systems

Often, I wake up wondering whether there will be sunny blue skies to enjoy, even though sophisticated algorithms have been used to predict the weather the day before. This level of uncertainty arises not from inaccuracies in the measurements or predictive models but from the underlying nonlinear behaviour of the weather system, known as chaos or "the butterfly effect." Chaotic instabilities (or chaos following classical instability) represent a fascinating and important nonlinear behaviour that can occur in engineering systems. They are characterised by random-like motion, sensitivity to initial conditions and long-term unpredictability. These phenomena manifest across a diverse array of deterministic applications, highlighting the ubiquitous nature of chaos in various engineering domains. The reason for its broad application is that chaotic dynamics is not restricted to a specific system in one area of science but rather it emerges from the underlying mathematics, namely, nonlinearity. This chapter will cover a wide range of these but is by no means exhaustive.

An example is the occurrence of chaos following railway wheel and brake squeal. Here, the periodic vibrations break into larger erratic ones, causing higher noise generation and degradation. These chaotic behaviours pose significant challenges to the design and operation of railway and automotive systems, necessitating a deep exploration of their underlying dynamics to develop mitigation techniques. Beyond railway vehicles and cars, chaotic instabilities manifest in flutter, aeroelastic phenomena affecting aircraft wings and structures, leading to potentially catastrophic consequences if not comprehensively understood and managed. In dragline motion, used extensively in mining operations, chaotic dynamics can affect the stability and control of these very large robots, influencing the safety and efficiency of the mining process. Moreover, chaos finds its way into spacecraft attitude motion, impacting the stability and orientation of satellites and spacecraft during mission-critical manoeuvres and communications. The human body itself, a complex biomechanical system, seems to not be immune to chaotic instabilities and indeed utilises them for healthy functioning. From irregular heart rhythms to the intricate dynamics of neuronal firing, chaos emerges as a crucial aspect of understanding physiological processes.

Mathematical modelling and nonlinear dynamics analysis tools have become indispensable in unravelling the complexities associated with these chaotic instabilities in engineering systems. Through the application of these tools, engineers and scientists strive to predict, control and harness chaotic behaviour for enhanced system performance and reliability. This exploration into chaotic instabilities underscores the interdisciplinary nature of engineering, where mathematics and dynamics

DOI: 10.1201/9781003496533-4

converge to decipher the intricate dance of chaos within complex systems. Hence, we will use these tools to look at the prediction of chaotic phenomena in this range of systems, following the onset of classic instability detailed in Chapter 3. By systematically examining chaotic instabilities across a spectrum of engineering scenarios, this chapter provides insights into predictive methodologies crucial for predicting when and how engineering chaos occurs.

4.1 RAILWAY WHEEL SQUEAL CHAOS

Wheel squeal arises from a friction-induced instability, where a wheel mode of vibration grows into a nonlinear limit cycle. Several mechanisms have been studied, such as the falling friction theory, where an unstable slope in the traction–creep curve excites the wheel [1–4]. More recently, a modal coupling mechanism has been investigated, where multiple interacting modes cause a phase shift between the frictional force and displacement, leading to instability [5–14]. Studies [10,11] clarified the effects of both mechanisms, demonstrating that modal coupling can produce large nonlinear limit cycles. Further research [12] identified chaotic vibrations in wheel squeal, a behaviour also observed in other friction-coupled systems (e.g., [15]). Feeny and Moon [16] experimentally and numerically analysed chaos in a fundamental harmonically forced spring–mass system with dry friction showing, funnel-like chaotic attractors across different friction models. Shin et al. [17] studied a simplified two-degree-of-freedom brake squeal system, representing the brake pad and disc. Their numerical stability analysis showed that the proximity of modes and damping can either stabilise or destabilise the system, and they found large chaotic-like oscillations after period doubling [18]. Chaotic instability was later confirmed through numerical and experimental measurements of full-scale brake squeal, capturing the transition from a limit cycle to an unstable torus attractor [15]. More recently, Lin et al. [19] successfully controlled brake disc squeal at lower damping using state feedback control.

Research on railway wheel squeal chaos appears only to be beginning with previous works focussed on other friction-coupled systems. The next section details how we can predict chaotic instabilities in wheel squeal. The critical stability behaviour leading to the onset of chaotic instability is reviewed first. Then, chaos is subsequently confirmed and quantified under "falling friction" and mode-coupled conditions using the tools discussed in Chapter 2 based on [12].

Prediction of Wheel Squeal Chaos

In Chapter 3, a wheel squeal model, including mode-coupled dynamics and a nonlinear creep model, was detailed based on [11,20] and [12]. The analytical methods for predicting squeal local instability and limit cycle amplitude were then described [11]. We will now use this to derive a conservative condition for the prediction of chaotic instability following wheel squeal.

For convenience, we review the limit cycle analysis results provided in Chapter 3. Squeal instability initially grows exponentially, as predicted by local stability analysis, before stabilising into a nonlinear limit cycle. This transition occurs when creep oscillations reach the highly damped slip region below the critical creep level, effectively limiting further growth. At this steady state, the wheel limit cycle amplitude, $A_w\omega_i$, can be determined by balancing the energy input and dissipation over the cycle, as described in [11]

$$A_w\omega_i = \left((V_c/\cos(\theta_c)) - \zeta_c V_0\right)\cos(\theta)/\cos(\omega_i t_c)$$
$$\text{where} \quad \sin(2\omega_i t_c) - 2\omega_i t_c = 2\pi\left(V_0 c_{si} + k_{Fi}k_{\zeta 2}k_{\zeta 3}\right)/\left(k_{Fi}k_{\zeta 3}\left(k_{\zeta 1} - k_{\zeta 2}\right)\right), \tag{4.1}$$

where ς_c and t_c represent the critical creep and corresponding time in the squeal cycle, respectively, occurring at the peak of the creep curve. Equation (4.1) provides a closed-form analytical solution for the wheel squeal limit cycle amplitude, incorporating both mode coupling and falling friction mechanisms, under the assumption of bounded positive creep oscillations. This limit cycle solution is valid within the bilinear creep approximation, where the system remains stable due to a balance between negative damping and the steeply rising creep slope. However, if the effective negative damping exceeds this balance, the oscillations can continue growing to highly nonlinear amplitudes. At extreme creep oscillations, the system may enter chaotic instability, where oscillations fluctuate unpredictably between large positive and negative creepage and forces. In this case, the creep curve behaviour shifts from a bilinear to a trilinear model, making it impossible to derive closed-form solutions for all conditions. This behaviour resembles the chaotic friction oscillator studied in [16]. The transition from a predictable limit cycle to chaotic dynamics can be approximated by a necessary condition for squeal amplitude, as

$$A_w\omega_i \geq V_c\cos(\theta)/\cos(\theta_c) \quad \text{for} \quad V_c > \zeta_c V_o. \tag{4.2}$$

This condition can be analytically solved using equation (4.1); however, for large creep oscillations under mode coupling, the impact of parametric excitation should also be considered. Parametric excitation occurs when system parameters vary at specific frequencies, amplifying motion. A simple example is pumping a swing without an external force by shifting the centre of mass or point of suspension at twice the natural frequency [21,22]. At zero creep ($\zeta=0$), the friction coefficient vanishes, making the modal equation of motion non-autonomous if the creep curve is treated as an external variable. Since the creep curve is influenced by external factors, such as humidity, rainfall, contact geometry and third-body debris or lubricant, large negative creep oscillations under mode coupling lead to significant parametric variations in system modal parameters. This can introduce parametric instability, further amplifying the limit cycle oscillations and causing squeal to increase until the negative critical creep threshold is exceeded. For wheel squeal induced by mode coupling, the dominant vibration mode (derived in Chapter 3) can experience substantial parametric oscillations in key system parameters, including the effective system damping term $c_{si} + k_{Fi}k_{\varsigma 1,2}k_{\varsigma 3}/V_0$, the stiffness, k_i, and the mass, m_i. These parameters depend on the friction coefficient, $\mu(\zeta)$, which varies with lateral creepage as a function of time. The resulting parametric excitation can introduce additional negative damping, further increasing squeal amplitude. To quantify this effect, the modal equation is reformulated into a fundamental form, aligning with the framework in [23] as

$$\ddot{y}_i(t) + p_{1i}(t)\dot{y}_i(t) + p_{2i}(t)y_i(t) = 0, \tag{4.3}$$

where

$$p_{1i}(t) = \left(c_{si}(t) + k_{Fi}k_{\zeta 1,2}k_{\zeta 3}/V_0\right)/m_i(t), \quad p_{2i}(t) = \text{Re}[k_i(t)/m_i(t)], \tag{4.4}$$

and the parametric coefficients p_{ni} are periodic functions as the creep oscillates during squeal limit cycles. Equation (4.3) accounts for complex roots and includes subscripts $i=1,2$ representing the two degrees of freedom in the system, as illustrated in Figure 3.2. Based on Floquet theory and the approach in [23], equation (4.3) can be transformed into

$$\ddot{x}_i(t) + p_i(t)x_i(t) = 0 \tag{4.5}$$

where

$$x_i(t) = y_i(t)\exp\left[\tfrac{1}{2}\int p_{1i}(t)dt\right] \quad \text{and} \quad p_i(t) = p_{2i}(t) - \tfrac{1}{4}p_{1i}^2(t) - \tfrac{1}{2}\dot{p}_{1i}(t). \tag{4.6}$$

The bilinear shape of the creep curve introduces a significant oscillation component at twice the limit cycle frequency, ω_{pi}, due to the presence of two distinct slopes on either side of the critical creep point. As a result, frictional creep oscillations can be linearised around their nominal creep conditions, treating them as small sinusoidal perturbations. These perturbations occur at twice the squeal frequency, reflecting the bilinear nature of the positive side of the creep curve, and can be mathematically expressed as

$$p_i(t) = \omega_{pi}^2\left(1 + f_{oi}\cos(2\omega_{pi}t)\right), \tag{4.7}$$

$$p_{1i}(t) \approx \omega_{oi}b_i(1 + \Delta p_{1i}\cos(2\omega_{pi}t)), \quad \Delta p_{1i} = \int_{-\pi}^{\pi} p_{1i}\cos(2\omega_{pi}t)d(\omega_{pi}t), \tag{4.8}$$

$$p_{2i}(t) \approx \omega_{oi}{}^2(1 + \Delta p_{2i}\cos(2\omega_{pi}t)), \quad \Delta p_{2i} = \int_{-\pi}^{\pi} p_{2i}\cos(2\omega_{pi}t)d(\omega_{pi}t), \tag{4.9}$$

where the parametric excitation amplitude, f_{oi}, is

$$\begin{aligned} f_{oi} &= \max\left[\sqrt{f_{o1i}^2 + f_{o2i}^2}, f_{o3i}\right] \quad \text{where} \\ f_{o1i} &= \left(\Delta p_{2i} - \tfrac{1}{2}b_i^2\Delta p_{1i}\right)/\left(1 - \tfrac{1}{4}b_i^2\right), \\ f_{o2i} &= (-b_i\omega_{pi}\Delta p_{1i})/\left(\omega_{oi}\left(1 - \tfrac{1}{4}b_i^2\right)\right), \\ f_{o3i} &= \left(b_i^2\Delta p_{1i}^2\right)/\left(8\left(1 - \tfrac{1}{4}b_i^2\right)\right). \end{aligned} \tag{4.10}$$

Equations (4.5) and (4.7) correspond to the well-known Mathieu's equation, which exhibits maximum parametric resonance at twice the limit cycle frequency. This resonance grows exponentially at a rate of $f_{oi}\omega_{pi}/4$ for the transformed variable $x_i(t)$, as outlined in [21]. By transforming back to the modal variable, $y_i(t)$, using equation (4.6), the system's modal damping, now incorporating parametric excitation, can be expressed as

$$c_{\text{spi}} = \left(\frac{1}{\tau_i} \int_0^{\tau_i} p_{1i}(t)\,dt - f_{oi}\omega_p/2 \right) m_i, \quad \tau_i = 2\pi/\omega_{pi}. \tag{4.11}$$

The magnitude of parametric excitation is directly linked to the size of the creep oscillation. Chaos may occur if the effective system damping – which accounts for mode coupling, falling friction and parametric excitation – becomes sufficiently negative to push squeal vibrations beyond the critical amplitude of negative creep oscillations, as defined in equation (4.2). However, solutions to equations (4.8)–(4.11) are only approximations as they assume small perturbations (following [21]). Additionally, they require multiple integrations, making them computationally complex and likely necessitating use of numerical methods, which reduces the efficiency of an analytical solution compared to directly integrating the equations of motion.

A simpler, more efficient analytical approach can be achieved by assuming that negative creep oscillations first arise due to stiffness mode coupling, where average creep is highest. A simplified criterion for chaos can then be derived based on stiffness mode coupling, using the maximum nonlinear creep at the angle of attack and the average value over a critical creep oscillation, including the critical creep level, ζ_{crit},

$$\Delta(\mu_{\text{ave}}) = 0,$$

$$\mu_{\text{ave}} \approx \max\left[\mu\left(\frac{\theta_A}{\cos(\theta)}\right), \int_0^{2\pi} \mu\left(\frac{\theta_A \sin(\tau)\left(1-\left(1-\zeta_{\text{crit}}\cos(\theta)/\theta_A\right)\right)}{\cos(\theta)}\right) d\tau \Big/ 2\pi \right]. \tag{4.12}$$

These conservative analytical criteria for wheel squeal chaos of (4.8)–(4.11) and (4.12) are compared with numerical solutions of the wheel squeal equations of motion in the subsequent section.

Results of Wheel Squeal Chaos Predictions

Chaotic instability in railway wheel squeal was investigated by simulations of the full nonlinear time-domain model (as described in section 3.1). Time history analysis, phase space representations, bifurcation diagrams, Poincaré maps and Lyapunov exponents were used to characterise the nonlinear behaviour. In particular, squeal was examined across a range of contact angles and angles of attack to determine the conditions for chaotic instability onset. The analytical criteria were validated against numerical simulations, helping identify and quantify the conditions leading to chaos. The system parameters used were based on previous squeal studies [11,24].

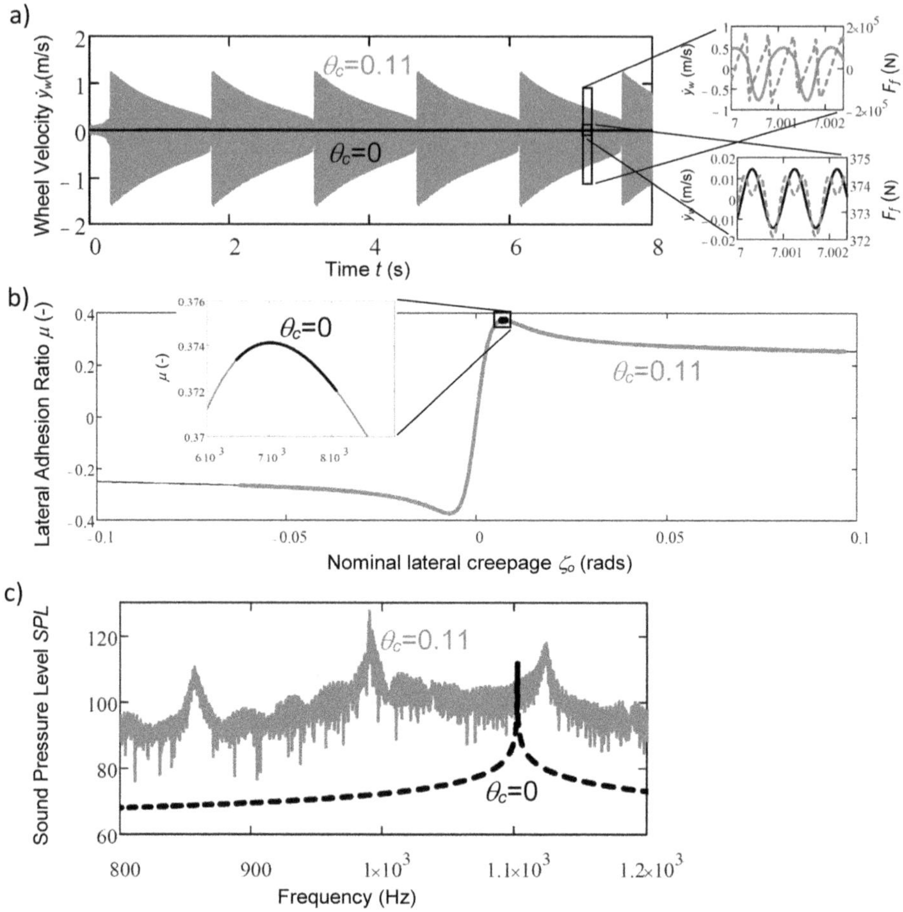

FIGURE 4.1 Wheel squeal simulations under no mode coupling $\theta_c=0$ (-) and stiffness mode coupling $\theta_c=0.11$ (-); (a) wheel vibration with zoom-ups at 7 seconds including frictional force (--) (b) creep oscillation with zoom-up at 0.006 rad. (c) Sound pressure level spectrums of noise; no mode coupling (- -), stiffness mode coupling (-) [12].

Figure 4.1 shows the time histories and frequency spectra of numerical simulations under no mode coupling (contact angle = 0) and with mode coupling (contact angle = 0.11).

The time histories in Figure 4.1a and b illustrate the differences in squeal behaviour with and without mode coupling:

- Without mode coupling, squeal occurs at lower amplitudes, with oscillations confined to the positively and negatively sloped regions near the critical creep on the positive side of the creep curve. The velocity oscillations appear approximately sinusoidal (as seen in the zoomed-in view) and remain in phase with the frictional force, supplying energy input to the system.

- With mode coupling, significantly larger, nonsinusoidal creep oscillations occur across all slopes of the creep curve. As shown in Figure 4.1b, these oscillations are self-excited, amplifying both squeal and creep oscillations beyond the negative critical creep. Once reverse (negative) creep is reached, oscillations grow further, as predicted in Section 2.3.

Additionally, the large wheel velocity and creep oscillations under mode coupling are asymmetric due to the nonlinear influence of friction/lateral adhesion. The extreme nonsinusoidal oscillations suggest that linear assumptions used in limit cycle amplitude prediction equations are no longer valid. A Fast Fourier Transform (FFT) over 8-second intervals at a sampling rate of 21.2 kHz was used to obtain the squeal frequency spectra for both cases shown in Figure 4.1c. Without mode coupling, squeal exhibits a pure tone at 1102 Hz, with an average noise level of 112 dB. With mode coupling, the noise level increases to 127 dB, with a broader frequency band and multiple tonal components at 856, 990 and 1122 Hz, which are approximately multiples of 133 Hz. This broader frequency spectrum, generated from a simple deterministic system, is indicative of chaotic motion. Although the amplitude of the creep oscillations may be unrealistic in practice, the broader spectral response could better match experimental and field measurements of squeal, as suggested in [12].

To further investigate the potential onset of chaotic behaviour, phase space and Poincaré map analyses were performed, and the impact of contact angle on squeal was examined, as shown in Figure 4.2.

Figure 4.2 illustrates how wheel squeal behaviour evolves as the contact angle, θ_c, increases, transitioning from a periodic limit cycle to quasiperiodic or possibly chaotic motion, before eventually settling into a higher-order periodic limit cycle:

- Figure 4.2a ($\theta_c = 0.02$) shows a simple closed limit cycle, indicating sinusoidal squeal behaviour with a stable periodic motion.
- Figure 4.2b ($\theta_c = 0.08$) reveals a wandering limit cycle that fills a circular region, suggesting multiple incommensurate periodic solutions, characteristic of quasiperiodic motion [26].
- Figure 4.2c ($\theta_c = 0.11$) shows the phase space becoming increasingly irregular, forming a large, nonclosed orbit. This suggests the onset of chaotic instability, which requires confirmation through Poincaré map analysis.
- Figure 4.2d demonstrates that for even higher contact angles, the system returns to a closed orbit, but with a more complex structure compared to the original sinusoidal limit cycle in Figure 4.2a.

To verify these behaviours, corresponding Poincaré maps were generated and analysed within a bifurcation diagram, as shown in Figure 4.3.

Figure 4.3 confirms the nonlinear behaviour observed in the phase space plots of Figure 4.2. In particular, the bifurcation diagram in Figure 4.3a initially shows a period-2 limit cycle, which transitions into quasiperiodic motion (in Figure 4.2) between contact angles $\theta_c = 0.06\ and\ 0.08$ validated by the Poincaré map in Figure 4.3b as a closed set of nonfractal points. At $\theta_c \approx 0.087$, the bifurcation diagram shows a nonlinear jump in the steady-state solution, indicating a transition from quasiperiodic

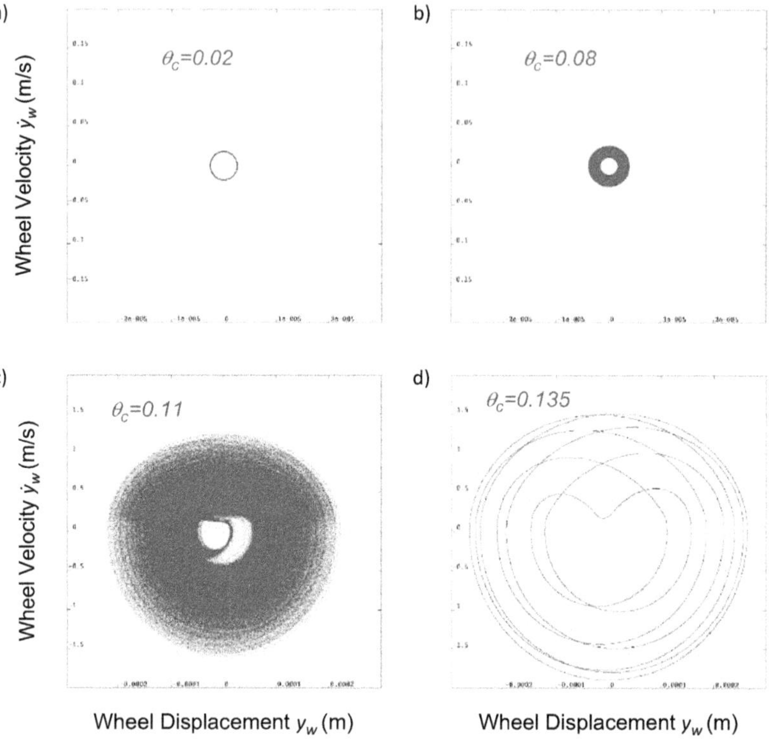

FIGURE 4.2 Simulated phase spaces of squeal oscillations under increasing contact angles; (a) $\theta_c = 0.02$ (b) $\theta_c = 0.08$ (c) $\theta_c = 0.11$ and (d) $\theta_c = 0.135$ [12].

to chaotic motion as the wheel squeal velocity amplitude exceeds the crabbing velocity. The chaotic motion is confirmed at $\theta_c = 0.11$, where the nonperiodic, random-like behaviour observed in Figure 4.2c appears as a complex fractal strange attractor in the Poincaré map of Figure 4.3c. A non-integer Lyapunov dimension for this strange attractor was calculated as 2.1, confirming its fractal nature from a Lyapunov spectrum of [3.1, 0.0, −38.9, −226.9], with the positive maximal exponent confirming exponential sensitivity to initial conditions, a hallmark of chaotic motion. For higher contact angles (Figure 4.3d), the system exhibits a window of nearly periodic behaviour, where multiple small line segments indicate a slightly quasiperiodic motion. This observed sequence of transitions – from a stable limit cycle to quasiperiodic motion, then chaotic behaviour, followed by periodic windows – is characteristic of the quasiperiodic route to chaos, as detailed in [27] and [28].

A key finding is that chaotic motion emerges after a nonlinear increase in wheel squeal amplitude, exceeding the crabbing velocity and inducing negative creep oscillations. This aligns with the analytical criteria (4.8)–(4.11) and (4.12). To systematically identify the conditions for chaotic behaviour, jumps in bifurcation diagrams (such as Figure 4.3a at $\theta_c = 0.087$) were analysed to determine the critical sound pressure levels and contact angles where chaos first appears across a range of angles of attack. These results were then quantitatively compared with the analytical

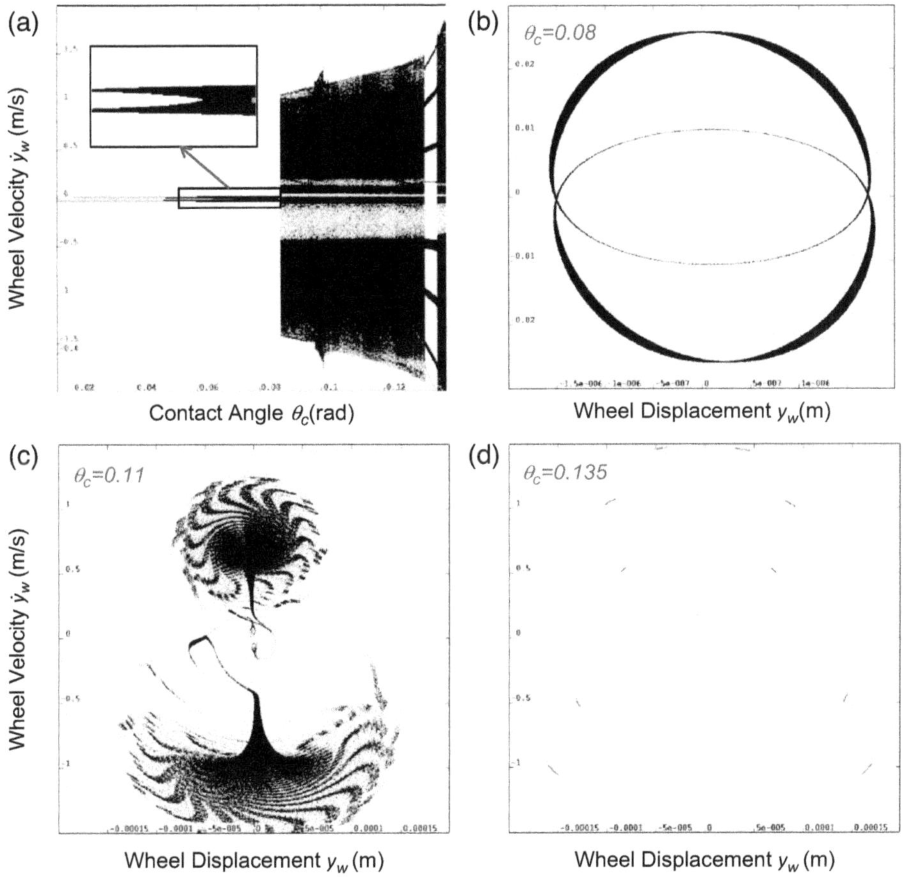

FIGURE 4.3 (a) Bifurcation diagram of Poincare maps for contact angle. Poincare maps for increasing contact angles; (b) $\theta_c=0.08$ (c) $\theta_c=0.11$ and (d) $\theta_{c=}0.135$ [12].

predictions of chaotic instability using equations (4.2)–(4.12). For this analysis, substantial falling friction was incorporated by reducing the rail mass and stiffness by 10% and 4%, respectively. The findings are summarised in Figure 4.4.

The steady-state squeal sound pressure levels, immediately before and after the transition to chaotic oscillations over a range of angles of attack (nominal lateral creepage), are shown in Figure 4.4a. These results compare well with the necessary analytical criteria from equation (4.2) with the numerically simulated squeal amplitudes being slightly lower than the analytical threshold (by less than 1 dB) for angles of attack above 10 mrad. Chaotic squeal is identified by extremely high sound pressure levels (exceeding 130dB) across a broad range of angles of attack (up to 25 mrad). The trend of chaos onset across different angles of attack is well captured by the analytical model, with a 20-dB offset between numerical and analytical results due to the nonlinear increase in squeal amplitude once negative critical creep oscillations develop. Figure 4.4b further compares the numerical simulations and analytical

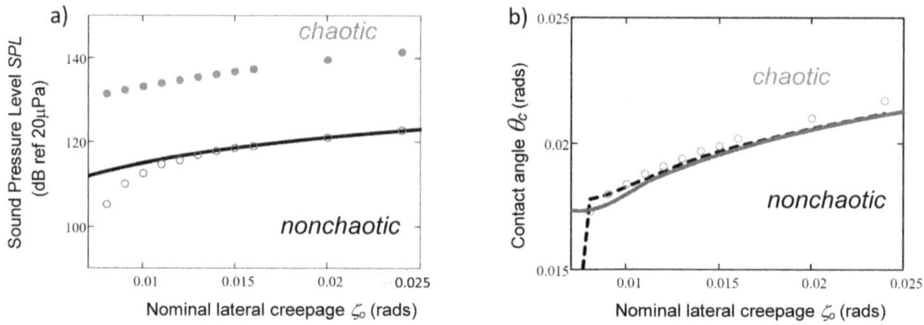

FIGURE 4.4 (a) Critical sound pressure level vs nominal lateral creepage at a rolling speed of 17.8 m/s; conservative analytical (4.2) (-) and numerically simulated results just before (o) and after (o) chaotic instability. (b) Critical contact angle θ_c at which chaotic instability first occurs; analytical; full solution to equation (4.2)–(4.9) (--) simplified solution for stiffness mode coupling occurrence equation (4.12) (-), and numerical simulations (o) [12].

predictions for critical contact angles at which chaotic instability first occurs over various angles of attack. The results show a close match between the analytical solutions (4.2)–(4.12) and numerical simulations, with discrepancies within 0.001 rad. The simplified analytical criterion (4.12) efficiently and conservatively predicts the critical contact angle in an instantaneous calculation, unlike equations (4.2)–(4.12), which require complex numerical integrations. Hence, the analytical criteria provide an efficient and conservative method for predicting steady-state chaotic instability, making them useful for avoiding or mitigating squeal. In this case, chaotic squeal instability is primarily driven by stiffness mode coupling, with parametric excitations further amplifying squeal vibrations into large positive and negative creep oscillations.

Hence, a very efficient model for prediction of wheel squeal chaos has been developed and validated using mathematical modelling and stability analysis to develop predictive criteria. A similar approach for other chaotic instabilities in engineering is demonstrated subsequently.

4.2 BRAKE SQUEAL CHAOS

Brake squeal is an undesirable, high-pitched tonal noise produced when slowing a vehicle using brakes. It occurs due to a friction-induced instability of one or more brake system modes of vibration that increases to a nonlinear limit cycle and/or bounded chaotic oscillations. Hence, an understanding of the occurrence of brake squeal chaos requires insights into the underlying mechanisms of its instability. Oberst and Lai [15] provided a comprehensive review of research into chaotic instabilities in brake squeal and related friction-coupled systems. Experimental and numerical studies have also demonstrated chaotic behaviour in friction-induced instabilities, when externally forced. Feeny and Moon [16] identified chaos in a fundamental dry friction oscillator, finding that chaotic attractors shared common characteristics despite variations in friction laws. In a simplified two-degree-of-freedom brake pad and disc system, Shin et al. [17] conducted a numerical stability analysis, revealing squeal limit cycle attractors in phase space, influenced by mode proximity and damping effects (which could either stabilise or destabilise the system). Later, nonlinear chaotic-like motion was observed following a period-doubling route, though it was not conclusively identified [18]. A nonlinear time-domain analysis of a brake model [29] showed intermittent bursts of high-amplitude vibrations, following an intermittency route to chaos. Feeny and Moon [30] also explored stick–slip oscillations, demonstrating that chaotic motion could be suppressed using large-amplitude dither. Brake disc squeal was identified and controlled using state feedback control in [19], while chaotic instability and its transition in a full-scale brake system was confirmed both numerically and experimentally in [15]. Additionally, [31] introduced a novel method for estimating Lyapunov exponents from brake squeal measurements and also identified quasi-periodic or slightly chaotic behaviour, while further advancing the study of chaotic behaviour in friction-induced vibrations. Similarly, limit cycles, quasi-periodic motion, weakly chaotic attractors and different types of stick–slip vibrations were measured experimentally in friction excited vibrations with transitions characterised by beating phenomena, sudden energy exchange and intermittent dynamics in [32]. Numerical work also confirmed observations of complex dynamical behaviour and chimera-alike dynamics in a self-excited frictional oscillator with a weak stiffness nonlinearity [33]. Although this previous research numerically or experimentally identified chaotic instability in brake squeal, chaos could only be predicted for a limited set of parameters under restricted computational or experimental conditions. In addition, combined mechanisms were not specifically investigated. To address this, [34] provided analytical identification and quantification of chaotic instabilities in brake squeal under "falling friction," mode-coupled and spragging conditions and hence will be a focus in the following.

Prediction of Brake Squeal Chaos

Chapter 3 reviewed and developed closed-form analyses to predict the local instability, growth and limit cycle amplitude of brake squeal. Building on these results, we now establish conservative analytical criteria to determine the onset of chaotic brake squeal. A key question addressed is: What happens when the brake squeal limit cycle destabilises due to excessive instability, leading to phase space expansion? To answer this, two analytical criteria are developed to predict chaotic vibrations in brake squeal:

1. Lyapunov Exponent Test for Chaos – This method quantifies the sensitivity of the system to initial conditions. The Lyapunov exponent (λ) determines whether an initial error in the system grows exponentially, indicating chaos.
2. Nonlinear Instability Analysis – Based on the previous chapter's findings, this approach assesses whether instability levels reach a critical nonlinear threshold that disrupts the squeal limit cycle.

Hence, we first define the Lyapunov exponent as the rate of expansion of an initial error in the system, following a power law growth, as described in [26] as

$$d = d_o 2^{\lambda(t-t_o)}, \tag{4.13}$$

where d is the maximum diameter of the ellipsoid in phase space that grows from the initial conditions at time t, and d_o is the diameter of the sphere of initial conditions at time t_o. A positive Lyapunov exponent implies chaos, i.e.,

$$\text{Chaos possible if } \lambda > 0. \tag{4.14}$$

Note this Lyapunov exponent analysis is merely a necessary condition for chaos. A positive maximal Lyapunov exponent characterises sensitivity to initial conditions, which on its own does not imply any sort of complicated behaviour. In a chaotic system, motion is bounded by nonlinearities, so divergence, λ, is measured locally over small time increments $(t - t_0)$ or linearly using partial derivatives (via the Jacobian matrix). This divergence is continuously reset to calculate an average local divergence over an extended time period. The eigenvalues of the Jacobian matrix determine local stability, creating a direct relationship between the maximum Lyapunov exponent and the largest system eigenvalue, evaluated dynamically. In general, the maximum Lyapunov exponent can be expressed as the time-averaged real part of the largest dynamic eigenvalue over a sufficiently long duration [35]. Chaos is most likely when this largest eigenvalue is most positive, indicating maximum local divergence. For the brake squeal case, stability and eigenvalue analysis from the previous chapter reveals that chaotic instability is most pronounced when the effective system damping, c_{eff}, is most negative. This condition arises under stiffness mode coupling, which is typical in brake squeal scenarios. The instability is driven by insufficient modal damping, allowing complex stiffness effects to dominate over damping influences. Based on these findings, a simple and conservative analytical criterion for predicting the critical friction coefficient at which brake squeal chaos occurs is formulated, using stiffness mode coupling as [34]

$$\text{Chaos possible if } \mu_s > \mu_{\text{crit}}, \quad \Delta(\mu_{\text{crit}}) = 0, \tag{4.15}$$

where the critical friction for chaos may be obtained as

$$\mu_{\text{crit}} = \frac{\left\{ \dfrac{m_a}{k_H} \left[2\sqrt{\dfrac{k_H}{m_b}\left(\dfrac{k_a}{m_a} - \dfrac{k_b}{m_b}\right)} - \left(\dfrac{k_H}{m_b} + \dfrac{k_a}{m_a} - \dfrac{k_b}{m_b}\right)\right] - \sin^2(\theta) \right\}}{\left(\delta_a \cos(\theta)\sin(\theta)\right)}. \tag{4.16}$$

Equations (4.14) and (4.15) provide a conservative criterion for predicting brake squeal chaos as they focus solely on high local phase space expansion caused by stiffness mode coupling. Additionally, for bounded chaos, sufficient nonlinearities must be present to allow for folding and contraction in the phase space. To account for this, an amplitude criterion should also be established to determine when significant expansion and nonlinear effects arise. For this purpose, the closed-form limit cycle solution for brake squeal can be used to identify the point at which the bilinear friction law is exceeded, leading to large nonlinear creep oscillations [36] as

$$A_a\omega_i \geq V_{c,o}/\cos(\theta) \quad \text{for} \quad V_c > \zeta_c V_o \tag{4.17}$$

This conservative condition may be simplified by solving for the critical system modal damping as

$$\text{Chaos if } \mu_s > \mu_{\text{crit}},$$

$$c_{si}\left(\mu_{\text{crit}}\right) = k_{Fi}N\left[\left(\frac{(\sin(2\omega_i t_c) - 2\omega_i t_c)}{\left(k_{\mu 1} - k_{\mu 2}\left(\mu_{\text{crit}}\right)\right)/2\pi}\right) - k_{\mu 2}\left(\mu_{\text{crit}}\right)\right]/\left(\zeta_c V\right) \tag{4.18}$$

$$\omega_i t_c = \text{acos}\left(1 - \zeta_c V/V_c\right).$$

Equation (4.18) provides a less conservative criterion for predicting brake squeal chaos compared to equations (4.15) and (4.16). Unlike the previous criteria, it considers both large local expansion and the requirement for negative sliding to introduce sufficient nonlinearities, ensuring bounded phase space behaviour in brake squeal. Since the analytical solution is not a fully closed form, solving for the critical friction coefficient, μ_{crit}, requires a numerical root-finding function using the equations for c_{si}. However, in practice, this computation is nearly instantaneous. The effects of parametric excitation from varying friction levels [36] are neglected, primarily due to the steep positive slope of the friction curve and to avoid complex integrations. To validate these predictive criteria for brake squeal chaotic instability, numerical solutions of the full nonlinear equations of motion are compared against the conservative analytical predictions from equations (4.15), (4.16) and (4.18). Finally, it is noted that the energy provision/generation for chaotic motion comes from the same frictional dynamics of the squeal instability. In particular, any chaotic attractor that develops is powered by individual or combined frictional instability mechanisms identified in the previous chapter, i.e., falling friction, mode coupling and spragging.

Results of Brake Squeal Chaos Predictions

The occurrence of chaotic instability in brake squeal was investigated for a real brake system case study [37] based on the numerical integration of the full nonlinear equations of motion, described in the previous chapter. The nonlinear phenomena were investigated using various phase space tools and the Lyapunov spectrum of exponents using algorithms described in Chapter 2. Comparisons between the full numerical solutions and the conservative analytical criteria developed in the previous section were conducted to determine the conditions for chaotic instability in brake squeal. The nominal system parameters were selected based on real-world brake squeal studies [6,37–39] and [40] with the squeal mode frequency at 3 kHz [39], aligning with finite-element model results presented in Figure 4.5a [37].

The dynamic transformations in Figure 4.5b were used to determine the parameters for modes a and b in Figure 4.5a, based on the coupled rotor and brake calliper dynamics, as detailed in [37]. The dependence of real brake squeal on brake pressure [39] was also modelled following the approach in section 3.2, where contact, modal stiffness and mass parameters are expressed as functions of pressure, as described

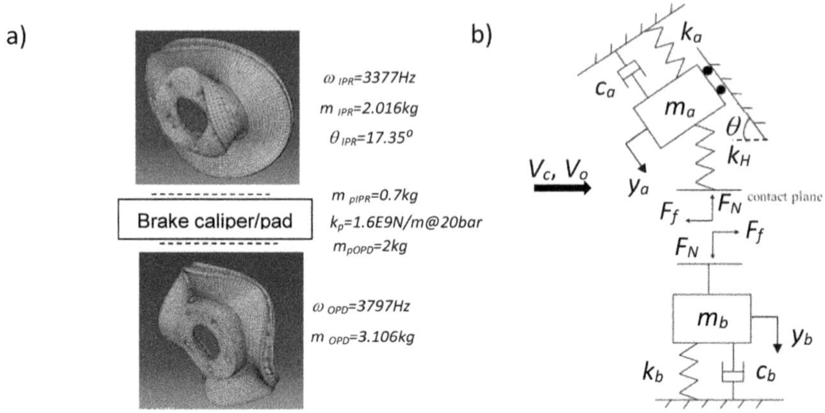

FIGURE 4.5 Real brake system squeal model (a) coupled in-plane radial (IPR) and out-of-plane diametral (OPD) modes, FEA performed in [39,37] (b) coupled two-mode brake squeal model [34,37].

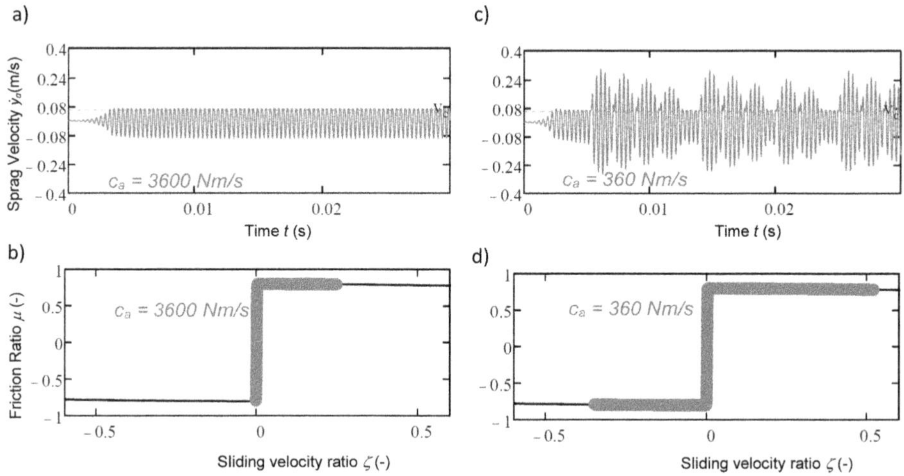

FIGURE 4.6 Simulated squeal oscillations for the real brake system; (a) sprag mode vibration velocity (-) and (b) corresponding friction oscillation range (bolder/thicker) under static friction $c_a = 3600$ Ns/m and (c) sprag mode vibration velocity (-) and (d) corresponding friction oscillation range (bolder/thicker) under static friction $c_a = 360$ Ns/m. In all cases, $c_b = 3.613\ c_a$ [34].

in [37]. In the subsequent analysis, damping levels were varied from their nominal values of 3.6% (mode a) and 4.2% (mode b), while maintaining the scaling ratio $c_b = 3.613\ c_a$. First, the time history and friction curve behaviour was investigated, as shown in Figure 4.6.

Figure 4.6 illustrates brake squeal behaviour under different damping conditions. In particular, Figure 4.6a shows a periodic limit cycle caused by stiffness mode coupling with nominal damping, while Figure 4.6c displays larger, irregular amplitude oscillations (later shown to be chaotic) when modal damping is significantly reduced (10%). The corresponding friction oscillations in Figure 4.6b and d reveal that in the low damping case, the sliding behaviour becomes highly nonlinear, with large positive and negative sliding that deviates from a sinusoidal pattern. A comparison of the time histories in Figure 4.6a and c (up to 0.05 seconds) suggests that the oscillations grow into nonlinear, irregular amplitudes once they enter the negative (reverse) sliding regime, consistent with [36]. To further examine the onset of chaotic brake squeal, phase space analysis under progressively lower damping levels is presented in Figure 4.7.

Figure 4.7 illustrates how brake squeal behaviour evolves from a simple limit cycle to increasingly complex, possibly chaotic motion as damping is reduced. In particular, Figure 4.7a shows a single closed-loop limit cycle for $c_a=3600$ Ns/m, indicating sinusoidal-like squeal behaviour. When damping is reduced to 50% of nominal, Figure 4.7b reveals a more complex periodic motion, with the limit cycle beginning to break apart. At 25% nominal damping, the system becomes even more irregular and intricate, and when damping is further reduced to 10%, the system exhibits a

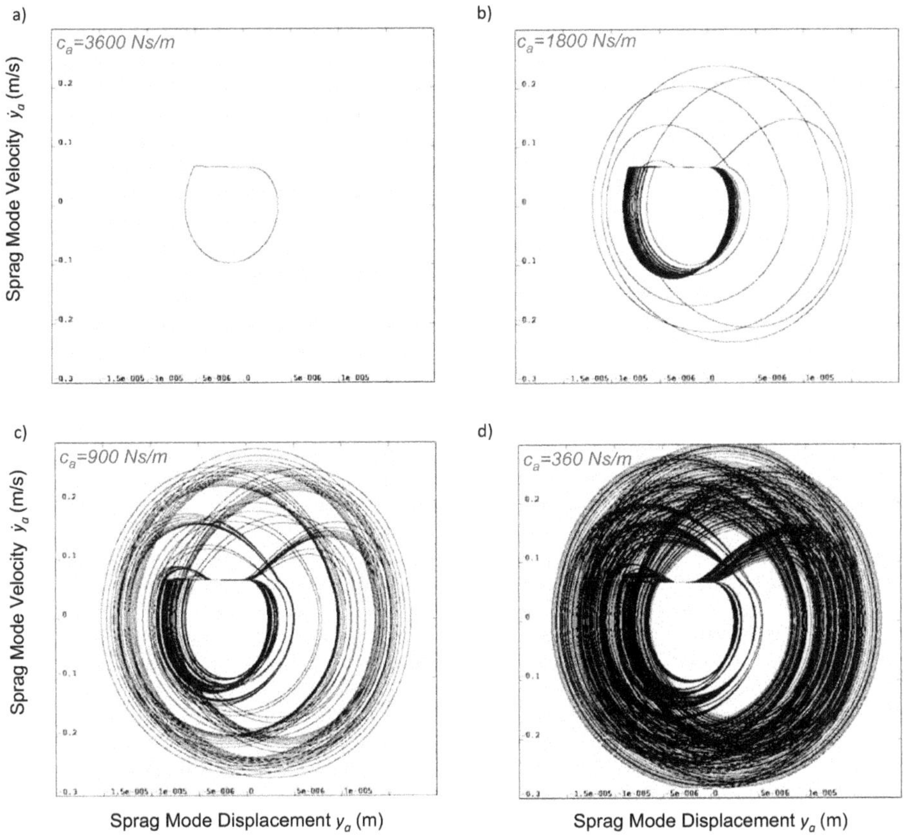

FIGURE 4.7 Simulated phase spaces of squeal oscillations in a real brake system under decreasing damping; (a) $c_a=3600$ Ns/m (b) $c_a=1800$ Ns/m (c) $c_a=900$ Ns/m and (d) $c_a=360$ Ns/m. In all cases, $c_b=3.613\ c_a$ [34].

wandering, non-closed orbit, suggesting the potential onset of chaotic behaviour. To confirm this transition, the corresponding bifurcation diagram and Poincaré maps (with a Poincaré plane at $y_b = 0.0$) are provided in Figure 4.8.

Figure 4.8 confirms that the nonlinear behaviour observed in Figure 4.7 corresponds to brake squeal chaos. The bifurcation diagram in Figure 4.8a reveals chaotic behaviour at 10% nominal damping, which gradually transitions to more ordered motion as the damping increases. At nominal damping, the single-loop sinusoidal squeal seen in Figure 4.7a is confirmed as period-2 motion in the Poincaré map (Figure 4.8b). When damping is reduced to 50% of the nominal one, the complex periodic motion in Figure 4.7b) is identified as an approximate period-16 limit cycle in Figure 4.8c. The breakdown into chaos occurs through a sudden crisis phenomenon rather than through period-doubling bifurcations, as evidenced by the bifurcation diagram in Figure 4.8a. As damping further decreases to 10% of the nominal one, the Poincaré map points become more smeared, forming a fractal attractor (Figure 4.8d),

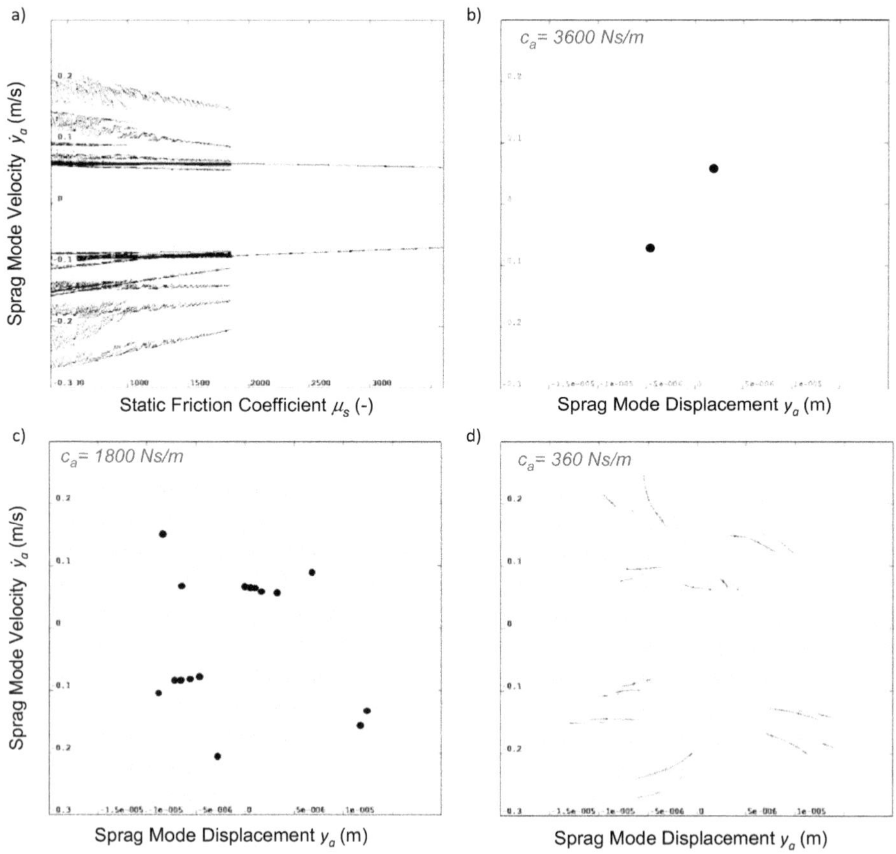

FIGURE 4.8 (a) Bifurcation diagram of Poincare maps for increasing damping in the real brake system. Poincare maps for decreasing damping; (b) $c_a = 3600$ Ns/m (c) $c_a = 1800$ Ns/m and (d) $c_a = 360$ Ns/m. In all cases, $c_b = 3.613\ c_a$ [34].

which corresponds to the nonperiodic, random-like motion in Figure 4.7d. The confirmation of brake squeal chaos is provided by a positive maximal Lyapunov exponent ([71, 0.0, −1062, -1.61×10^5]), indicating chaotic instability as well as the fractal strange attractor with a non-integer Lyapunov dimension of 2.07.

These numerical predictions can now be compared with the analytical criteria developed in the previous section. Nonclosed limit cycles were identified using phase space and Poincaré map analyses, similar to those in Figures 4.7 and 4.8, to determine the critical friction coefficient for a range of contact angles and brake pressures. These numerical solutions were then compared with the conservative analytical predictions of chaotic instability from equations (4.15)–(4.18). Two cases of chaotic behaviour were analysed for different brake model conditions. In the fundamental brake model, one case used nominal parameters [34], while another reduced the damping and normal force to 30% and 0.8% of nominal values, respectively. In the real brake model conditions, brake pressure was varied, while damping was held at 10% of nominal values. Figure 4.9 summarises these results, highlighting the conservative nature of the analytical predictions for the onset of steady-state brake squeal chaos.

Figure 4.9a compares the analytically predicted onset of chaotic instability with results from full nonlinear numerical simulations for the fundamental brake model. The negative sliding criterion provides a conservative estimate of chaos onset via critical static friction level across various sprag angles, showing errors within 15% for the first scenario (with reduced damping and normal force). A similar but more conservative trend is provided by the simpler stiffness mode coupling criterion, albeit with somewhat larger errors. In the nominal case, the negative sliding criterion does not apply, as the normal load is too large to reach negative sliding conditions. Here, chaos likely arises from other frictional nonlinearities, suggesting further

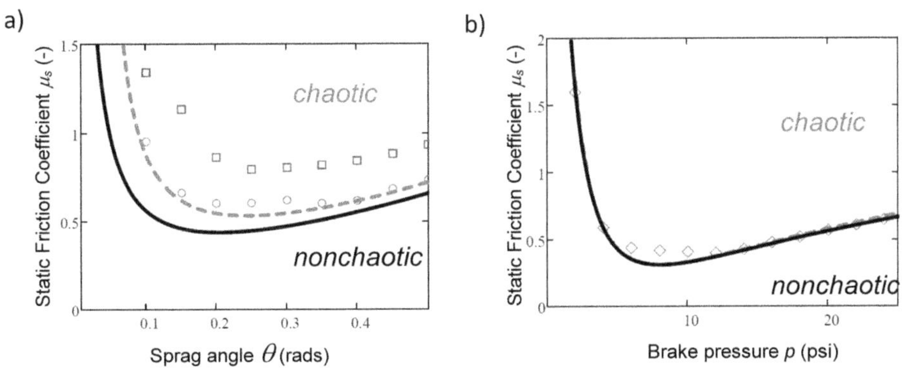

FIGURE 4.9 (a) Fundamental model, critical static friction coefficient vs sprag angle at which chaos first occurs; conservative analytical criteria; stiffness mode coupling (-), negative sliding (- -) and numerically simulated results for nominal conditions (□) and 30% and 0.8% nominal damping and normal force (o). (b) Real brake model, critical static friction coefficient at which chaos first occurs; conservative analytical criteria; stiffness mode coupling (-), negative sliding (mainly overlapping) (- -) and numerical simulations (◊) 10% nominal damping [34].

analysis using methods outlined in [41]. However, the stiffness mode coupling criterion remains valid and provides a conservative prediction for chaos initiation, accurately representing the relationship between critical static friction and sprag angle. Here, chaotic squeal typically occurs at high static friction coefficients (>0.6) across various sprag angles, due to stiffness mode coupling.

For the real brake model (Figure 4.9b), the analytical predictions and numerical simulations match closely, with errors under 25%, for critical friction levels at the onset of chaotic instability across a range of brake pressures. Both stiffness mode coupling and negative sliding criteria from equations (4.15)–(4.18) closely align/overlap, possibly due to the high sensitivity of brake squeal amplitude to static friction near stiffness mode coupling conditions. These results suggest brake squeal chaos primarily arises from stiffness mode coupling, with further amplification due to friction nonlinearities during negative sliding. This understanding provides a foundation for investigating strategies to suppress chaotic instability, explored in Chapter 6.

4.3 AERODYNAMIC CHAOS FOLLOWING FLUTTER

Chaotic motion of an elastic structure in a fluid, such as an airfoil section of an aircraft or wind turbine blade, can occur following flutter instability that was explored in Chapter 3. The mode coupling mechanism causes instability and limit cycle flutter due to a Hopf bifurcation. Following the critical flutter speed, the response can transition to chaos through successive other bifurcations like period doubling.

Chaos was first numerically found in an airfoil with cubic pitching stiffness [42] and then coupled cubic stiffnesses [43] under incompressible flow. This chaos following flutter (termed throughout this book) is characterised by a transition to chaos following the instability of a Hopf bifurcation to self-sustained flutter limit-cycle oscillations. Chaos following flutter has also been identified in a piezo-aeroelastic airfoil energy harvester supported by nonlinear springs [44] as well as with nonlinear motion coupling, inertial and aerodynamic nonlinearity [45]. Other instances of chaotic behaviour include panel flutter under thermal loads, supersonic flow [46] and experimental turbulent flow [47]. In supersonic conditions, nonlinear flutter in variable stiffness composite plates has exhibited limit cycle oscillations, quasi-periodic motion or chaotic responses, depending on the fibre path orientation [48]. Additionally, chaos has been observed following stall flutter, driven by a combination of structural, aerodynamic and kinematic nonlinearities [49,50]. Similarly, fin vibration chaos has been shown to occur under aerodynamic, thermal and structural nonlinearities in hypersonic flow [51]. Recent numerical simulation research on the transitions to chaos in general aeroelastic systems include [45,52,53]. This includes numerical work on freeplay in the pitch degree of freedom [54] and a flapping aerofoil [55]. Also, recently, chaos following flutter was identified in an airfoil without the need for structural or thermal nonlinearity [56]. This was identified based on interesting analogies between bilinear lift aerodynamics and friction under mode-coupled structural dynamics causing chaos in railway wheel squeal and brake squeal [11–25,27–34,36–40,42,43,46,47,49,50,51,57–135]. This research was then extended to analytical closed-form predictions of chaos following flutter, in optimally designed wind turbines. This provided computational acceleration in the order of 10^7 times compared to nonlinear numerical simulation [41]. This analysis forms the basis of the following.

Prediction of Flutter Chaos

As shown in Chapter 3, flutter is a local instability that typically grows to a stable limit cycle, due to a balance of energy between pre and post stall motion [56]. When this instability is combined with significant nonlinearities, it can lead to chaos following flutter, characterised by bounded motion that has local phase space expansion. Specifically, in [56], the Lyapunov exponent, λ, was used to define a condition for expansion or sensitivity to initial conditions as [26]

$$\text{Chaos possible if } \lambda > 0 \text{ where } d = d_o 2^{\lambda(t-t_o)}. \tag{4.19}$$

Here, d_o and d represent the initial and final spherical diameters at, t_o, and t, respectively. In [35], it was demonstrated that the maximal Lyapunov exponent can be expressed as the time-averaged real part of the maximum dynamic eigenvalue integrated over a sufficiently long duration. Consequently, chaos is most likely to occur under stiffness mode-coupled flutter conditions as this scenario produces a larger maximum dynamic eigenvalue due to the influence of complex stiffness negative damping, in contrast to the lower instability effects observed under viscous mode coupling (assuming realistic damping values remain subcritical). This observation forms the basis of the first necessary condition for chaos following flutter as [26]

$$\text{Chaos possible if } W > W_f, \quad \Delta\left(W_f\right) = 0. \tag{4.20}$$

Here, W_f represents the critical flutter speed, determined in closed form under zero damping (as discussed in Chapter 3). Equation (4.20) serves as a conservative criterion since it incorporates the continuous averaging of local trajectory expansion in phase space. However, it remains a necessary condition as it only assesses the sensitivity to initial conditions rather than fully capturing the nonlinear folding and bounding of chaos. In the case of flutter, nonlinearity arises due to the bilinear lift curve. Specifically, it was proposed and tested in [56] that an offset in the angle of attack induces asymmetry in the bilinear lift curve response, introducing the necessary nonlinearity for chaotic behaviour, as described by

$$\alpha_o \neq 0. \tag{4.21}$$

The necessary criterion in equation (4.21) provides only a discrete measure of the nonlinear conditions leading to chaos. To refine this approach, it is important to recognise that the asymmetrical lift curve behaviour functions as a parametric bilinear stiffness excitation. It is well established that such parametric excitation can induce subharmonic and higher harmonic instabilities, which, in cases of insufficient damping, evolve into interacting limit cycles in phase space [136–138]. If the excitation level is sufficiently high, the instability zones of different resonances may overlap, leading to bifurcations and ultimately chaotic behaviour in the system [137]. This chaotic response is primarily influenced by the magnitude of stiffness variation and the level of damping. Therefore, less conservative conditions for chaos following flutter are proposed based on the existence of all parametric harmonic instability regions, as outlined in [139] as

$$\zeta_{\text{ti}} < \zeta_{\text{tic1}} = \zeta_{\text{crit}} 2\sqrt{K_f / M_f} / \left(1 + 1/\sqrt{k_{\text{stall}}}\right), \tag{4.22}$$

where

$$\begin{gathered}
\log(k_{\text{stall}}) = 4\zeta_{\text{crit}} U^{-1}\left[\operatorname{atan}\left(-U/(2\zeta_{\text{crit}})\right) + \pi\right] + 4\zeta_{\text{crit}} V^{-1} \operatorname{atan}\left(-V/(2\zeta_{\text{crit}})\right), \\
U^2 = \left(1 + \sqrt{k_{\text{stall}}}\right)^2 - 4\zeta_{\text{crit}}^2, \quad V^2 = \left(1 + 1/\sqrt{k_{\text{stall}}}\right)^2 - 4\zeta_{\text{crit}}^2, \\
K_f = 1 - \left(k_{22}(S_x/m)^2 / k_\alpha\right), \quad M_f = m_\alpha (S_x/m)^2 / (J_\alpha + J_{ae}),
\end{gathered} \tag{4.23}$$

and ζ_{tic1} is the critical damping (with respect to the torsional degree of freedom) above which the first subharmonic instability region disappears and U and V are non-dimensional intermediate parameters. The derivation of equations (4.22) and (4.23) based on [139] involves modification from the critical damping ζ_{crit} using factors M_f and K_f of the mass moment of inertia and stiffness terms, respectively, to represent the parametrically excited mass-decoupled flutter equation, see [41]. The additional constraint that

$$0 < \zeta_{\text{crit}} < \left(1 + 1/\sqrt{k_{\text{stall}}}\right)/2, \tag{4.24}$$

This condition is always met if $k_{\text{stall}} > 1$. Based on this, the least conservative criterion for chaos can be proposed by requiring at least the first two instability regions to overlap, leading to bifurcational chaos. The damping threshold for the second instability region is one-third lower than that of the first region. Consequently, the least conservative condition for the onset of chaos is defined as [56,41]:

$$\zeta_{\text{ti}} < \zeta_{\text{tic1}}/3. \tag{4.25}$$

Equations (4.20), (4.21) and (either (4.22) or (4.25)) provide conservative and necessary conditions for chaos following flutter. When damping and non-angle-of-attack flutter derivatives are small, these conditions can be evaluated in closed form using the critical flutter speed solution (see previous chapter). Alternatively, the analytical solutions for c_{effi} can be used. The criteria highlight that chaos following flutter is driven by local expansion due to (1) high dynamic divergence caused by flutter instability from stiffness mode coupling following a Hopf bifurcation (4.20) and (2) nonlinear parametric excitation due to the bilinear lift aerodynamics causing interacting limit cycles if damping is insufficient according to (4.21) and ((4.22) or (4.25)). It should be noted that the generality of the flutter model and analysis of (4.19)–(4.25) suggest that the criteria for chaos following flutter, in an airfoil section, can be applied to many structures including aircraft and wind turbines. Subsequently, predictions of chaos for a wind turbine blade are compared with full numerical solutions of the nonlinear equations of motion (in Chapter 3).

Results of Flutter Chaos Predictions

The nonlinear numerical and analytical predictions were compared for the onset and chaotic behaviour of flutter in an optimised wind turbine blade [140]; a growing area of research due to important restrictions on turbine speed, power and size ie [141–145]. In particular, the accuracy of the analytical predictions was validated through an optimal wind turbine case study, using full nonlinear numerical simulations and experimental aerodynamic design and parameters from the literature ie [146–151]. The system parameters were selected based on optimal conditions for a large three-bladed composite wind turbine. The study utilised NACA 4412 airfoil blade sections, operating at an optimal lift coefficient, with an offset angle of attack fixed at 3° and a constant tip speed ratio of 6, as detailed in [41].

Figure 4.10a and b present the time history of the full nonlinear equations of motion, showing the blade tip section angle of attack oscillations under nominal conditions with a freestream wind speed of $U_\infty = 19.3$ m/s (i.e., local tip wind speed of 117.2 m/s). The response for modified conditions with a sharper stall slope is also included. The corresponding lift curve oscillations are shown in Figure 4.10c and d. Note that the variation in stall slope remains consistent with experimentally observed variations at high angles of attack near or within the stall region [142].

Figure 4.10a shows a periodic stable limit cycle occurring at a speed above the critical flutter speed, after a transition through a supercritical Hopf bifurcation [56]. Comparing Figure 4.10a and b, it is evident that the limit cycle oscillation amplitudes become irregular when the stall slope of the lift curve is steeper, suggesting nonperiodic motion. The

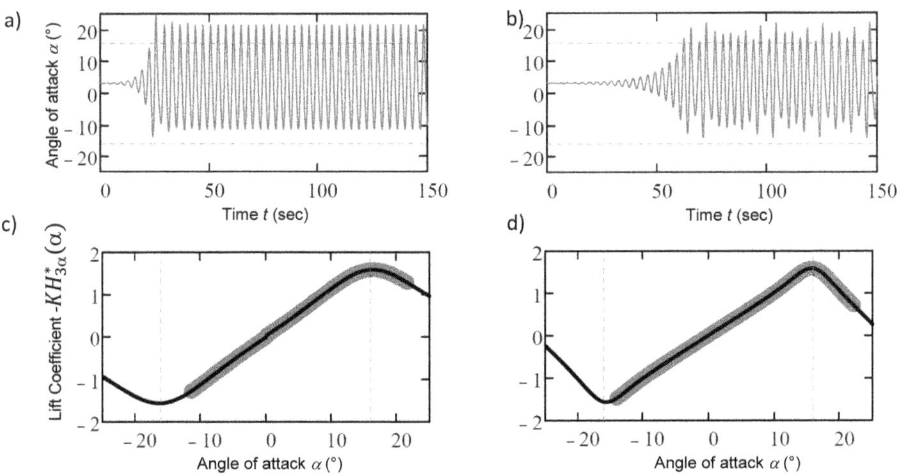

FIGURE 4.10 Simulated time histories of flutter at wind speed 19.3 m/s (local tip wind speed 117.2 m/s) for blade tip section with (a,c) nominal lift curve [152] and for (b,d) a sharper stall slope (smoothing factor $\beta = 20$, $k_{stall} = 1.5$) [41].

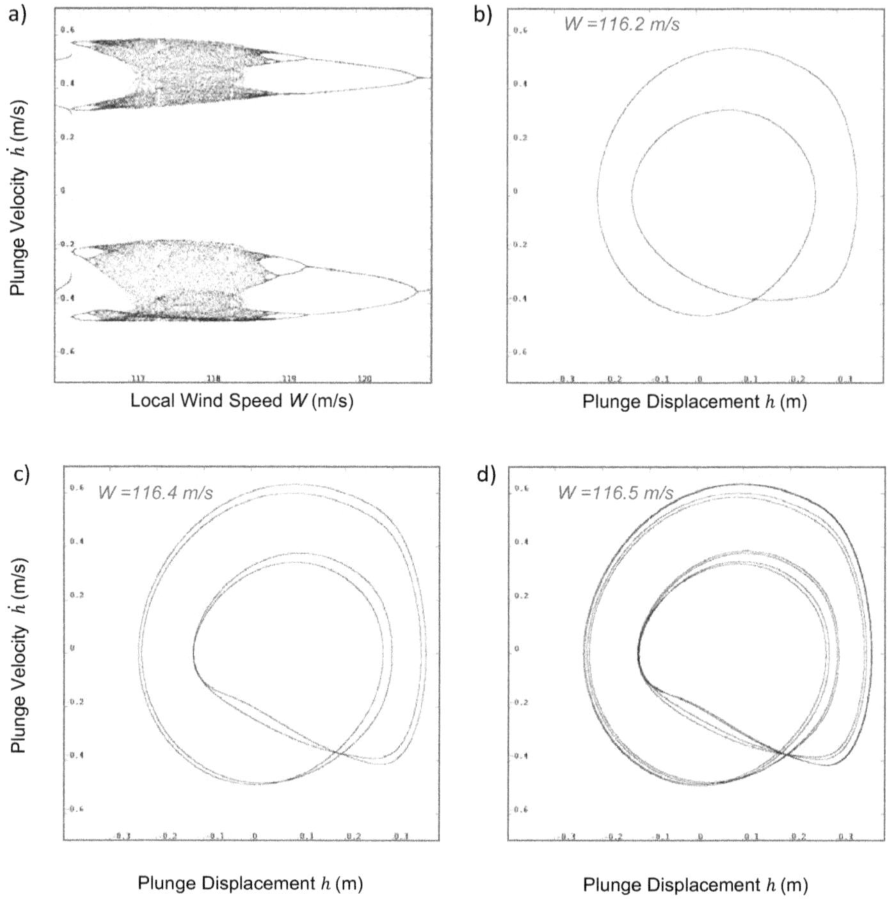

FIGURE 4.11 Bifurcation diagram (a) and phase spaces of flutter oscillations under increasing local wind speed; (b) $W = 116.2$ m/s (c) $W = 116.4$ m/s and (d) $W = 116.5$ m/s [41].

corresponding lift curve oscillations in Figure 4.10c and d show a steeper stall slope when possible chaos occurs, consistent with analytical predictions. Additionally, the results show that this irregular behaviour remains confined within a bilinear lift curve, influenced by the angle of attack offset and wind speed. To further investigate the effect of wind speed on nonlinear behaviour, a bifurcation and phase space analysis was conducted, as shown in Figure 4.11. The solutions were obtained after initial transients had decayed, and the Poincaré plane at $\alpha = 0.0$ was selected for convenience.

Figure 4.11a presents a bifurcation diagram that follows a well-known period-doubling route to chaos as the local tip wind speed surpasses the critical flutter speed. The system undergoes a transition to chaotic behaviour between approximately 116.6 and 118.8 m/s, before reverting to a more ordered state, similar to the airfoil study in [56]. Figure 4.11b–d further confirm this progression with (Figure 4.11b) illustrating a two-loop limit cycle at 116.2 m/s, represented by a closed set of four points in the bifurcation diagram, due to the Poincaré plane at $\alpha = 0.0$ being crossed twice per loop. As the local tip wind speed increases, the system undergoes successive bifurcations, leading to increasingly complex motion in Figures 4.11c and d. With further increases in wind speed, the system exhibits progressively compressed bifurcations, eventually forming a strange attractor at $W = 117.2$ m/s, as depicted in Figure 4.12. In particular, the phase space plot in Figure 4.12a shows a wandering, nonclosed orbit, which is confirmed by the fractal distribution of nonrepeating points in the Poincaré map (Figure 4.12b), characteristic of chaotic motion.

The chaotic state is confirmed in Figure 4.12b as it shows a strange attractor fractal quantified by a non-integer Lyapunov dimension of 2.16 and a Lyapunov spectrum of [0.084 0.000 −0.392 −0.710] [153,154]. Here, the first (maximum) Lyapunov exponent is greater than 0, meaning there is exponential sensitivity to initial conditions. The lowest local tip wind speed that chaos occurs in the time domain numerical simulations is $W = 116.6$ m/s (see Figure 4.11a), which compares very closely to $W_f = 116.5$ m/s (theoretical prediction) as part of the analytical criteria for chaos following flutter of (4.19)–(4.25). In agreement with a bilinear nonsymmetric criterion [56], bifurcation chaos was also found for offset angles of attack of 1°, 4° and 6°. A deeper parametric investigation and comparison of the numerical and analytical predictions of the onset of chaos was also performed for two wind turbine cases, as shown in Figure 4.13.

Figure 4.13 shows a strong correlation between the (conservative) analytically predicted chaos onset and the numerical bifurcation diagram simulations. In both cases, the critical torsional damping levels are plotted over a range of stall slope factors, k_{stall}. The bifurcation diagrams were used repeatedly to numerically identify the onset of bifurcation chaos for different stall slope factors (similarly to Figure 4.11). The analysis included both the nominal blade case and a lighter blade case, where the blade section wall thickness was reduced to 20% of the nominal

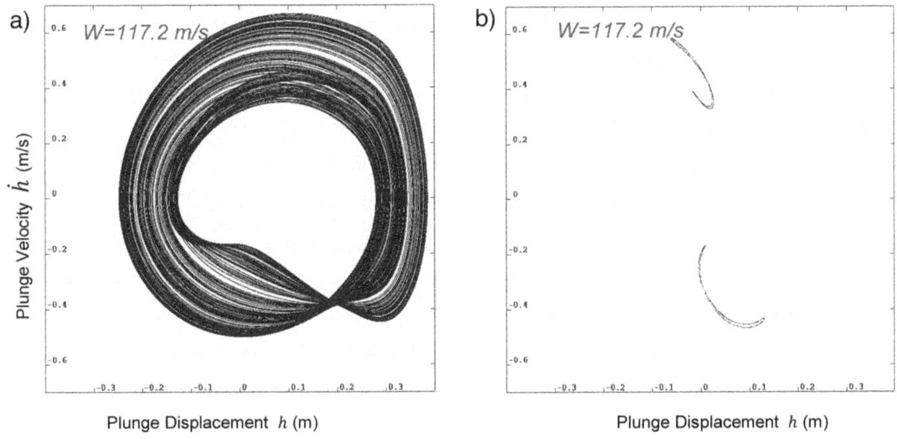

FIGURE 4.12 Phase space (a) and Poincare map (b) of wind turbine flutter oscillations for flow speed 117.2 m/s [41].

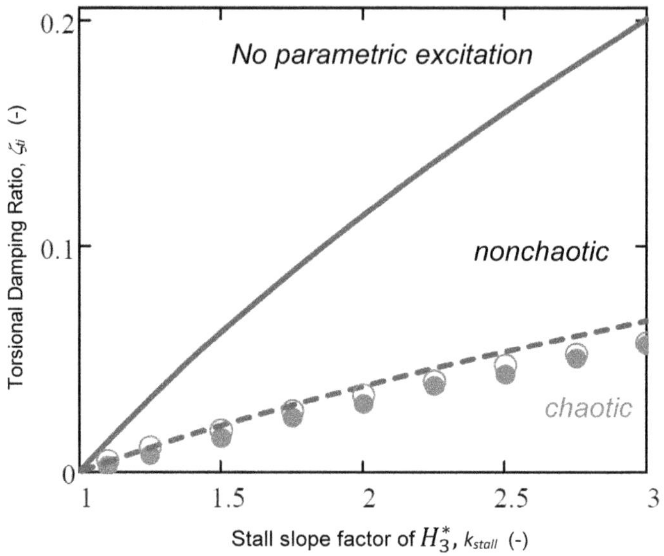

FIGURE 4.13 Critical torsional damping vs stall slope factor for chaos onset; conservative analytical criteria; no parametric excitation (-) (4.22), less conservative analytical criteria; no overlapping harmonics (- -) (4.25) and numerically simulated results for nominal case (•), thinner wall blade section case (○) [41].

value. This modification reduced the local critical flutter speed to 52.1 m/s (by a factor of $1/\sqrt{5}$, based on modal mass reduction) [41]. The computational time for numerical simulations was approximately 4 hours for 9 data points, compared to just 50 ms for 400 points using analytical predictions, resulting in a computational speedup of approximately 10^7 times. The critical torsional damping at chaos onset, predicted using the non-overlapping harmonics criterion, is accurate within 1% damping error for both wind turbine blade cases. The more conservative criterion, based on the nonparametric excitation, has larger errors, but follows a similar trend. This trend indicates that chaos following flutter is more likely to occur under low damping conditions and high stall slope factors. In any case, the stiffness mode coupling flutter criterion provides a useful and efficient conservative analytical prediction of chaos following blade flutter onset, as identified in [56]. However, the additional refinement based on overlapping parametric resonances allows for a more accurate quantification of the effects of damping and stall characteristics on the onset of chaos after flutter. These findings confirm that chaos following flutter is driven by conventional (stiffness mode coupled) flutter, combined with parametric excitation nonlinearities associated with lower damping and steeper stall slopes. Additionally, the results highlight that the analytical predictions can be generalised across different parameter sets. They also demonstrate that the analytical approach can replace thousands of numerical simulations, achieving computational acceleration by a factor of 10^7. These validated analytical criteria provide quantitative insights into the suppression of chaotic instability in wind turbine blades, a topic further explored in Chapter 6.

4.4 DRAGLINE BUCKET SWING CHAOS

It has been discovered that chaotic instabilities can occur in the swing motion of a dragline bucket during operation cycles. A dragline is a large, powerful, rotating multibody system utilised in the mining industry for removal of overburden and is the practical application of a tilted pendulum on a rotating arm known as an inclined Furuta pendulum, as shown in Figure 4.14.

The system comprises a slewing (rotating) assembly that includes a house, which contains essential components in the operator and engine rooms, along with a boom and bucket, as shown in Figure 4.14a). The dragline's operational cycle consists of three main phases:

1. Digging – The bucket collects overburden.
2. Slew – The house and boom rotate around a vertical axis while the bucket is hoisted.
3. Dump and Return – The bucket releases the overburden, and the house and boom return to the original digging position.

The bucket position is controlled within a plane by the hoist and drag rope lengths but remains free to swing perpendicular to this plane, requiring operator expertise to manage this unwanted motion.

As a fundamental rotating multibody system, the dragline offers insights into the dynamic behaviour of a variety of complex mechanical systems, including lifting equipment, spinning spacecraft and satellites, as studied in [155–162]. Additionally, the Furuta pendulum is widely used as a benchmark nonlinear system for the development and tuning of nonlinear control strategies [163–165]. Nonlinear dynamics and stability research on the non-inclined Furuta pendulum (dragline with loose dragrope) [166] has identified pitchfork and Hopf bifurcations, providing insights for inverted position control. Further research includes [164], where new branching point bifurcations were discovered, and [160], where similar methodologies were applied on a real-world Furuta pendulum system – the dragline. The nonlinear swing

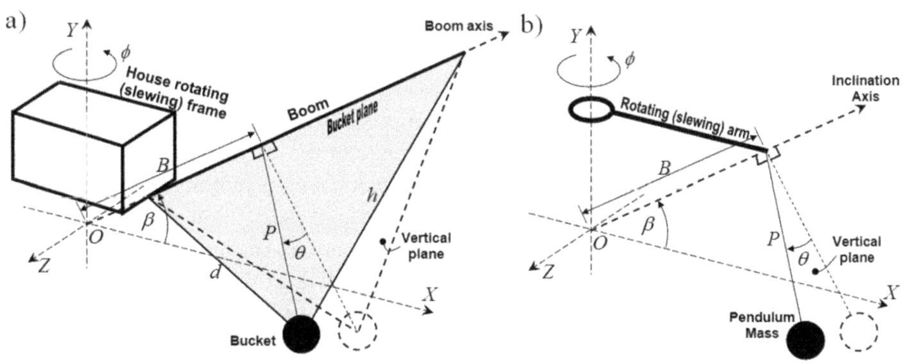

FIGURE 4.14 Equivalent models of an inclined Furuta pendulum; (a) slewing dragline with tight dragrope; (b) generic model [168].

dynamics of the dragline bucket during house slewing are particularly significant as excessive oscillations lead to fatigue damage and make control more challenging [167]. The study in [160] derived an analytical closed-form Melnikov solution to predict chaotic instability in the non-inclined Furuta pendulum/dragline system. While these predictive results provide valuable insights for avoiding and controlling chaos, they assume a slack drag rope that only occurs sometimes during real dragline cycles [168]. Here, we detail more general results for more typical dragline operation.

Prediction of Dragline Bucket Swing Chaos

To analyse the nonlinear behaviour of the dragline system, a model of an inclined Furuta pendulum is developed based on Figure 4.14. The house and boom structure (modelled as a rigid slewing arm) has moment of inertia, I_h, rotating about the vertical slew axis with boom inclination, β, relative to the horizontal X-axis. The bucket is represented as a pendulum mass, with its position determined by (massless) drag and hoist ropes of lengths, d and h, within the bucket plane. The bucket's 3D position is described by four independent degrees of freedom: slew angle, ϕ, that measures the rotation of the house and boom assembly about the slew axis; out-of-plane angle, θ, representing the angle between the vertical plane of the boom axis and the bucket plane; in-plane positions, B and P, parallel and perpendicular to the boom axis.

Due to the tight drag rope, the bucket swings about the boom axis rather than a horizontal axis, which would occur with a loose drag rope. This configuration transforms the system into an inclined Furuta pendulum, rather than a non-inclined one. The bucket is subjected to gravity, g, acting in the $-Y$ direction, and experiences a damping torque proportional to its out-of-plane angular velocity, with damping constant, c_t. Additionally, a net slew torque M is applied about the Y-axis, controlled by the operator through the slew transmission. The equations of motion for the system for the two rotational degrees of freedom, φ and θ, may be derived using Lagrange's equations in a similar manner to [160] as

$$\ddot{\phi}\underset{\sim}{I} + m\left(\ddot{\theta}P\underset{\sim}{D}_2 - \dot{\theta}^2 PB\sin\theta\cos\beta + 2\dot{\phi}\dot{\theta}P\sin\theta\left(P\cos\theta - \underset{\sim}{D}_1\sin\beta\right)\right) = M, \tag{4.27}$$

$$m\left(\ddot{\theta}P^2 + gP\cos\beta\sin\theta + \ddot{\phi}P\underset{\sim}{D}_2 - \dot{\phi}^2 P\sin\theta\left(P\cos\theta - \underset{\sim}{D}_1\sin\beta\right)\right) + c_t\dot{\theta} = 0, \tag{4.28}$$

where the underscore ~ indicates varying functions defined as follows; $\underset{\sim}{D}_1$ is the horizontal distance from the slew axis to bucket image in the vertical boom plane, $\underset{\sim}{D}_2$ is the perpendicular horizontal distance from the bucket swing velocity to the slew axis and $\underset{\sim}{I}$ represents the instantaneous moment of inertia of the system about the Y axis.

Note the inclination angle, β, influences the equations of motion in a couple of ways [160,168]. Gravity now acts at an angle, β, relative to the bucket swing plane, which reduces the natural frequency of the bucket swing (without slewing) to $\omega_o = \sqrt{g\cos\beta/P}$. Also, the bucket swing plane is not parallel to the slew axis, and therefore, a centrifugal force component emerges due to slewing, acting within the bucket's swing plane. This centrifugal stiffening effect plays a critical role in system stability and its perturbed nonlinear behaviour. The energy, E, and angular momentum, H_{Oy}, are defined as they are constant under unperturbed conditions as

$$E = \tfrac{1}{2}\underset{\sim}{I}\dot{\phi}^2 + \tfrac{1}{2}mP^2\dot{\theta}^2 + m\underset{\sim}{D}_2P\dot{\phi}\dot{\theta} + mgP\cos\beta(1-\cos\theta), \tag{4.29}$$

$$H_{Oy} = \underset{\sim}{I}\dot{\phi} + mP\dot{\theta}\underset{\sim}{D}_2. \tag{4.30}$$

The phase space of the unperturbed system, where both the energy, E, and angular momentum, H_{Oy}, remain constant, was analysed to examine its stability and nonlinear behaviour. Therefore, Eqs. (4.29) and (4.30) lead to the phase space representation described by

$$\dot{\theta}^2 = \frac{2\underset{\sim}{I}\left(\mathrm{E} - mgP\cos\beta(1-\cos\theta)\right) - H_{\mathrm{Oy}}^2}{mP^2(I_h + m(P^2+B^2)\cos^2\beta\sin^2\theta)}. \tag{4.31}$$

Equation (4.31) is in a similar (but distinct) form to that of the previous analyses (e.g. [160]) for the non-inverted Furuta pendulum and may also be shown to be integrable, a necessary condition for application of Melnikov's method. The corresponding set of phase curves under different energy and angular momentum conditions for different values of $C = 2EI - H_{\mathrm{Oy}}^2$ are plotted in Figure 4.15. These phase space plots are representative of the unperturbed dynamics of the inclined Furuta pendulum for out-of-plane bucket angles of magnitude less than π.[1] Figure 4.15a corresponds to the two equilibrium conditions described by

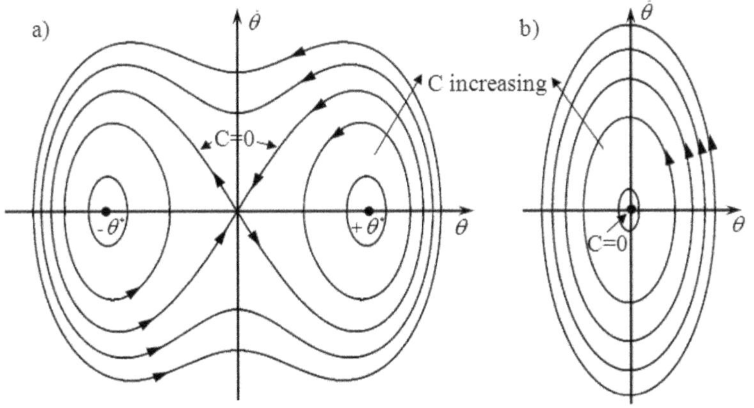

FIGURE 4.15 Phase space curves for an inclined Furuta pendulum without external forcing or dissipation; (a) $H_{\mathrm{Oy}}^2 > \left(I\dot{\phi}_{\mathrm{crit}}\right)^2$, (b) $H_{\mathrm{Oy}}^2 < \left(I\dot{\varphi}_{\mathrm{crit}}\right)^2$ [168].

$$\begin{bmatrix}\theta & \dot{\phi}\end{bmatrix}^{\mathrm{T}} = \left[\pm\theta^{*}\sqrt{g\cos\beta/(P\cos\theta^{*} - \underset{\sim}{D}_{1}^{*}\sin\beta)}\right]^{T}, \quad \text{if } H_{\mathrm{Oy}}^{2} \leq \left(I\dot{\phi}_{crit}\right)^{2}, \tag{4.32}$$

while Figure 4.15b corresponds to that described by

$$\begin{bmatrix}\theta & \dot{\phi}\end{bmatrix}^{T} = \begin{bmatrix}0 & H_{\mathrm{Oy}}/I\end{bmatrix}^{T}, \text{ if } H_{\mathrm{Oy}}^{2} \leq \left(I\dot{\phi}_{\mathrm{crit}}\right)^{2}, \tag{4.33}$$

where θ^{*} is a function of the parameters, H_{Oy} and I, as defined by the roots of a fourth-order polynomial

$$\theta^{*} = \cos^{-1}\left[\text{root}\begin{pmatrix}\left(I_{h} + m\left[P^{2} + B^{2} - (B\sin\beta - xP\cos\beta)^{2}\right]\right)^{2} \\ +H_{\mathrm{Oy}}^{2}(B\sin\beta - xP\cos\beta)/g = 0\end{pmatrix}\right]. \tag{4.34}$$

The stability of these equilibrium points may be derived using Lyapunov's direct method. Physically, the symmetrical properties of the bucket swing allow the possibility of two stable equilibrium points, corresponding to positive and negative swing displacements of the bucket/pendulum mass, $\pm\theta^{*}$. This pair of offset stable equilibrium points (also occurring in a variety of analogous systems, e.g. [158]) has physical significance, in that they serve to limit the slew rate to a constant critical value

$$\dot{\phi}_{\mathrm{crit}} = \sqrt{g/h_{Y}}, \quad h_{Y} = P\cos\beta - B\sin\beta. \tag{4.35}$$

where h_Y is the vertical distance of the bucket below the X-axis at zero swing angle. Note the inclined configuration significantly delays the formation of the two stable equilibrium points to bucket positions below the horizontal (i.e., at longer rope lengths) and shifts the critical slew rate to a higher value, $\dot{\phi}_{\mathrm{crit}} = \sqrt{g/h_{Y}} \geq \sqrt{g/P}$, before stability conditions are met. The existence of a pair of homoclinic orbits in Figure 4.15a encircling the two equilibrium points is important as the system will be attracted to either one of the equilibrium points depending on the initial conditions. There will also be a region in phase space where this dependence will be highly sensitive to small changes. By solving for the origin point using equation (4.31), the set of homoclinic orbits is found to occur under the conditions

$$C = 2EI - H_{\mathrm{Oy}}^{2} = 0 \quad \text{and} \quad \left\{H_{\mathrm{Oy}}^{2} > I^{2}g/h_{Y} \quad \text{where} \quad h_{Y} > 0\right\}. \tag{4.36}$$

It can be inferred that the pitchfork bifurcation leading to homoclinic orbits in Figure 4.15a occurs when the angular momentum first surpasses the critical value given by equation (4.32). Regarding chaotic instability, it is expected to emerge initially for large enough perturbations and small bucket swing angles once the slew velocity exceeds a critical threshold of $\sqrt{g/h_Y}$. To investigate this under the assumption of small bucket swing angles, the harmonic terms in equations (4.27) and (4.28) were approximated as

$$\sin\theta \approx \theta, \quad 1-\cos\theta \approx \theta^2/2, \tag{4.37}$$

and the system dynamics are reduced to a Lagrangian form [159] of

$$\hat{L} = \left[\hat{H}_{Oy}^2 + (\hat{I} - 1 + y^2)y'^2\right]/(\hat{I} + y^2) - \hat{k}\hat{y}^2/2, \tag{4.38}$$

where

$$\tau = \Omega t, \quad \hat{y} = \frac{\theta}{\hat{d}}, \quad \hat{d}^2 = \frac{I}{mP\cos\beta h_Y} - \frac{I_h}{m\cos^2\beta(D^2 + h_Y{}^2)}, \quad \hat{I} = \frac{I}{mP\cos\beta h_Y\hat{d}^2},$$

$$\hat{k} = \frac{\dot{\phi}_{\text{crit}}^2}{\Omega^2\left[1+(D/h_Y)^2\right]}, \quad \hat{H}_{Oy} = \frac{H_{Oy}}{mP\cos\beta h_Y\hat{d}^2\Omega\sqrt{1+(D/h_Y)^2}},$$

$$\left\{\hat{L},\hat{E}\right\} = \frac{\{L,E\}}{mP\cos\beta h_Y\hat{d}^2\Omega^2\left[1+(D/h_Y)^2\right]}, \tag{4.39}$$

and the perturbation parameters are

$$\hat{M} = \frac{M}{mP\cos\beta h_Y\hat{d}^2\Omega^2\left[1+(D/h_Y)^2\right]}, \quad \hat{c} = \frac{c_t}{mP\cos\beta h_Y\Omega\left[1+(D/h_Y)^2\right]}, \tag{4.40}$$

Note the static radial distance of the bucket from the slew axis is defined as

$$D = \underset{\sim}{D}_1(\theta = 0) = \underset{\sim}{D}_2(\theta = 0) = B\cos\beta + P\sin\beta. \tag{4.41}$$

The transformed Lagrangian (4.38) is now in a form suitable for Melnikov analysis to conservatively predict the required perturbations to cause chaotic instability [158].

Prediction under Perturbed Steady Slewing Conditions

Melnikov's method, as introduced in Chapter 2, is used for analytically predicting the onset of chaotic instability in the perturbed dragline system, using the unperturbed phase space as a reference. First, the system is expressed in the form:

$$\dot{\mathbf{x}} = \mathbf{f}(\mathbf{x}) + \varepsilon\mathbf{g}(\mathbf{x},t); \quad \mathbf{x} = \begin{pmatrix} u & v \end{pmatrix}^T \in \Re^2 \tag{4.42}$$

where $\mathbf{f}(\mathbf{x})$ is considered to be the unperturbed Hamiltonian system of state equations defined on $\Re^2$, and $\varepsilon\mathbf{g}(\mathbf{x},t)$ is a small periodic perturbation (which does not have to be Hamiltonian). The equations of motion for the dragline (4.27) and (4.28) can be expressed in the form of (4.42), using (4.39) and (4.40) as

$$
\begin{aligned}
d\hat{y}/d\tau_\Omega &= \hat{p}_y\left(\hat{\mathrm{I}}+\hat{y}^2\right)\left(\hat{\mathrm{I}}-1+\hat{y}^2\right)^{-1} \\
d\hat{p}_y/d\tau_\Omega &= \frac{\hat{y}}{\left(\hat{\mathrm{I}}+\hat{y}^2\right)^2}\left\{\hat{H}_{\mathrm{Oy}}^2-\hat{k}\left(\hat{\mathrm{I}}+\hat{y}^2\right)^2+\left[\hat{p}_y\left(\hat{\mathrm{I}}+\hat{y}^2\right)\beta(\hat{y})\right]^2\right\} \\
&\quad +\varepsilon\begin{bmatrix}\hat{M}_E\cos\tau+2\hat{y}\hat{M}_E\left(\hat{H}_{\mathrm{Oy}}/\hat{I}\right)\sin\tau \\ +\hat{y}\left(\hat{M}_E^2/\hat{I}\right)\sin^2\tau-\hat{c}\hat{p}_y\left(\hat{\mathrm{I}}+\hat{y}^2\right)^2\beta(\hat{y})\end{bmatrix}+O\left(\varepsilon^2\right)
\end{aligned}
\tag{4.43}
$$

where the generalised momentum, $\hat{p}_y$, is defined by

$$
\hat{p}_y=\hat{y}'\left(\hat{\mathrm{I}}-1+\hat{y}^2\right)/\left(\hat{\mathrm{I}}+\hat{y}^2\right), \tag{4.44}
$$

and $\hat{M}=\hat{M}_E\cos\tau$ is the perturbational slew torque. Note only first-order terms from a Taylor series expansion have been shown in equation (4.43), with more details in [159]. The Melnikov function, denoted as $M(t_0)$, is expressed as an integral

$$
M(t_0)=\int_{-\infty}^{\infty}\mathbf{f}(\mathbf{q}_0(t))\wedge\mathbf{g}(\mathbf{q}_0(t),t+t_0)\,dt, \tag{4.45}
$$

where the symbol $\wedge$ is the wedge product defined by $a\wedge b=a_1b_2-a_2b_1$, and $\mathbf{q}_0(t)$ is the solution for the set of homoclinic orbits in the unperturbed system defined in equation (4.42). The equations describing the homoclinic orbits depicted in Figure 4.15a may be obtained via solution to equations (4.31), subject to (4.36), (4.37) and (4.40), and for small angle, θ, as

$$
\left[\hat{y}_0\ \ d\hat{y}_0/d\tau_\Omega\right]=\left[\sqrt{\eta}\,\mathrm{sech}(\vartheta\tau_\Omega)-\vartheta\sqrt{\eta}\,\mathrm{sech}(\vartheta\tau_\Omega)\tanh(\vartheta\tau_\Omega)\right], \tag{4.46}
$$

where

$$
\eta=2\hat{E}/\hat{k}-\hat{I},\quad \vartheta=\sqrt{\left(2\hat{E}-\hat{I}\hat{k}\right)/\left(\hat{I}-1\right)}, \tag{4.47}
$$

and the subscript $(\ )_0$ denotes the homoclinic orbit solutions. Using equations (4.46), (4.42) and (4.43), the Melnikov function may be obtained explicitly. In particular, under the assumption that $\hat{M}_E \ll \hat{I}\sqrt{\hat{k}/\left(\hat{I}-1\right)}$ or $\hat{M}_E \ll 2\hat{H}_{\mathrm{Oy}}\cosh\left(\pi/2\vartheta\right)$, a relationship for the critical torque amplitude may be simplified to the form

$$
\hat{M}_E>2\hat{c}\vartheta^3\left(5\hat{I}+2\eta\right)\Bigg/\left(15\pi\sqrt{\frac{\vartheta^2}{\eta}\mathrm{sech}^2\left(\frac{\pi}{2\vartheta}\right)+\frac{\hat{H}_{\mathrm{Oy}}^2}{\hat{I}^2}\ \mathrm{cosech}^2\left(\frac{\pi}{2\vartheta}\right)}\right). \tag{4.48}
$$

Equation (4.48) provides a conservative analytical condition for the onset of chaotic instability in a dragline (or Furuta pendulum) subjected to slew torque perturbations, assuming small bucket swing angles.

Prediction under Unsteady Slewing Conditions

The normal dig, dump and return operation of a dragline is driven by an approximate cyclic (unsteady) slew torque from stationary initial conditions (at dig) of the form

$$M = M_{\text{Eu}} \cos(\Omega t) - c_h \dot{\phi}, \tag{4.49}$$

where c_h is the house slew damping constant that arises due to rotational dissipation and/or derivative motor control. The sufficient critical torque amplitude under these unsteady slewing conditions was found to reduce to the condition for the existence of the pitchfork homoclinic orbits that may be expressed as the dimensional inequality

$$M_{\text{Eu}} > \dot{\phi}_{\text{crit}} \sqrt{(I\Omega)^2 + c_h^2} \tag{4.50}$$

The reason for the simplification of this criterion is that the required torque amplitude, to achieve the critical slew rate for pitchfork bifurcation, was found to always exceed the critical torque required for chaos under the Melnikov condition (4.48). Hence, under normal cyclic slew operation, once the slew rate is instantaneously large enough for the pair of stable equilibrium bucket positions to exist, the criterion for chaotic instability (4.48) will also be instantaneously satisfied. This resultant sufficient criterion (4.50) is identical to the previous non-inclined Furuta pendulum analysis, except for a substantially larger critical slew rate $\dot{\phi}_{\text{crit}} = \sqrt{g/h_Y} \geq \sqrt{g/P}$ due to the inclination, as discussed previously. In particular, the critical slew rate is determined by the smaller vertical displacement of the bucket below the horizontal, as opposed to that relative to the boom tip suspension point. It is, therefore, expected that chaotic instability will be delayed to higher slew torque amplitudes and will only occur for bucket positions below the horizontal as compared to non-inclined (slack dragrope) conditions. However, it is uncertain whether the predicted transient chaotic behaviour related to instantaneous pitchfork bifurcation existence will persist throughout the entire cycle. These key issues are investigated numerically and using field measured data of dragline in operation, in the subsequent sections.

Results of Dragline Bucket Swing Chaos Predictions

Predictions of the occurrence of chaotic instability in an inclined (tight dragrope) and non-inclined (loose dragrope) Furuta pendulum under both steady and unsteady slewing motions were investigated both analytically and numerically for realistic dragline parameters. The mechanical parameters for a standard 1370 BE dragline were used to compare with previous research results: $m = 66{,}000\,\text{kg}$, $c_t = 4.1$ MNms, $I_h = 1.588 \times 10^9$ kgm^2 and $\Omega = 0.05 \rightarrow 0.2\,\text{rad/s}$. Two different bucket positions and inclinations were investigated, as shown in Figure 4.16. Case 1a represents the non-inclined case and is included so that direct comparison with previous results [160] can be made. The inclined Cases 1b and 2 represent typical quasistatic bucket positions encountered during normal operation with a taut dragrope. For these parameters, the critical slew rate at which a pitchfork bifurcation first occurs under inclined and non-inclined conditions is described in Table 4.1. The slew torque

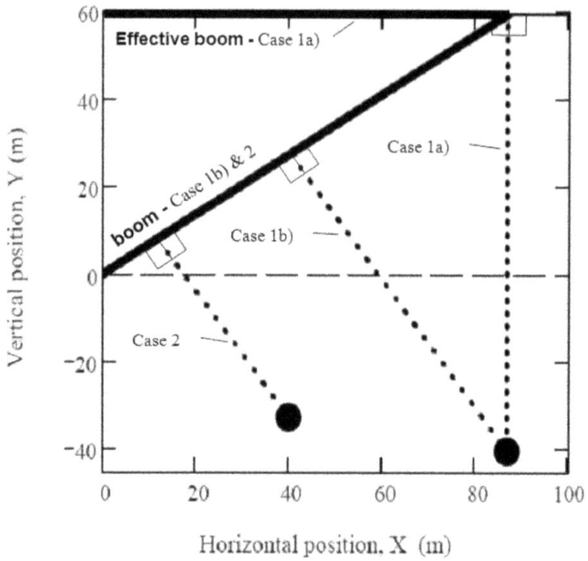

FIGURE 4.16 Pendulum mass/bucket positions and inclinations for test Cases 1a, b & 2 [168].

TABLE 4.1
Case Configuration Parameters and Corresponding Critical Slew Rates [168]

	β (rad)	D (m)	h_Y (m)	$\dot{\phi}_{crit}$ (rad/s)
Case 1a non-inclined	0	87	100	0.3132
Case 1b inclined	0.602	87	40.224	0.4938
Case 2 inclined	0.602	40	32.517	0.5492

frequency was chosen to be of the same order as the actual swing cycle used in dragline operation. Numerical integration of equations (4.27) and (4.28) was performed using a fourth-order Runge–Kutta method with at least 200 timesteps over the bucket response cycle, ensuring convergence of the numerical solution. Plots were generated after a number of preiterates (no recording) of forcing periods, in order to ensure transients had subsided.

Results under Perturbed Steady Slewing Conditions

According to the criterion in equation (4.48), initial slew rates slightly exceeding the critical value were examined to ensure that the critical torque amplitude remained within realistic limits for draglines (on the order of 100 MNm) while keeping bucket swing angles small and practical. Unless otherwise specified, the initial slew rate was set to 2% above the critical slew torques listed in Table 4.1. Figure 4.17 presents

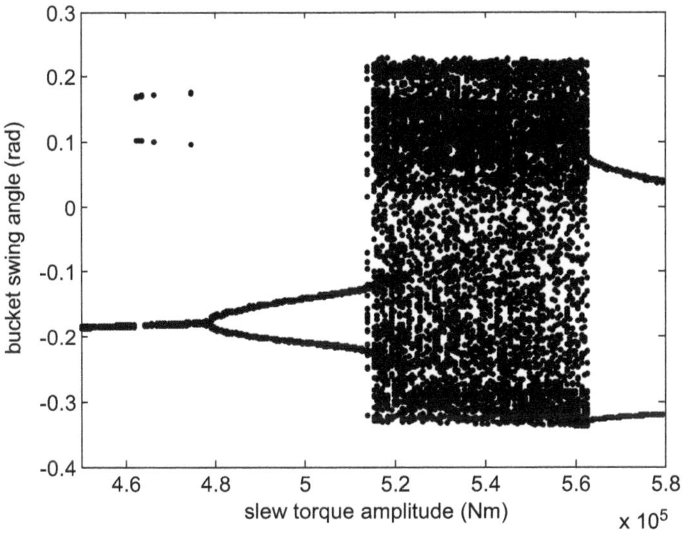

FIGURE 4.17 Bifurcation diagram for the inclined Furuta pendulum/dragline Case 1b under perturbed, steady slew conditions ($\dot{\phi}_i = 0.505$ rad/s, $\Omega = 0.05$ rad/s) [168].

a bifurcation diagram for the dragline/inclined Furuta Case 1b using 200 preiterates. The diagram reveals the presence of steady-state chaotic instability across a range of torque amplitudes, with its initial onset occurring at approximately M_E=515 kNm. Notably, this threshold is significantly higher than M_E=390 kNm [160], found in the loose dragrope/non-inclined Case 1a. Following the same approach, critical torque amplitudes were determined for multiple bifurcation diagrams at different slew torque frequencies, allowing for a quantitative comparison with the analytical Melnikov predictions from equation (4.48). The compiled results for the three cases in Table 4.1 are summarised in Figure 4.18.

There is a strong correlation between the analytical and numerical results shown in Figure 4.18, confirming the conservative Melnikov predictions for chaotic instability. Comparing Cases 1a and 1b reveals that the inclined Furuta pendulum (dragline system) requires significantly higher critical torques (more than 3.5 times greater) to reach chaotic instability. This increase is primarily due to centrifugal stiffening of the bucket swing motion, which requires stronger perturbations to overcome in the inclined configuration. Similarly, comparing Cases 1b and 2 suggests that the degree of centrifugal stiffening is influenced by the bucket's position relative to the horizontal *X*-axis, with higher critical torques needed as the bucket moves closer to the horizontal *X*-axis. This trend is further validated by equation (4.28), which indicates that the total centrifugal torque acting on the bucket swing motion is proportional to this distance, $\underset{\sim}{h}_Y$. These findings, observed under steady slewing conditions, suggest that a dragline system may encounter chaotic instability under larger perturbations, particularly those associated with a typical (periodic/unsteady slewing) operational cycle.

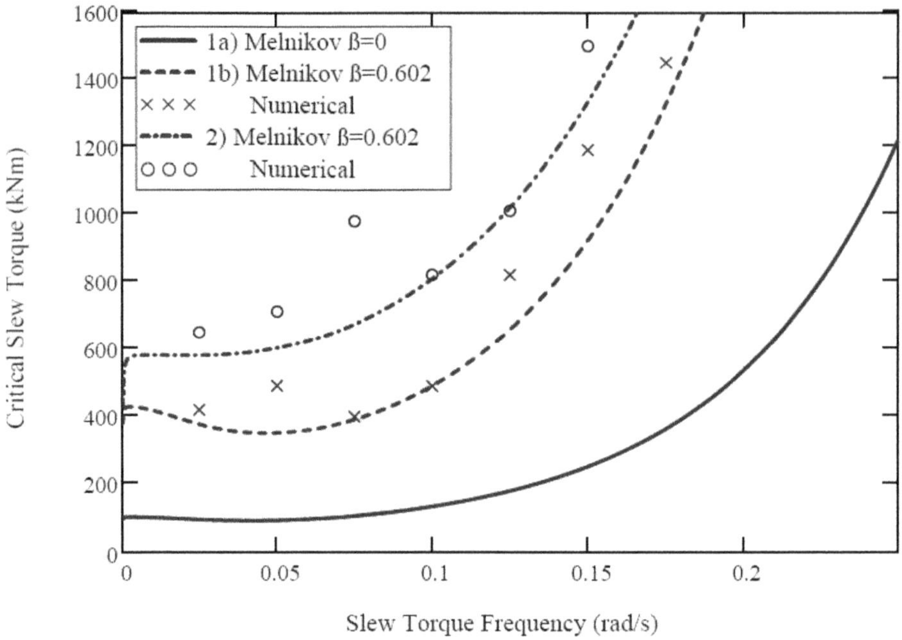

FIGURE 4.18 Comparison of analytical and numerical results for the onset of chaotic instability for the inclined Furuta pendulum/dragline Cases 1a, b & 2 under perturbed, steady slew conditions ($\dot{\phi}_i = 0.505$ rad/s) [168].

Results under Unsteady Slewing Conditions

The validity of the analytical criterion (4.50) was also investigated via numerical simulations of typical unsteady slewing conditions. In particular, a nominal slew torque frequency of $\Omega = 0.0623$ rad/s representing a cycle time of 100 seconds was used. All other parameters were set values as specified previously, with exception to house slew damping, which was set to a value of $c_h = 8.97 \times 10^7$ Nms; obtained from field measurements. Trajectory simulations and the Poincare map of Figure 4.19 show chaotic instability at a marginally higher torque amplitude of 84 MNm compared to the onset of 78 MNm predicted by the analytical criterion.

The complexity observed in the time history, phase space trajectories and the strange attractor in the Poincaré map confirms chaotic instability. Note the steady-state chaotic behaviour persists, even though the boom maintains the necessary slew rate for pitchfork homoclinic orbits for only a small portion of the cycle time. These results were confirmed over a wider range of conditions using a bifurcation diagram generated for all Cases 1a, b & 2 as shown in Figure 4.20.

Figure 4.20 shows that chaotic instability first emerges through bifurcations at a torque amplitude of approximately 82 MNm, which closely aligns with the analytical prediction of 78 MNm. A comparison between Cases 1a and 1b further confirms that the dragline/inclined Furuta pendulum requires significantly higher critical torques,

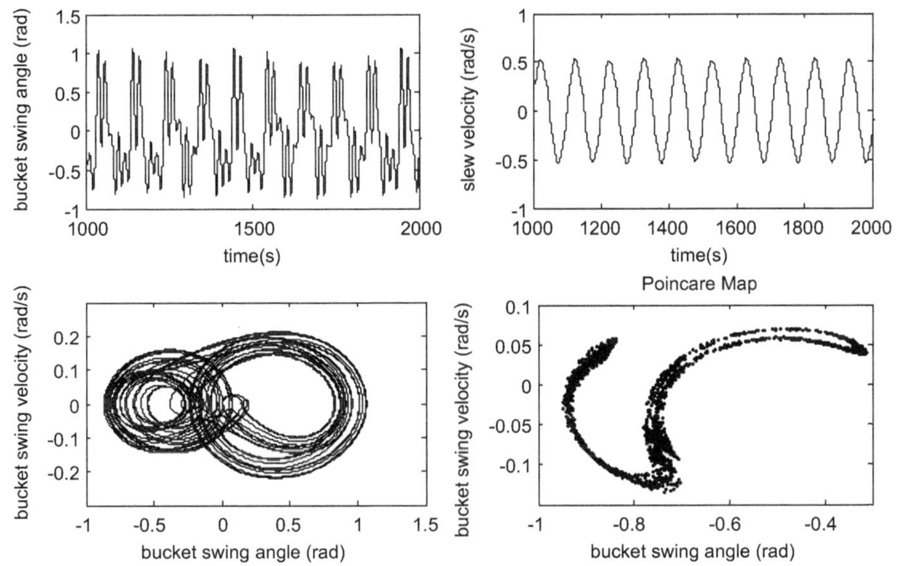

FIGURE 4.19 Time histories and phase spaces for the inclined Furuta pendulum/dragline Case 1b under large unsteady slew conditions ($\dot{\phi}_i = 0$ rad/s, $\Omega = 0.0623$ rad/s, $M = 84 \times 10^6$ Nm) [168].

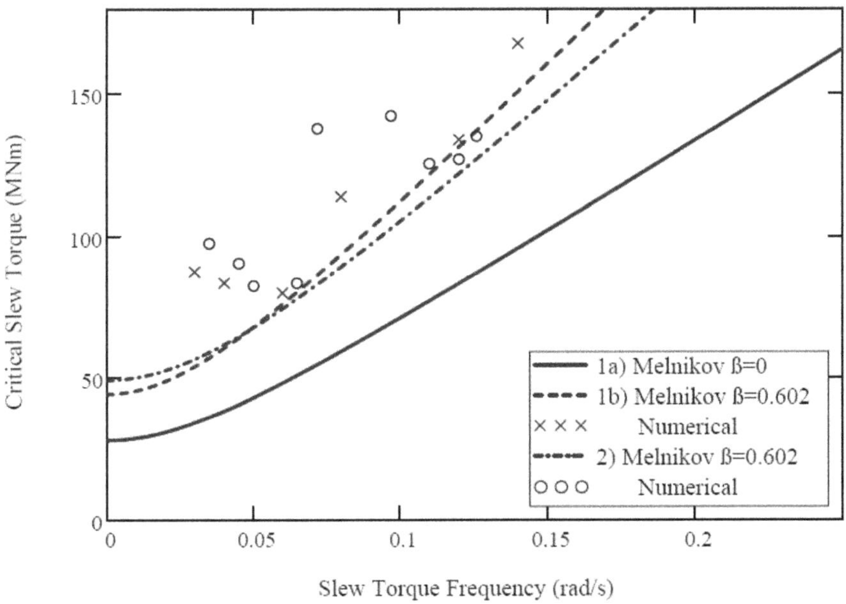

FIGURE 4.20 Comparison of analytical and numerical results for the onset of chaotic instability for the dragline/inclined Furuta pendulum/dragline Cases 1a, b & 2 under unsteady slew conditions.

compared to the non-inclined case, due to its higher critical slew rate, which is a direct consequence of the system's inclination. Comparing the inclined Cases 1b and 2, it is interesting to note that the criteria are relatively close and crossover at approximately 0.05 rad/s. By inspection of (4.50) and Table 4.1 it may be deduced that the closeness of these cases is associated with that of the critical slew rates. Also, crossover occurs due to the substantial difference in horizontal bucket locations from the slew axis and hence rotational inertias, I.

This analysis confirms that steady-state chaotic instability can occur during high slew rate conditions in dragline operations, regardless of whether the drag rope is tight or loose. The presence of chaos negatively impacts productivity and maintenance and may contribute to challenges in automation. The findings confirm that Melnikov's method offers a useful (and extremely computationally efficient) conservative estimate for predicting the onset of steady-state chaotic instability in both steady and unsteady slewing conditions. Additionally, chaotic instability is observed to be delayed to higher torque amplitudes in the tight dragrope or inclined Furuta pendulum configuration, due to the centrifugal stiffening effect on the pendulum swing motion. Further research shows this persists under real operation cycles where the rope lengths and mass varies [169]. From mining machinery, we now move to spacecraft where we find the same fundamental nonlinear dynamics.

4.5 SPACECRAFT CHAOS

A number of investigations on spinning spacecraft stability have been performed in the past after abnormalities were observed in practice. Many of these observed instabilities have been found to arise from excitations from structural imperfections in combination with the inherent nonlinearity within the dynamics of the system. For instance, many researchers have investigated the attitude stability and behaviour of a dual-spin spacecraft due to parametric fluctuations resulting from structural imperfections. However, attention has also been concentrated on investigating chaotic instabilities that can arise in a spacecraft system. Satellite examples of spinning spacecraft and their dynamic modelling are illustrated in Figure 4.21.

Bearing axis wobble arising from both static and dynamic rotor imbalances has been studied by Tonkin [171] as well as McIntyre and Gianelli [172]. Cronin [173] and Scher [174], amongst others, have developed models incorporating different types of joint flexibility between platform and rotor, which also revealed parametric instabilities. Instabilities caused by the asymmetry of rotor and platform have also been found by Tseng [175], Tsuchiya [176] and Lukich and Mingori [177]. A number of authors have incorporated some or all of these effects, such as Agrawal [178,179] and Konstantopoulos [180]. Van Doorn and Asokanthan [181] have studied these forms of instabilities using the method of averaging. More recently, Chang et al. [182] and Ross [183] have found further stability results for a spacecraft containing a rotor and nutation damper, respectively. The results obtained in the above studies are of concern to designers. This is because a spacecraft may pass through these instability regions during rotor spin-up or during a malfunction such as run-down of the momentum wheel during temporary power loss. However, in many previous investigations on stability, some form of linearisation of the system equations of motion has been performed;

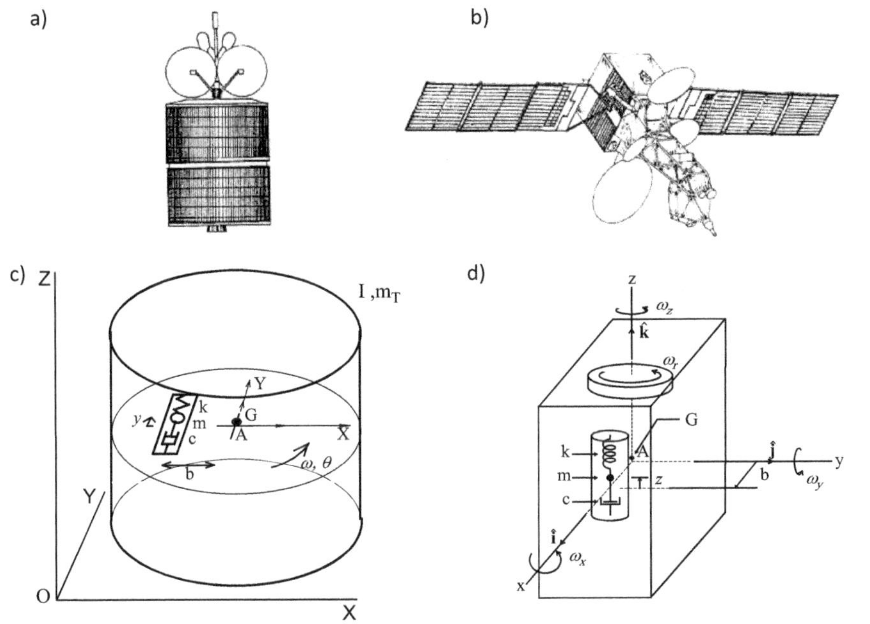

FIGURE 4.21 (a) External rotor spacecraft: Intelsat IV [170], (b) body stabilised spacecraft: Intelsat V [170], (c) rotating body with internal spring-mass-damper, (d) dual-spin spacecraft with an axial nutational damper; bias momentum type [194], ch.5 [10].

thus, other forms of nonlinear behaviour such as chaotic instabilities were not considered. However, fundamental rotating body systems similar to spinning spacecraft have been studied for chaos. Zaslavsky [184] has presented results for a kicked rotor subjected to both a steady torque and a periodic series of pulsed torques. The problem of a kicked double rotor has also been investigated by Grebogi [185]. An analysis on a rigid body with an attached imperfect flywheel has been performed by Holmes et al. [186] and Koiller [187] using Melnikov's method to show the existence of horseshoes. Other research works on chaos due to spacecraft attitude instability include [188–192].

We, here, first describe the equations of motion governing the dynamics of a simplified spinning spacecraft system. The stability configuration of the system is then determined so that a physical description of the dynamics of the system can be given. The dynamics of the system under the external influence of a periodically varying torque is then investigated numerically by integrating the equations of motion. Results are shown using several criteria for the purpose of identifying nonlinear behaviour such as chaos. These are then developed further for the more complex dual-spin spacecraft.

Prediction of Spacecraft Chaos

For this analysis, a simple model of spinning spacecraft with a circumferential nutational damper is first chosen, and no linearisation of the equations of motion has been performed. Such a model represents basic satellite systems as well as a simplification of a dual spin spacecraft in that platform and non-principal axis rotations are not considered. A circumferential damper was chosen over an axial damper because of its effectiveness in operation over a wider range of nutation angles [193]. Also, it allows simplification of the problem to one plane. The problem involves the study of a body with internal moving parts such that the resulting dynamics involves a coupling of the motion of the damper mass and the spinning rotor like the slewing dragline discussed in the previous section. Chaotic motion has been found in this simplified model when the spacecraft is subjected to an external periodically varying torque. A similar situation may arise practically under malfunction of the control system or during spin-up of an unbalanced rotor of a dual-spin spacecraft.

Figure 4.21a,c shows the rotating multibody system being analysed and consisting of a rigid body rotating at an angular velocity, ω, incorporating a spring–mass–dashpot system for internal energy dissipation. The degrees of freedom of the system, y and θ describe the displacement of the damper mass and rotation of the system about A (or G), respectively. The damper is centred on the body fixed X-axis and has a point mass of m that moves along an axis parallel to the Y-axis at a distance b from A. The spring and the damping constants of the circumferential damper are denoted by k and c, respectively. The system rotates about its centre of mass at point G, which coincides with A when $y = 0$. Note that the position of A with respect to the motionless centre of mass G and the instantaneous moment of inertia of the system about A vary with the position of the damper mass. The system is considered to have a total mass m_T and moment of inertia I about the Z-axis when $y = 0$. The system is also considered to be subjected to a net external torque M, composed of an excitational time-varying torque M_E such that $M = M_E$. In practical applications, such a disturbance torque may occur in a dual-spin spacecraft, particularly if a control system malfunction leads to rotor driver fluctuations, resulting in periodic torquing of the platform. Similar disturbances can also arise during the spin-up of an unbalanced rotor, from vibrations in spacecraft appendages, or due to environmental and orbital variations.

The equations of motion for this system can be formulated using Lagrange's equations with a generalisation for multibody systems or other means [194,195]. They may then be expressed in nondimensionalised form as

$$\left(\hat{I} + \hat{y}^2\right)\hat{\omega}' + 2\hat{y}\hat{y}'\hat{\omega} - \hat{y}'' = \hat{M}(\tau), \tag{4.51}$$

$$\hat{y}'' + \hat{c}\hat{y}' + \hat{k}\hat{y} - \hat{\omega}^2\hat{y} - \hat{\omega}' = 0, \tag{4.52}$$

using the following dimensionless quantities:

$$\tau = \Omega t, \quad \hat{y} = \frac{(1-\mu)y}{b}, \quad \hat{\omega} = \frac{\omega}{\Omega}, \quad \hat{I} = \frac{(1-\mu)I}{mb^2}, \quad \hat{c} = \frac{c}{m(1-\mu)\Omega},$$
$$\hat{k} = \frac{k}{m(1-\mu)\Omega^2}, \quad \hat{M} = \frac{(1-\mu)M}{mb^2\Omega^2}, \quad \mu = \frac{m}{m_T}. \tag{4.53}$$

Here, Ω and τ represent the frequency of the time-varying torque and the nondimensional time, respectively, while the superscript, ^, indicates a nondimensional parameter. Additionally, differentiation with respect to the dimensionless time is denoted,′. For stability and chaotic analysis, it is useful to define energy and angular momentum quantities, which remain conserved under unperturbed conditions as [194]

$$\hat{E} = \frac{1}{2}\left[\hat{\mathrm{I}} + \hat{y}^2\right]\hat{\omega}^2 + \frac{1}{2}\hat{y}'^2 + \frac{1}{2}\hat{k}\hat{y}^2 - \hat{y}'\hat{\omega}, \quad \text{where} \quad \hat{E} = \frac{(1-\mu)E}{mb^2\Omega^2}, \tag{4.54}$$

$$\hat{h}_G = \left(\hat{\mathrm{I}} + \hat{y}^2\right)\hat{\omega} - \hat{y}', \quad \text{where} \quad \hat{h}_G = \frac{(1-\mu)h_G}{mb^2\Omega}, \tag{4.55}$$

Equations (4.51) and (4.52) can be considered to be the nondimensional, simplified Euler's equation representing the dynamics of a spinning spacecraft with circumferential nutational damper. The coefficient of $\hat{\omega}'$ in equation (4.51) represents the instantaneous moment of inertia of the system about G, while the terms involving $\hat{y}'$ and $\hat{y}''$ represent the moments arising from the Coriolis and other inertia forces with respect to *G, respectively*. Equation (4.52) represents the acceleration–force balance of the spring–mass–damper system where the last two terms characterise the effects of the centrifugal and angular accelerations, respectively. It may be noted that these two equations completely describe the system dynamics. In this case, the system is of order 5 since $\hat{M}$ is assumed to vary periodically with time and state equations described by $\hat{\theta}' = \hat{\omega}$ as trivial. It may also be observed that equations (4.51) and (4.52) are coupled through three nonlinear terms resulting from the dynamic effects of the damper point mass. The subsequent analysis will investigate the dynamical response of this system to an external disturbance torque $\hat{M} = \hat{M}_E$. The stability analysis performed using Lyapunov's direct method in [196] identified a pitchfork bifurcation to two equilibrium states as found for the previous dragline analysis. This forms the basis of the following phase space analysis, leading to chaos prediction.

Phase Space Analysis

To derive analytical criteria for chaotic motion using Melnikov's method, the system is initially treated as a perturbed integrable system. This requires the unperturbed system to possess a homoclinic orbit and be integrable. Using equations (4.54) and (4.55), the unperturbed phase space of constant energy and angular momentum can be represented by a set of corresponding phase curves. In particular, Figure 4.22 shows the phase spaces for particular values of C in configurational state space (X_1, X_2) where the following nondimensional terms have been introduced for simplicity [197],

$$C = \frac{2\hat{E}\hat{I} - \hat{h}_G^2}{\hat{k}}, \quad \gamma = \hat{I} - 1, \quad \eta = \frac{2\hat{E} - \hat{k}\hat{I}}{\hat{k}}, \quad r_{1,2} = \left(\eta \pm \sqrt{\eta^2 + 4C}\right)/2, \tag{4.56}$$

Note the analogy with dragline phase space curves in the previous section, i.e., a pitchfork bifurcation from one to two equilibrium points as energy of the unperturbed system changes. Figure 4.22a shows the pairs of homoclinic orbits about the

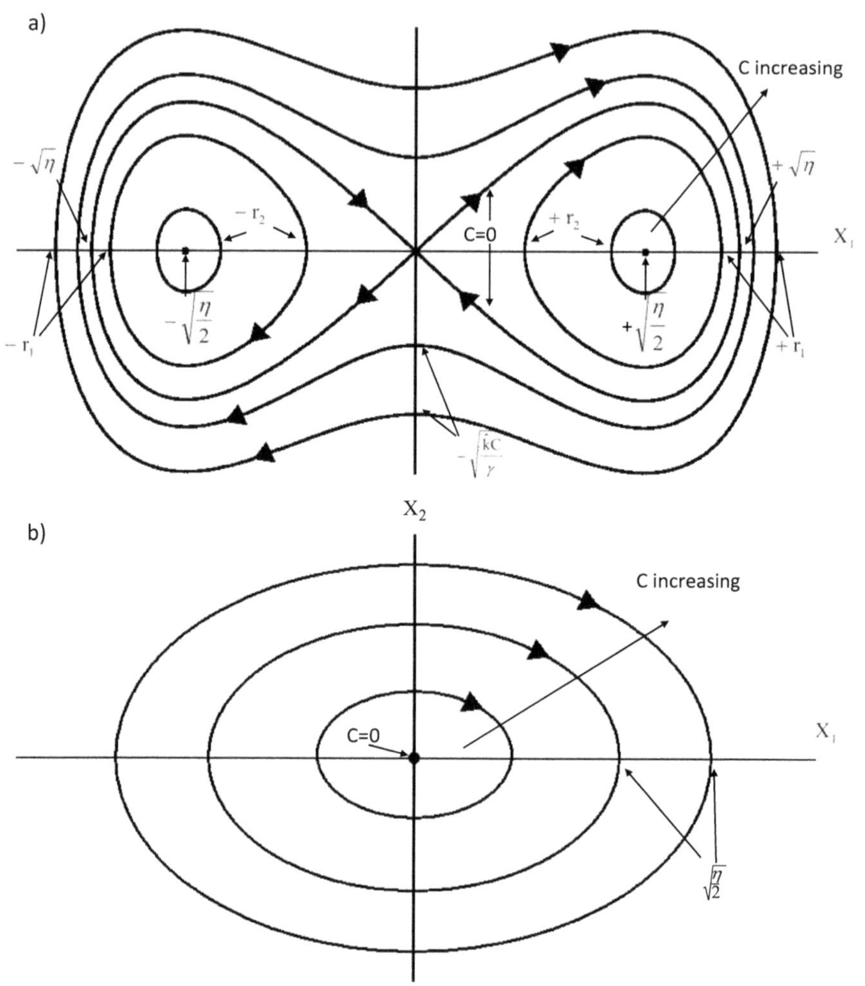

FIGURE 4.22 Phase space curves for the unperturbed rotating body with (a) $\eta \geq 0$, (b) $\eta \leq 0$ (ch.5 [8]).

two equilibrium points for consideration of Melnikov's method, which may be determined when $C = 0$ as

$$\left[X_1^0 \quad X_2^0\right] = \left[\sqrt{\eta}\,\text{sech}\left(\sqrt{\frac{\eta\hat{k}}{\gamma}}\tau\right) \quad -\eta\sqrt{\frac{\hat{k}}{\gamma}}\,\text{sech}\left(\sqrt{\frac{\eta\hat{k}}{\gamma}}\tau\right)\tanh\left(\sqrt{\frac{\eta\hat{k}}{\gamma}}\tau\right)\right], \tag{4.57}$$

where X_i^0 for $i = 1,2$ denote the closed solutions for the homoclinic orbits. Analytical prediction of chaotic instabilities in this system was then done considering the damping and external torque as perturbations using Melnikov's method.

Melnikov's Method

Noting the analogy with the dragline system described in the previous section, the nondimensionalised Hamiltonian equations of motion and Melnikov analysis described by equations (4.42)–(4.47) are also pertinent for the spacecraft system described here and so are not repeated. However, more detailed solutions to the non-trivial integrals [197] are provided here for interest, i.e., the Melnikov function can be simplified to the form

$$m(t_0) = \frac{\hat{M}_E \pi \eta \Lambda_0}{\vartheta^2} - \frac{2\hat{c}\vartheta\eta\left(5\hat{I}+2\eta\right)}{15}, \tag{4.58}$$

where

$$\Lambda_0 = \frac{\vartheta}{\sqrt{\eta}}\operatorname{sech}\left(\frac{\pi}{2\vartheta}\right)\sin\tau_0 - \frac{\hat{h}_G}{\hat{I}}\operatorname{cosech}\left(\frac{\pi}{2\vartheta}\right)\cos\tau_0 - \frac{\hat{M}_E}{\hat{I}}\operatorname{cosech}\left(\frac{\pi}{\vartheta}\right)\sin(2\tau_0). \quad 4.59)$$

Solving for the root of the Melnikov function (4.58) provides the criterion for the onset of chaotic instability as

$$\hat{M}_E \Lambda_{max} = \frac{2\hat{c}\vartheta^3\left(5\hat{I}+2\eta\right)}{15\pi}. \tag{4.60}$$

where Λ_{max} is the maximum amplitude of the periodic function Λ_0. In order to analytically solve for $\Lambda_{max,}$ two assumptions may be made based on the physical parameters of the system in (4.59) as follows:

1. In the case where $\hat{M}_E << \hat{I}\sqrt{\hat{k}/\left(\hat{I}-1\right)}$ is order $O\left(\hat{h}_G\right)$, the amplitude of the third term in equation (4.59) has a negligible effect on the amplitude $\Lambda_{max,}$ which then reduces to

$$\Lambda_{max} = \sqrt{\frac{\vartheta^2}{\eta}\operatorname{sech}^2\left(\frac{\pi}{2\vartheta}\right) + \frac{\hat{h}_G^2}{\hat{I}^2}\operatorname{cosech}^2\left(\frac{\pi}{2\vartheta}\right)}$$

2. In the case where $\hat{I}\sqrt{\hat{k}/\left(\hat{I}-1\right)} << \hat{M}_E$ is order $O\left(\hat{h}_G\right)$, the amplitude of the first term in equation (4.59) has a negligible effect on the amplitude $\Lambda_{max,}$ which in turn reduces to

$$\Lambda_{max} = \frac{1}{4}\left(3A+\sqrt{A^2+32B^2}\right)\sqrt{\frac{1}{2}-\frac{A}{32B^2}\left(A-\sqrt{A^2+32B^2}\right)},$$

where $A = \left(\hat{h}_G/\hat{I}\right)\operatorname{cosech}\left(\pi/(2\vartheta)\right)$ and $B = \left(\hat{M}_E/\hat{I}\right)\operatorname{cosech}\left(\pi/\vartheta\right)$.

Using this criterion, it is of interest to verify, identify and explore how the occurrence of chaotic instability is affected by system parameters.

Results of Spacecraft Chaos Predictions

To verify the analytical criterion for chaos, numerical simulations were conducted by integrating the full nonlinear equations (4.51) and (4.52), while applying an excitational time-varying torque. To ensure numerical accuracy, both the fourth-order Runge–Kutta method and the Huen predictor-corrector routine were used for integration. A range of tools were employed to analyse the nonlinear behaviour, including time history plots, bifurcation diagrams, Poincaré maps and Lyapunov exponents. The Lyapunov spectrum was computed using the Wolf et al. [198] algorithm, iterating over at least 1000 forcing periods, with renormalisation performed at half the forcing period. The parameters of the spinning spacecraft with a circumferential nutational damper were chosen to be similar to those of INTELSAT II or III, with numerical integrations starting from the specified initial conditions. Further details on the numerical onset of chaotic instability through bifurcations can be found in [194] and are summarised in the bifurcation diagram of Figure 4.23.

The bifurcation diagram shows that the onset of chaotic motion is clearly defined by period doubling, with a later crises phenomena (where the attractor suddenly expands at approximately $M_E = 7.9\text{e-}4$ Nm). Intermittency was also identified in other cases. Notably, the entire bifurcation process occurred over a very narrow range of torque magnitudes, which is a critical concern for spacecraft designers as it highlights the system's sensitivity to small changes in control parameters. Fully developed chaotic instability is observed at $M_E = 9\text{e-}4$ Nm, for which the Lyapunov exponents were computed as $[0.57 \;\; 0 \approx 0 - 3.0] \times 10^{-2}$, confirming chaotic motion

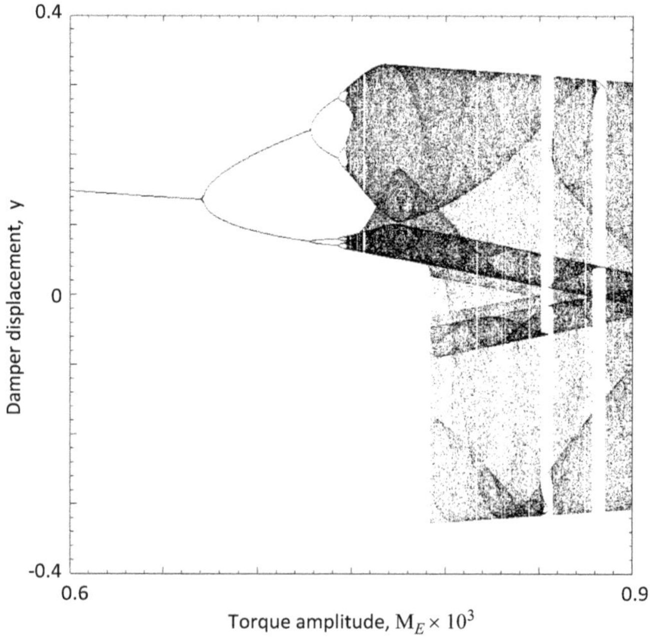

FIGURE 4.23 Bifurcation diagram showing a period-doubling route to chaotic motion (ch.5 [8]).

due to the presence of a positive exponent. Furthermore, the correlation dimension of the system, calculated from the Poincaré map, yielded a noninteger value of 2.093, verifying chaos and suggesting that the system is highly dissipative (being close to 2). For comparison, Figure 4.24 depicts the analytic criteria of (4.60) in 3D parameter space.

The figure shows that chaotic instability emerges at very low torque amplitudes when the torque frequency and damping is small. To compare these analytical predictions with numerical results, the first occurrence of steady-state chaotic instability is observed at approximately $M_E = 7.5e\text{-}4$ Nm in the bifurcation diagram of Figure 4.23. This method was also applied to other parameter configurations of the same model, as summarised in Table 4.2: Comparison of Analytical and *Numerical* results *for the onset of Chaotic Instability.*

When comparing the analytical and numerical results in Table 4.2, note that Melnikov's analysis is conservative as it provides only a condition for the occurrence of transient chaotic motion (or sensitivity to initial conditions), whereas the numerical results confirm the presence of steady-state chaotic motion. A more detailed investigation was conducted over a range of forcing frequencies for the first two parameter configurations of Table 4.2, as shown in Figures 4.25 and 4.26, respectively.

These figures demonstrate that Melnikov's prediction provides a good estimate of the trend and a conservative lower bound for the onset of steady-state chaotic instability. However, the level of agreement between the analytical and numerical results varies, which may be attributed to several factors such as the levels of damping [190], type of onset, overlapping attractors and perturbation magnitude [197]. Nevertheless, based on the comparisons in Table 4.2, it can be concluded that, for conservative and practical design purposes, Melnikov's analysis serves as a valuable analytical tool for predicting and avoiding chaotic instability in a simplified spacecraft system.

It was of interest to determine whether chaotic vibrations can exist in more complex satellite systems. Hence, we also show evidence for a complex dual-spin

TABLE 4.2
Comparison of Analytical and Numerical Results for the Onset of Chaotic Instability (ch.5 [8])

Parameter Configuration	Critical Value for Torque Amplitude, *M* (Nm)	
	Analytical	Numerical
(a) INTELSAT II or III $c = 0.005$ Ns/m, $\Omega = 0.03$ rad/s (see Meehan and Asokanthan [194])	7.3×10^{-4}	7.4×10^{-4}
(b) INTELSAT II or III $c = 0.002$ Ns/m, $\Omega = 0.05$ rad/s (see Meehan and Asokanthan [194])	6.3×10^{-4}	10.1×10^{-4}
(c) Rotating body with internal energy dissipation (see Meehan and Asokanthan [195])	0.70	1.67

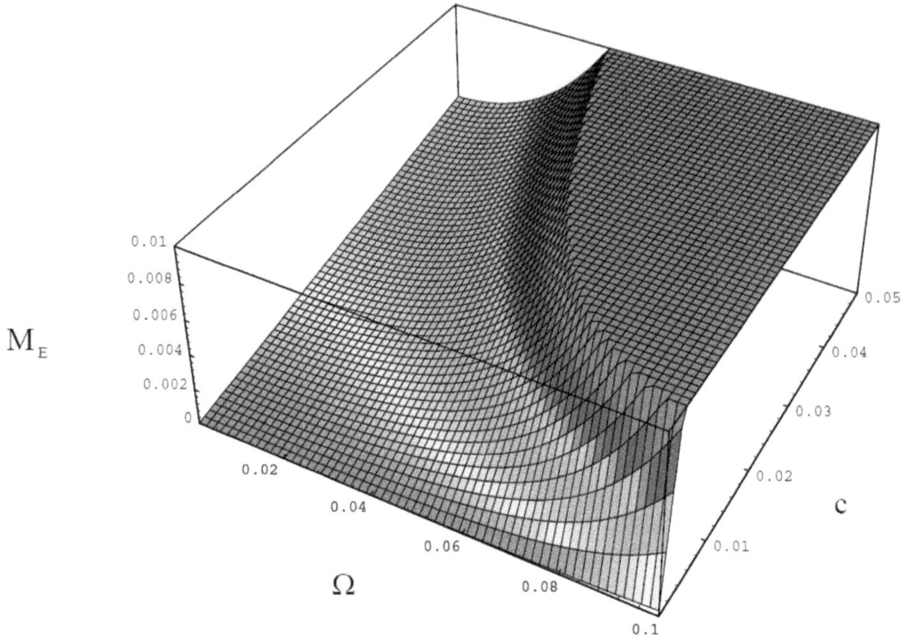

FIGURE 4.24 Analytical surface separating chaotic from nonchaotic motion for parameter space M_E - Ω - c for rotating body with satellite parameters (ch.5 [8]).

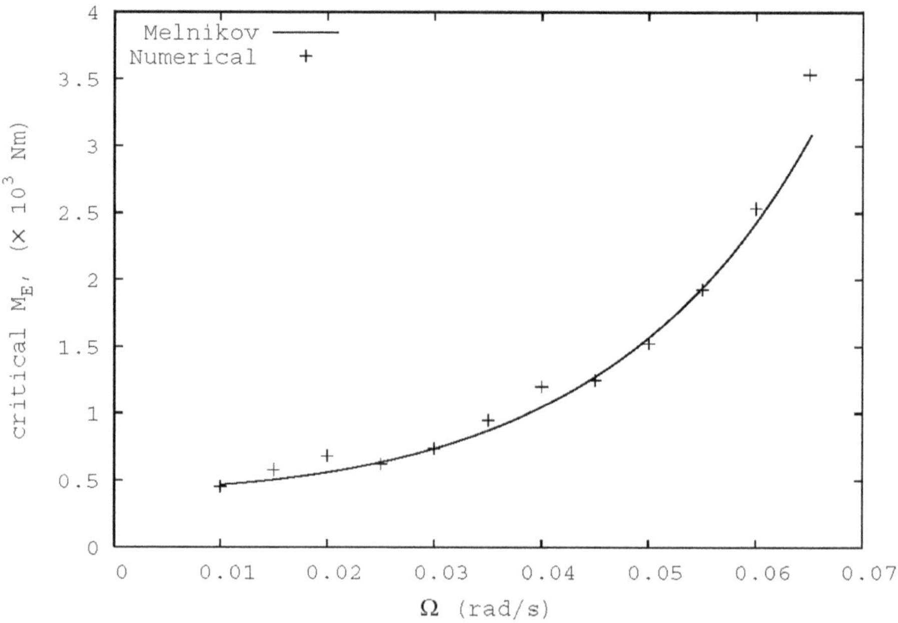

FIGURE 4.25 Chaos diagram of Melnikov and numerical results for a rotating body with internal energy dissipation with parameter configuration of Table 4.2.

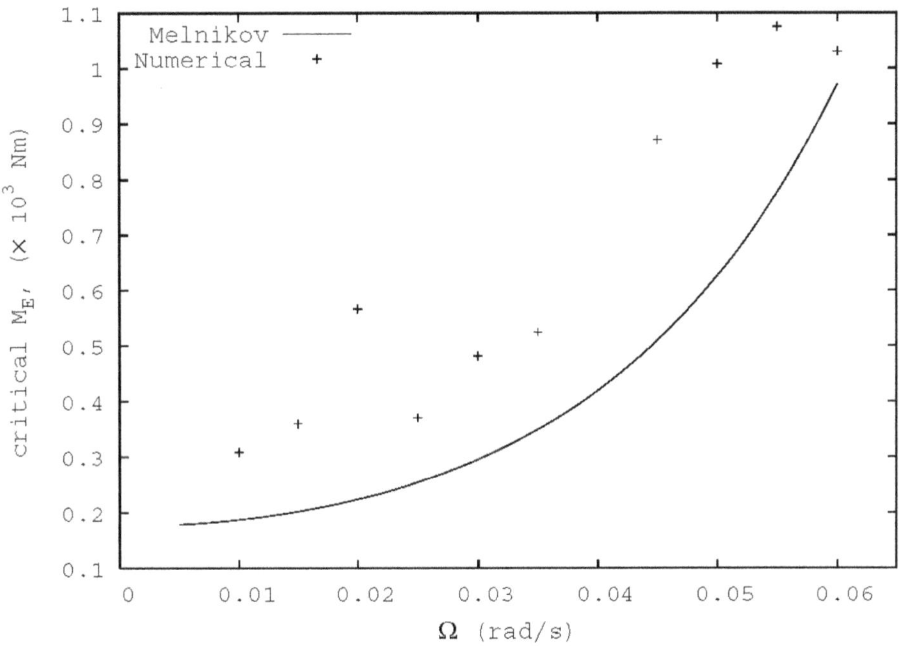

FIGURE 4.26 Chaos diagram of Melnikov and numerical results for a rotating body with internal energy dissipation with parameter configuration of Table 4.2.

spacecraft depicted in Figure 4.21b. Figures 4.27a–d show various views of the resultant four-dimensional strange attractor of chaotic behaviour in different combinations of three-dimensional spaces. The Lyapunov spectrum in this case was calculated to be $[13.7\ \ 0\ \ -0.12\ \ -2.69\ \ -17.4\ \ -30.7\ \ -64.6]\times 10^{-3}$. The fractal dimensions of this strange attractor were calculated as Lyapunov dimension of 2.62, correlation dimension of 2.32 and box dimension estimate of 2.57. The noninteger values confirm chaotic motion. It is of note that the fractal nature of the strange attractor of this higher-dimensional system was able to be viewed clearly in two-dimensional Poincáre maps since the Fractal dimension value is less than 3.

The results highlight that spacecraft designers should design to avoid regions of chaotic instability through careful parameter selection. Alternatively, implementing an effective automatic control strategy would enable the use of spacecraft parameters optimised for other requirements, even if they would otherwise make the system vulnerable to chaotic instabilities. Hence, a comprehensive analysis of this fundamental spacecraft system is provided in this book: from predicting to controlling chaotic instability that is covered in Chapter 6. In the meantime, some other interesting occurrences of chaotic instability in engineering systems are provided in the subsequent section.

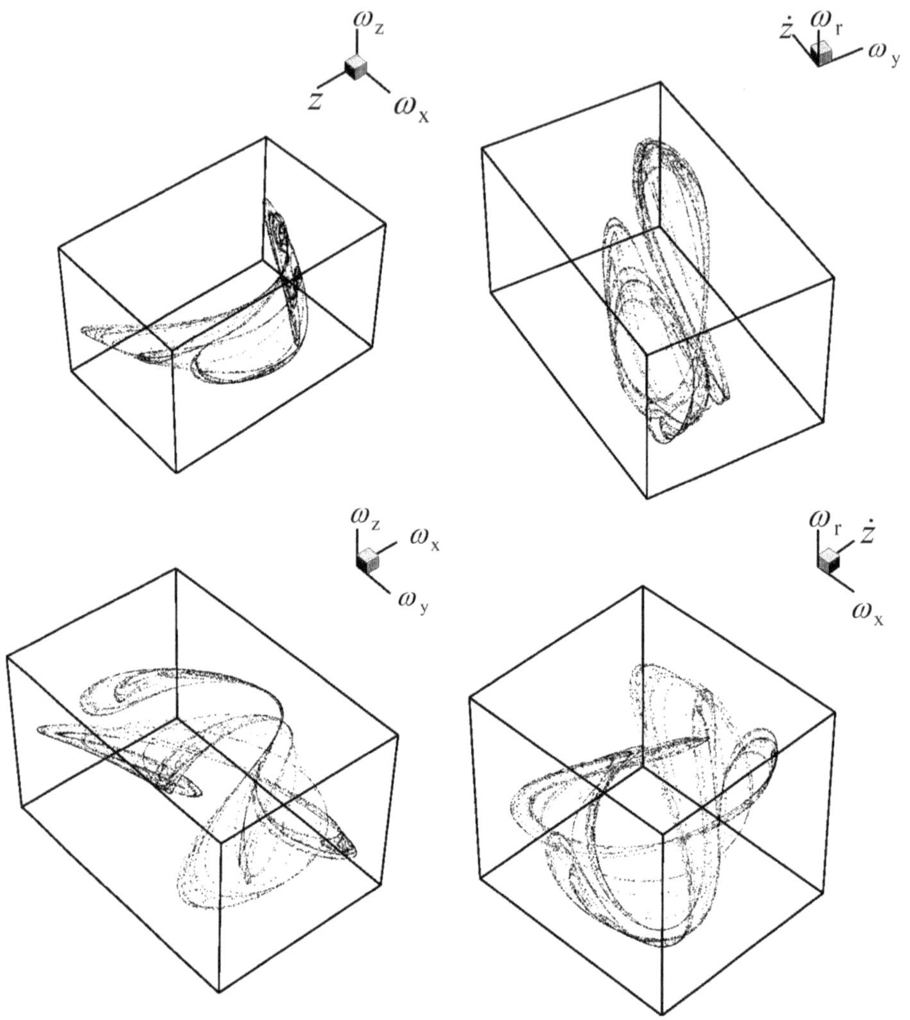

FIGURE 4.27 3D Poincaré map views showing chaotic motion for bias momentum satellite with $T = 1.3$ Nm and $\Omega = 0.21$ rad/s (ch.5 [8]).

4.6 OTHER CHAOTIC INSTABILITY PHENOMENA

Chaos has been mathematically predicted and shown to occur in many engineering systems due to the universal requirements of sensitivity to initial conditions arising from instability and nonlinearity. Here, we touch on a few additional processes to those detailed in the previous sections. In addition, the reader is directed to other more focussed texts on this phenomenon for a wider range of applications, such as Moon [26].

Transmission Driveline Chaos

In Chapter 3, we showed that a transmission driveline incorporating Hooke's (Universal) joint can exhibit twisting instabilities due to parametric excitations at running frequencies matching combinations of the natural frequencies of the shafts. The parametric excitation amplitude increases as the angle between the shafts enlarge. Therefore, it was interesting to investigate whether chaotic twisting vibrations occur due to the parametric excitation [199]. Figure 4.28a shows a bifurcation diagram of maximum Lyapunov exponents for the Hooke's joint driven system. For this case, the input frequency Ω remained constant at 55 rad/s, close to the primary combination resonance frequency $(\omega_1 + \omega_2)/2$ and the Hooke's joint angle is varied from 0° to 60°. From the diagram, it is evident that for Hooke's joint angles less than 14°, instabilities due to combination resonance are not excited. As the Hooke's joint

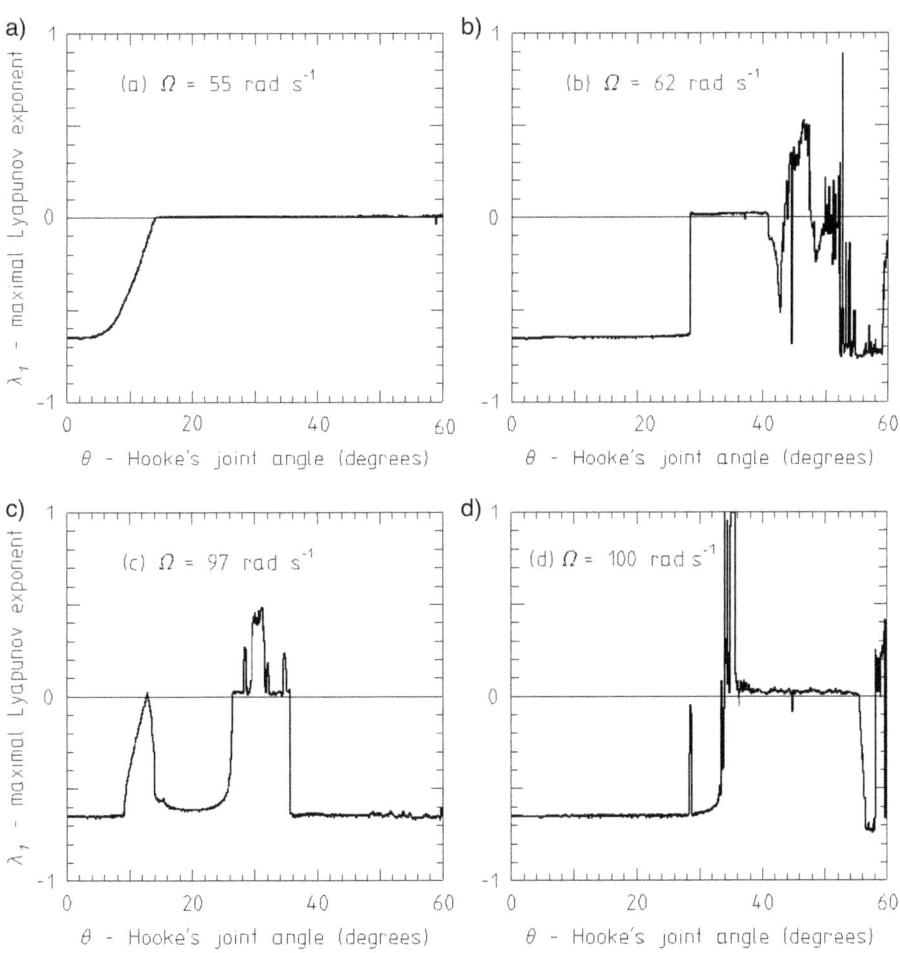

FIGURE 4.28 Bifurcation diagram of maximal Lyapunov exponent for range of Hooke's joint angles. (a) Ω = 55 rad/s; (b) Ω = 62 rad/s; (c) Ω = 97 rad/s; (d) Ω = 100 rad/s [199].

angle increases to 14°, it is seen that the largest Lyapunov exponent increases towards 0 but is still negative, indicating less stability and periodic motion. It is found that the transient time to stable period-1 motion in the range of Hooke's joint angles from 0° to 14° increases with the maximum Lyapunov exponent, which is 0 throughout the Hooke's joint angle range of 14°–60°, indicating quasiperiodic motion. Figure 4.28b shows a bifurcation diagram of maximum Lyapunov exponents for a constant input frequency Ω of 62 rad/s. In this case, quasiperiodic motion occurs at a Hooke's joint angle of 28.5° and continues for all values up to 41°, as indicated by the 0 maximum Lyapunov exponent.

For the interval of θ in the range of 41°–43°, the largest Lyapunov exponent becomes negative again, indicating periodic motion. For Hooke's joint angles greater than 43°, the maximum Lyapunov exponent becomes positive, indicating chaotic motion. This phenomenon persists until 47°, after which there are many subsequent periodic and chaotic intervals until at 53° the system settles to periodic motion again. This sequence of stable to quasiperiodic to periodic to chaotic motion is a typical example of the quasiperiodic route to chaos and has been documented by many other authors (see [200]). Figures 4.28c and d also depict similar bifurcation diagrams as Figure 4.28b, with quasiperiodic routes to chaotic motion for constant input frequencies Ω of 97 and 100 rad/s. More detailed analysis of the route to chaotic motion, as Hookes joint angle increases, for a constant input frequency of 62 rad/s, is shown in Figure 4.29, via a series of Poincaré map diagrams of state variables y vs dy/dt.

As illustrated in the Lyapunov bifurcation diagram of Figure 4.28b, for a Hooke's Joint angle of 20° and an input frequency of 62 rad/s, the system behaves in a stable manner. Figure 4.29a indicates this motion is a period-1 stable limit cycle,

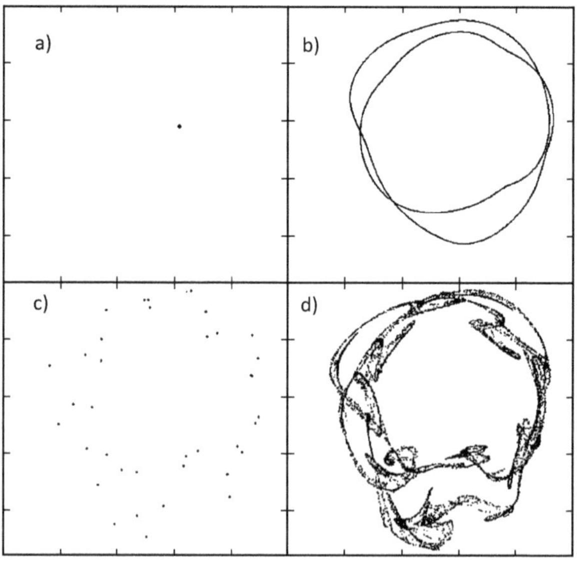

FIGURE 4.29 Poincaré map history for $\omega = 62$ rad/s. (a) $\theta = 20°$; (b) $\theta = 30°$; (c) $\theta = 42°$; (d) $\theta = 45°$ [199].

and the output frequency is equal to the Hooke's joint excitation frequency 2 $\Omega = 124$ rad/s. Figure 4.29b reveals the quasiperiodic attractor on the Poincaré map indicated by the closed curve for a Hooke's joint angle of 30°. The Lyapunov spectrum for these parameters was calculated to be [0 0 −0.826 −0.826 −1.444], giving a Lyapunov attractor dimension of 2. The integer value confirms quasiperiodic as opposed to chaotic motion. At a Hooke's joint angle of 42°, the periodic motion found in the Lyapunov bifurcation diagram of Figure 4.28b is period-37 in the Poincaré map of Figure 4.29c. Finally, at a Hooke's angle of 45°, chaotic motion is indicated in Figure 4.29d by the infinite set of highly organised points known as a strange attractor. The Lyapunov spectrum of exponents were calculated in this case to be [0.308 0 −0.282 −1.278 −1.844], with the positive value indicating chaotic motion.

Hence, chaotic torsional instabilities in a torsional driveline, due to a fluctuating angular velocity ratio across the joint, have been identified and quantified for particular input shaft frequencies and for sufficiently large Hooke's joint angles. Under these conditions, the response of the driveline becomes highly sensitive and unstable. These instability regions occur for ranges about critical input shaft frequencies. For operations close to critical input shaft frequencies, very small variations in input shaft frequency as well as Hooke's joint angle could strongly influence the type of torsional vibrations of the system. Increased damping in the system delays the onset of instability to higher angles of shaft misalignment. It has been shown that for an increasing angle of misalignment in shafts, the onset of chaotic motion was characterised by the quasiperiodic route. From a practical point of view, the results obtained have shown that parametric instabilities of the subharmonic type close to the higher natural frequency as well as the combination type can be a problem for significant Hooke's joint angles. Control of this phenomenon is investigated in the following chapters.

Chaos in Nanosystems

Chaos has also been found in applications across nanoscience. Nanosystems, characterised by their nanoscale size (1–100 nm) and properties, exhibit behaviours that often defy classical predictions and have also been found to exhibit chaos. At this scale, quantum effects become pronounced, leading to unconventional behaviours not observed in macroscopic systems. Chaos in nanosystems arises from the interplay of deterministic dynamics and nonlinearities typically from these quantum effects. Quantum chaos manifests as unpredictable energy level spacings, sensitive dependence on initial conditions and irregular spectral statistics. Nanoparticle systems, quantum dots, and quantum billiards serve as prototypical examples of quantum chaos in nanosystems. Chaotic trajectories of electrons arise due to the interplay of disorder, quantum interference and nonlinearity. This leads to phenomena such as weak localisation, quantum chaos-assisted tunnelling and chaotic scattering. There are also self-organisation and pattern formation processes observed in nanosystems. Examples include the spontaneous formation of nanostructures through self-assembly, chaotic mixing in microfluidic devices and the emergence of complex spatiotemporal patterns in reaction–diffusion systems on nanopatterned surfaces.

Presently, we focus on chaotic dynamics in nanomechanical oscillators used in atomic force microscopy, chemical sensing, biosensing and environmental monitoring at the nanoscale. Chaotic behaviour in these systems arises from the coupling between mechanical degrees of freedom, nonlinearities in the force-displacement relationship and external perturbations. This has implications for sensing, signal processing and mechanical computing. Micromechanical systems already have been shown to exhibit chaotic instability. As an example, in [201], a chaotic signal from a micromechanical resonator is generated using a disruptive approach, experimentally and numerically. This dynamical approach may be used on any kind of micro/nanomechanical resonator, such as accelerometers to microphones and in this case is used to generate a true random number generator. The experimental proof of chaotic vibrations is shown in the time histories and bifurcation diagrams of Figure 4.30.

A similar investigation is performed on a nanomechanical oscillator and its integrated capacitive actuation in [202] to be applied to the next generation of nano-electro-optomechanical systems. Research has also been focussed on chaos in carbon nanotubes (CNTs), such as [203,204]. CNTs have exceptional properties for applications in medical and nanoelectronic devices water purification, transportation of

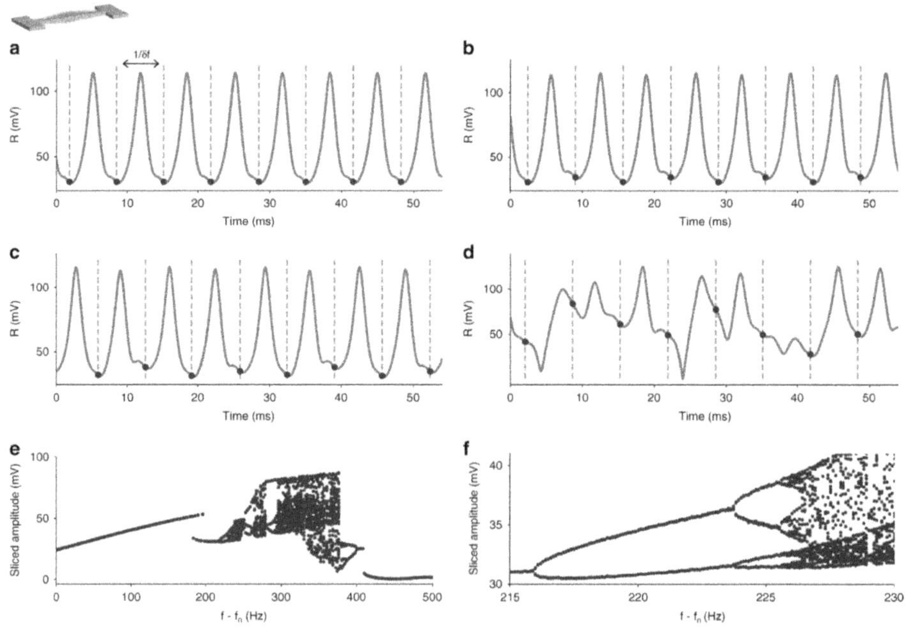

FIGURE 4.30 Time histories and bifurcation diagrams Poincaré map history for AFM beam. Typical mechanical responses are presented (from a to d) for detuning frequencies f–f0 of 215, 220, 225, 340 Hz, respectively, which highlight the presence of 1, 2 and 4 harmonics in the signal, until the chaotic regime is reached. The projection of the sliced data for each driving frequency forms a bifurcation diagram (e). Looking closer at its evolution before entering chaos (f), we observe the period-doubling route to chaos with period-doubling bifurcations [201].

nanoparticles, resonant sensors, switches and actuators and nanocomposites. The results indicate that the harmonic force and structural parameters have considerable effects on the bifurcation behaviour and can be used as effective control parameters for avoiding chaos.

Chaos in nanosystems is a rich interdisciplinary research area with important applications for nanoscience, engineering and technology. Understanding and harnessing chaotic dynamics offer opportunities for developing novel nanodevices and advancing applications in sensing, computing and beyond. As research in this field continues to evolve, interdisciplinary collaboration and innovative methodologies will be crucial for unlocking the full potential of chaos in nanosystems.

Human Body System Chaos

Chaos has been increasingly recognised as a fundamental aspect of physiological processes within the human body. Physiological processes often exhibit nonlinear dynamics, stochastic fluctuations and feedback interactions, leading to chaotic behaviour. Chaos in physiology arises from the intricate interplay of deterministic and stochastic factors, influencing phenomena ranging from heartbeat dynamics to neural activity and hormone regulation. Even the structure of our physiological systems manifests chaos in fractal forms [206], as shown in Figure 4.31.

In the cardiovascular system, heartbeat dynamics display chaotic behaviour, characterised by irregular intervals between successive heartbeats. Measures such as heart rate variability (HRV) and nonlinear analysis techniques have revealed chaotic dynamics in heart rhythm. This has implications for cardiovascular health, autonomic regulation and risk assessment of cardiac arrhythmias. The respiratory system exhibits complex dynamics, including irregular and chaotic breathing variability. These occur during conditions such as sleep apnoea, respiratory distress

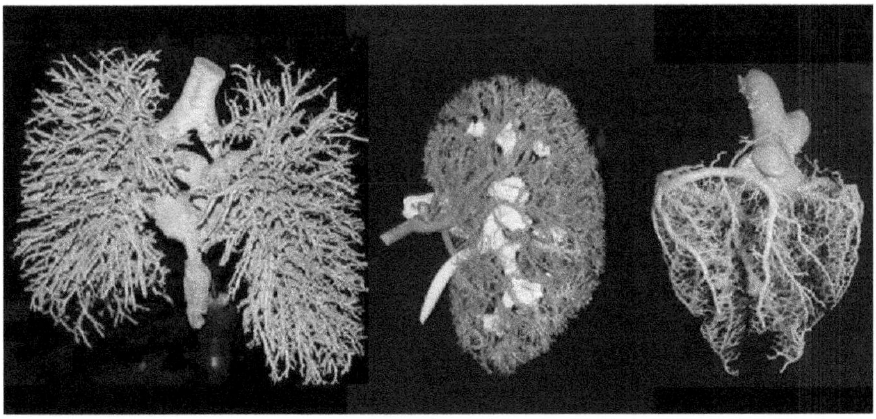

FIGURE 4.31 Fractal physiology examples of "tree-like" fractal anatomic systems. From left to right: bronchial tree, renal vascular and urinary systems and heart coronary system. By kind concession from Prof. Dr. Manfred Tschabitscher (Centre for Anatomy and Cell Biology, Medical University of Vienna, Vienna, Austria) [206].

and ventilatory instability, highlighting the importance of understanding chaotic behaviour in respiratory control. Brain activity displays chaotic features, with neural networks exhibiting spontaneous oscillations, synchronisation phenomena and complex spatiotemporal patterns. These underlie cognitive processes, consciousness and information processing, with implications for understanding brain disorders such as epilepsy, Parkinson's disease and schizophrenia. Hormone secretion and feedback loops within the endocrine system exhibit nonlinear dynamics. In particular, chaotic hormone secretion patterns have been observed in the hypothalamic–pituitary–adrenal (HPA) axis. This has implications for stress response, circadian rhythms and hormonal imbalances associated with diseases such as diabetes and adrenal disorders.

The complex and nonlinear behaviour of human body dynamics has motivated many researchers to consider the possibility of low-dimensional chaos as a paradigm for modelling such systems, i.e., [205]. To examine if a signal exhibits low-dimensional chaotic behaviour, chaotic measures such as the correlation dimension (CD) and maximum Lyapunov exponent (MLE) are required. The CD measures the fractal nature of the phase space of the signal, and the MLE tests for exponential divergence of neighbouring trajectories in phase space, indicating extreme sensitivity to initial conditions. In practical applications, real signals are not completely noise-free; however, careful application of these measures has yielded many interesting results about the behaviour of many systems. In human body systems, the apparent relationship between chaotic behaviour and proper biological functioning has been investigated, such as the existence of chaos and fractal behaviour in heart rate variability measurements [207–218], as shown in Figure 4.32.

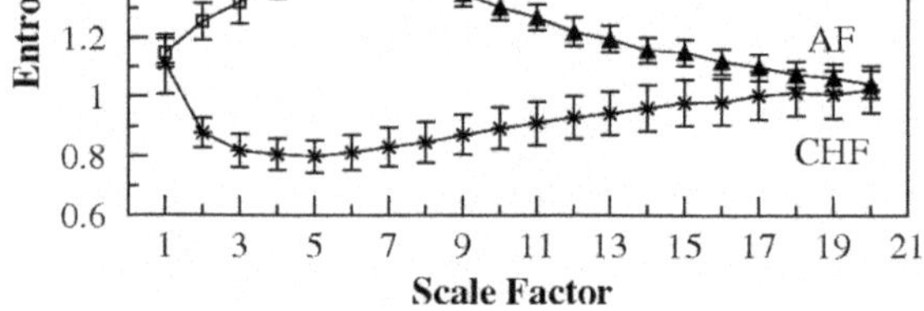

FIGURE 4.32 Multiscale entropy analysis of interbeat interval time series derived from healthy subjects, subjects with congestive heart failure (CHF) and subjects with atrial fibrillation (AF). Values are given as means ± standard error. Time series were filtered to remove outlier points due to artifacts and ventricular ectopic beats. The values of entropy depend on the scale factor. For scale 1, AF time series are assigned the highest value of entropy, and the values corresponding to healthy and CHF groups completely overlap. For larger scales, e.g., 20, the entropy value for the coarse-grained time series derived from healthy subjects is significantly higher than those for AF and CHF. At this scale, AF and CHF groups become indistinguishable [218].

This figure shows that healthy subjects appear to have a natural higher level of complexity (entropy) in their heart beat for all time scale factors greater than 7 (where only every seventh data point is analysed). Alternatively, the Hurst exponent, a method from statistical mechanics based upon detecting fractal behaviour, has been applied to a variety of real, complex systems [211,212] such as detecting unhealthy heart fluctuations [213]. Similarly, the Hurst exponent of spike interarrival times of the STN activity of Rhesus monkeys has been determined discriminating persistent and nonpersistent behaviour [210]. Another method that distinguishes noisy random and chaotic systems is the non-Markovity and relaxation parameters. This has been applied to Parkinson's disease gait and finger tremor measurements and epilepsy [214–217]. This showed a low NMP for untreated PD patients, i.e., regular behaviour, and an increase, i.e., more chaotic, when under DBS or medication. Chaotic signal analysis of deep brain signals of Parkinson's disease patients has also distinguished language task types in [217].

Hence, the understanding and identification of chaotic behaviour in the human body offer new innovations in health monitoring and diagnostics. Chaos analysis techniques offer insights into physiological states and disease conditions. Monitoring chaotic biomarkers such as HRV, respiratory variability and neural dynamics provides valuable diagnostic information for assessing health status and predicting disease risk. Chaos-based biomarkers may facilitate early detection, prognosis and personalised treatment strategies, leading to improved patient outcomes and healthcare management. Ultimately, understanding human body chaos can inform novel therapeutic approaches for managing health conditions and optimising treatment outcomes. Chaos-based interventions, including biofeedback training, neural stimulation techniques, and control strategies for stabilising chaotic dynamics, hold promise for treating disorders. These include cardiac arrhythmias, respiratory dysfunction, epilepsy and possibly many other neural disorders such as Parkinson's disease and depression. Hence, it appears that chaos is an inherent aspect of physiological processes within the human body, manifesting across various systems and influencing health and disease outcomes. As our understanding of this evolves, integrating chaos theory into medical practice holds promise for advancing precision medicine and improving patient care.

NOTE

1 For larger out-of-plane bucket angles, the phase space contains heteroclinic orbits that intersect at the inverted unstable equilibrium points, similar to the behaviour observed in a simple pendulum. However, these are not realistic for a dragline.

REFERENCES

Wheel Squeal Chaos

1. Rudd, M.J. (1976). Wheel/rail noise Part II: Wheel squeal. *Journal of Sound and Vibration*, 46(3), 381–394.
2. Remington, P.J. (1985). Wheel/Rail squeal and impact noise: What do we know? What don't we know? Where do we go from here? *Journal of Sound and Vibration*, 116(2), 339–353.
3. Thompson, D.J., Jones, C.J.C. (2000). A review of the modelling of wheel/rail noise generation. *Journal of Sound and Vibration*, 231(3), 519–536.

4. Thompson, D.J., Squicciarini, G., Ding, B., Baeza, L. (2018). A state-of-the-art review of curve squeal noise: Phenomena, mechanisms, modelling and mitigation. *Notes on Numerical Fluid Mechanics and Multidisciplinary Design*, 139, 3–41.
5. Hoffmann, N., Fischer, M., Allgaier, R., Gaul, L. (2002). A minimal model for studying properties of the mode-coupling type instability in friction induced oscillations. *Mechanics Research Communications*, 29, 197–205.
6. Hoffmann, N., Gaul, L. (2003). Effects of damping on mode-coupling instability in friction induced oscillations. *ZAMM-Zeitschrift fur Angewandte Mathematik und Mechanik*, 83(8), 524–534.
7. Pieringer, A. (2014). A numerical investigation of curve squeal in the case of constant wheel/rail friction. *Journal of Sound and Vibration*, 333(18), 4295–4313.
8. Zenzerovic, I., Kropp, W., Pieringer, A. (2018). Influence of spin creepage and contact angle on curve squeal: A numerical approach. *Journal of Sound and Vibration*, 419, 268–280.
9. Cui, X.L., Chen, G.X., Yang, H.G., et al. (2015). Effect of the wheel/rail contact angle and the direction of the saturated creep force on rail corrugation. *Wear*, 330–331, 554–562.
10. Ding, B., Squicciarini, G., Thompson, D., Corradi, R. (2018). An assessment of mode-coupling and falling-friction mechanisms in railway curve squeal through a simplified approach. *Journal of Sound and Vibration*, 423, 126–140.
11. Meehan, P.A. (2019). Prediction of wheel squeal noise under mode coupling. *Journal of Sound and Vibration*, 465, 115025.
12. Meehan, P.A. (2020). Investigation of chaotic instabilities in railway wheel squeal. *Nonlinear Dynamics*, 100, 159–172.
13. Moon, F.C. (1992). *Chaotic and Fractal Dynamics: An Introduction for Applied Scientists and Engineers*. Wiley Interscience, New York, USA.
14. Ding, B., Squicciarini, G., Thompson, D. (2019). Effect of rail dynamics on curve squeal under constant friction conditions. *Journal of Sound and Vibration*, 442, 183–199.
15. Oberst, S., Lai, J.C.S. (2011). Chaos in brake squeal noise. *Journal of Sound and Vibration*, 330(5), 955–975.
16. Feeny, B., Moon, F.C. (1994). Chaos in a forced dry-friction oscillator: Experiments and numerical modelling. *Journal of Sound and Vibration*, 170(3), 303–323.
17. Shin, K., Brennan, M.J., Oh, J.-E., Harris, C.J. (2002). Analysis of disc brake noise using a two-degree-of-freedom model. *Journal of Sound and Vibration*, 254(5), 837–848.
18. Shin, K., Oh, J., Brennan, M.J. (2002). Nonlinear analysis of friction induced vibrations of a two-degree-of-freedom model for disc brake squeal noise. *JSME International Journal Series C Mechanical Systems, Machine Elements and Manufacturing*, 45(2), 426–432.
19. Lin, B., Chang, S., Hu, J., Lue, Y. (2015). Controlling chaos for automotive disc brake squeal suppression. *Journal of Mechanical Science and Technology*, 29(6), 2313–2322.
20. Meehan, P.A., Liu, X. (2018). Modelling and mitigation of wheel squeal noise amplitude. *Journal of Sound and Vibration*, 413, 144–158.
21. Landau, L.D., Lifshitz, E.M. (1960). *Course of Theoretical Physics Vol. 1: Mechanics*. Pergamon Press, London.
22. Sagdeev, R.Z. (1988). *Nonlinear Physics: From the Pendulum to Turbulence and Chaos*. Harwood Academic Publishers, New York.
23. Nayfeh, H., Mook, D.T. (1979). *Nonlinear Oscillations*. Wiley, New York.
24. Liu, X., Meehan, P.A. (2016). Investigation of squeal noise under positive friction characteristics condition provided by friction modifiers. *Journal of Sound and Vibration*, 371, 393–405.
25. Eadie, D.T., Santoroa, M., Kalousek, J. (2005). Railway noise and the effect of top of rail liquid friction modifiers: Changes in sound and vibration spectral distributions in curves. *Wear*, 258, 1148–1155.

26. Moon, F.C. (1992). Chaotic and Fractal Dynamics: An Introduction for Applied Scientists and Engineers. Wiley Interscience, New York.
27. Hilborn, R.C. (1994). *Chaos and Nonlinear Dynamics: An Introduction for Scientists and Engineers*. Oxford University Press, New York.
28. Asokanthan, S.F., Meehan, P.A. (2000). Non-linear vibration of a torsional system driven by a Hooke's Joint. *Journal of Sound and Vibration*, 233(2), 297–310.

Brake Squeal Chaos

29. Oberst, S., Lai, J.C.S. (2015). Nonlinear transient and chaotic interactions in disc brake squeal. *Journal of Sound and Vibration*, 342, 272–289.
30. Feeny, B.F., Moon, F.C. (2000). Quenching stick–slip chaos with dither. *Journal of Sound and Vibration*, 237(1), 173–180.
31. Oberst, S., Lai, J.C.S. (2015). A statistical approach to estimate the Lyapunov spectrum in disc brake squeal. *Journal of Sound and Vibration*, 334, 120–135.
32. Stender, M., Di Bartolomeo, M., Massi, F., Hoffmann, N. (2019). Revealing transitions in friction-excited vibrations by nonlinear time-series analysis. *Nonlinear Dynamics*, 98(4), 2613–2630.
33. Stender, M., Jahn, M., Hoffmann, N., Wallaschek, J. (2020). Hyperchaos co-existing with periodic orbits in a frictional oscillator. *Journal of Sound and Vibration*, 472, 115203.
34. Meehan, P.A. (2022). Prediction and suppression of chaotic instability in brake squeal. *Nonlinear Dynamics*, 107, 205–225.
35. van der Kloet, P., Neerhoff, F. (2003). Dynamic eigenvalues and Lyapunov exponents for nonlinear circuits. In: *Proceedings of 2003 Workshop on Nonlinear Dynamics of Electronic Systems (NDES May 2003)*, 2003, 287–290.
36. Meehan, P.A. (2020). Investigation of chaotic instabilities in railway wheel squeal. *Nonlinear Dynamics*, 100, 159–172.
37. Meehan, P.A., Leslie, A.C. (2021). On the mechanisms, growth, amplitude and mitigation of brake squeal noise. *Mechanical Systems and Signal Processing*, 152, 107469.
38. Stender, M., Tiedemann, M., Hoffmann, N. (2019). Energy harvesting below the onset of flutter. *Journal of Sound and Vibration*, 458, 17–21.
39. Park, J., Lee, J.C., Cho, S., Yoon, K. (2012). Reduction of brake squeal analyzed in terms of coupling between in-plane and out-of-plane modes. *SAE International Journal of Passenger Cars - Mechanical Systems*, 5(4), 1230–1243.
40. Eriksson, M., Jacobson, S. (2001). Friction behaviour and squeal generation of disc brakes at low speeds. *Proceedings of the Institution of Mechanical Engineers*, 215(D), 1245–1256.
41. Meehan, P.A. (2023). Prediction and suppression of chaos following flutter in wind turbines. *Nonlinear Dynamics*, 111, 22153–22176.

Flutter Chaos

42. Zhao, L.C., Yang, Z.C. (1990). Chaotic motions of an airfoil with non-linear stiffness in incompressible flow. *Journal of Sound and Vibration*, 138(2), 245–254.
43. Dai, H., Yue, X., Xie, D., Atluri, S.N. (2014). Chaos and chaotic transients in an aeroelastic system. *Journal of Sound and Vibration*, 333(26), 7267–7285.
44. Abdelkefi, A., Nayfeh, A.H., Hajj, M.R. (2012). Design of piezoaeroelastic energy harvesters. *Nonlinear Dynamics*, 68(4), 519–530.
45. Wu, Y., Li, D., Xiang, J. (2018). Dimensionless modeling and nonlinear analysis of a coupled pitch–plunge–leadlag airfoil-based piezoaeroelastic energy harvesting system. *Nonlinear Dynamics*, 92(2), 153–167.

46. Xie, D., Xu, M., Dai, H., Dowell, E.H. (2015). Proper orthogonal decomposition method for analysis of nonlinear panel flutter with thermal effects in supersonic flow. *Journal of Sound and Vibration*, 337, 263–283.
47. Brouwer, K.R., Perez, R.A., Beberniss, T.J., et al. (2021). Investigation of aeroelastic instabilities for a thin panel in turbulent flow. *Nonlinear Dynamics*, 104(4), 3323–3346.
48. Akhavan, H., Ribeiro, P. (2021). Stability and bifurcations in oscillations of composite laminates with curvilinear fibres under a supersonic airflow. *Nonlinear Dynamics*, 103(4), 3037–3058.
49. dos Santos, L.G.P., Marques, F.D., Vasconcellos, R.M.G. (2022). Dynamical characterisation of fully nonlinear, nonsmooth, stall fluttering airfoil systems. *Nonlinear Dynamics*, 107(3), 2053–2074.
50. Esbati Lavasani, R., Shams, S. (2020). A new dynamic stall approach for investigating bifurcation and chaos in aeroelastic response of a blade section with flap free-play section. *International Journal of Bifurcation and Chaos*, 30(14), 2050200.
51. Tian, W., Yang, Z., Zhao, T. (2019). Nonlinear aeroelastic characteristics of an all-movable fin with freeplay and aerodynamic nonlinearities in hypersonic flow. *International Journal of Non-Linear Mechanics*, 116, 123–139.
52. Godavarthi, V., Kasthuri, P., Mondal, S., et al. (2020). Synchronisation transition from chaos to limit cycle oscillations when a locally coupled chaotic oscillator grid is coupled globally to another chaotic oscillator. *Chaos*, 30(3), 033121.
53. Tian, W., Zhao, T., Yang, Z. (2021). Nonlinear electro-thermo-mechanical dynamic behaviors of a supersonic functionally graded piezoelectric plate with general boundary conditions. *Composite Structures*, 261, 113326.
54. Vishal, S., Raaj, A., Bose, C., et al. (2022). Numerical investigation into discontinuity-induced bifurcations in an aeroelastic system with coupled non-smooth nonlinearities. *Nonlinear Dynamics*, 108(4), 3025–3051.
55. Bose, C., Gupta, S., Sarkar, S. (2019). Transition to chaos in the flow-induced vibration of a pitching–plunging airfoil at low Reynolds numbers: Ruelle–Takens–Newhouse scenario. *International Journal of Non-Linear Mechanics*, 109, 189–203.
56. Meehan, P.A. (2022). Flutter prediction of its occurrence, amplitude and nonlinear behaviour. *Journal of Sound and Vibration*, 535, 117117.
57. Heckl, M.A., Abrahams, I.D. (2000). Curve squeal of train wheels, Part 1: Mathematical model for its generation. *Journal of Sound and Vibration*, 229(3), 669–693.
58. Chiello, O., Ayasse, J.-B., Vincent, N., Koch, J.-R. (2006). Curve squeal of urban rolling stock—Part 3: Theoretical model. *Journal of Sound and Vibration*, 293(3–5), 710–727.
59. Brunel, J.F., Dufrénoy, P., Naït, M., et al. (2006). Transient models for curve squeal noise. *Journal of Sound and Vibration*, 293(3), 758–765.
60. Heckl, M.A. (2000). Curve squeal of train wheels, Part 2: Which wheel modes are prone to squeal? *Journal of Sound and Vibration*, 229(3), 695–707.
61. de Beer, F.G., Janssens, M.H.A., Kooijman, P.P. (2003). Squeal noise of rail-bound vehicles influenced by lateral contact position. *Journal of Sound and Vibration*, 267(3), 497–507.
62. Kinkaid, N.M., O'Reilly, O.M., Papadopoulos, P. (2003). Automotive disc brake squeal. *Journal of Sound and Vibration*, 267(1), 105–166.
63. Glocker, C., Cataldi-Spinola, E., Leine, R.I. (2009). Curve squealing of trains: Measurement, modelling and simulation. *The Journal of Sound and Vibration*, 324(1), 365–386.
64. Oberst, S., Lai, J.C.S. (2011). Chaos in brake squeal noise. *Journal of Sound and Vibration*, 330(5), 955–975.
65. Feeny, B., Moon, F.C. (1994). Chaos in a forced dry-friction oscillator: Experiments and numerical modelling. *Journal of Sound and Vibration*, 170(3), 303–323.

66. Fourie, D.J., Gräbe, P.J., Heyns, P.S., Fröhling, R.D. (2019). Frequency domain model for railway wheel squeal resulting from unsteady longitudinal creepage. *Journal of Sound and Vibration*, 445, 228–246.
67. Meehan, P.A., Liu, X. (2019). Modelling and mitigation of wheel squeal noise under friction modifiers. *Journal of Sound and Vibration*, 440, 147–160.
68. Liu, X., Meehan, P.A. (2015). Investigation of the effect of crabbing velocity on squeal noise based on a rolling contact two disc test rig. In: *Proceedings of the 10th International Conference of Contact Mechanics and Wear of Rail/wheel Systems CM2015*, September 2015, Colorado, USA.
69. Johnson, K.L. (1985). *Contact Mechanics*. Cambridge University Press, Cambridge.
70. Liu, X., Meehan, P.A. (2014). Investigation of the effect of relative humidity on lateral force in rolling contact and curve squeal. *Wear*, 310(1–2), 12–19.
71. Thompson, D.J. (2009). *Railway Noise and Vibration: Mechanisms, Modelling and Means of Control*. Elsevier, Oxford.
72. Oregui1, M., Núñez, A., Dollevoet, R., Li, Z. (2017). Sensitivity analysis of railpad parameters onvertical railway track dynamics. *Journal of Engineering Mechanics*, 143(5), 04017011.
73. Harris, C.M., Piersol, A.G. (2002). *Harris' Shock and Vibration Handbook*, 5th Ed. McGraw Hill, New York.
74. Fieldhouse, J., Ashraf, N., Talbot, C. (2008). The measurement and analysis of the disc/pad interface dynamic centre of pressure and its influence on brake noise. SAE Technical Paper 2008-01-0826.
75. Kinkaid, N.M., O'Reilly, O.M., Papadopoulos, P. (2003). Automotive disc brake squeal. *Journal of Sound and Vibration*, 267(1), 105–166.
76. Jarvis, R.P., Mills, B. (1963). Vibrations induced by friction. *Proceedings of the Institution of Mechanical Engineers*, 178(32), 857–864.
77. Spurr, R.T. (1961). A theory of brake squeal. *Proceedings of Automotive Division, Institution Mechanical Engineers (AD)*, 1, 33–40.
78. Spurr, R.T. (1971). Brake squeal. Paper No. C95/71, Vibration and Noise in Motor Vehicles, I.Mech.E., 13–16.
79. Earles, S.W.E. (1977). A mechanism of disc-brake squeal. Technical Report 770181, SAE, Warrendale, PA.
80. Hoffmann, N., Gaul, L. (2004). A sufficient criterion for the onset of sprag-slip oscillation. *Archive of Applied Mechanics*, 73, 650–660.
81. Fieldhouse, J., Ashraf, N., Talbot, C., et al. (2006). Measurement of the dynamic center of pressure of a brake pad during a braking operation. SAE Technical Paper 2006-01-3208.
82. North, M.R. (1976). Disc brake squeal. In: Institution of Mechanical Engineers, Automobile Division (ed) *Braking of Road Vehicles*. Mechanical Engineering Publications Limited, London, England, 169–176.
83. Hoffmann, N., Fischer, M., Allgaier, R., Gaul, L. (2002). A minimal model for studying properties of the mode-coupling type instability in friction induced oscillations. *Mechanics Research Communications*, 29, 197–205.
84. Cantone, F., Massi, F. (2011). A numerical investigation into the squeal instability: Effect of damping. *Mechanical Systems and Signal Processing*, 25(5), 1727–1737.
85. Grange, P., Clair, D., Baillet, L., Fogli, M. (2009). Brake squeal analysis by coupling spectral linearisation and modal identification methods. *Mechanical Systems and Signal Processing*, 23(8), 2575–2589.
86. Nack, W.V. (2000). Brake squeal analysis by finite elements. *International Journal of Vehicle Design*, 23(3–4), 263–275.

87. Chen, F., Mckillip, D., Luo, J., Wu, S. (2004). Measurement and analysis of rotor in-plane mode induced disc brake squeal and beyond. SAE Technical Paper 2004-01-2798. doi:10.4271/2004-01-2798.
88. Park, C., Han, M., Cho, S.S., et al. (2001). A study on the reduction of disc brake squeal using complex eigenvalue analysis. SAE Technical Paper 2001-01-3141.
89. Papinniemi, A., Zhao, J., Stanef, D., Ding, J. (2005). An investigation of in-plane vibration modes in disc brake squeal noise. SAE Technical Paper 2005-01-3923. doi:10.4271/2005-01-3923.
90. Yokoyama, T., Matsushima, T., Matsui, N., Misumi, R. (2012). A study of reduction for brake squeal in disc in-plane mode. SAE Technical Paper 2012-01-1825.
91. Matsuzaki, M. (1993). Brake noise caused by longitudinal vibration of the disc rotor. SAE Technical Paper 930804.
92. Earles, S., Soar, G. (1974). A vibrational analysis of a pin disc system with particular reference to squeal noise in disc-brakes. In: *1974 Institute of Physics, Stress Analysis Group, Annual Conference*, University of Aston, Birmingham, UK, January 1974, 237–251.
93. Kang, J. (2010). The onset of friction-induced vibration and sprigging. *Journal of Sound and Vibration*, 329, 3537–3549.
94. Massi, F., Baillet, L., Giannini, O., Sestieri, A. (2007). Brake squeal: Linear and nonlinear numerical approaches. *Mechanical Systems and Signal Processing*, 21(6), 2374–2393.
95. Stender, M., Tiedemann, M., Hoffmann, L., Hoffmann, N. (2019). Determining growth rates of instabilities from time-series vibration data: Methods and applications for brake squeal. *Mechanical Systems and Signal Processing*, 129, 250–264.
96. Oberst, S., Lai, J.C.S. (2015). Pad-mode-induced instantaneous mode instability for simple models of brake systems. *Mechanical Systems and Signal Processing*, 62–63, 490–505.
97. Nechak, L., Besset, S., Sinou, J.-J. (2018). Robustness of stochastic expansions for the stability of uncertain nonlinear dynamical systems – Application to brake squeal. *Mechanical Systems and Signal Processing*, 111, 194–209.
98. Nobari, A., Ouyang, H., Bannister, P. (2015). Uncertainty quantification of squeal instability via surrogate modelling. *Mechanical Systems and Signal Processing*, 60–61, 887–908.
99. Stender, M., Tiedemann, M., Hoffmann, N., Oberst, S. (2018). Impact of an irregular friction formulation on dynamics of a minimal model for brake squeal. *Mechanical Systems and Signal Processing*, 107, 439–451.
100. Stender, M., Tiedemann, M., Spieler, D., et al. (2021). Deep learning for brake squeal: Brake noise detection, characterisation and prediction. *Mechanical Systems and Signal Processing*, 149, 107181.
101. Denimal, E., Sinou, J.-J., Nacivet, S. (2019). Influence of structural modifications of automotive brake systems for squeal events with kriging meta-modelling method. *Journal of Sound and Vibration*, 463, 114938.
102. Murakami, H., Tsunada, N., Kitamura, T. (1984). A study concerned with a mechanism of disc brake squeal. SAE841233.
103. Kim, J., Kim, J., Kim, Y., et al. (2015). An improvement of brake squeal CAE model considering dynamic contact pressure distribution. *SAE International Journal of Passenger Cars - Mechanical Systems*, 8(4), 1188–1195 (SAE Technical Paper 2015-01-2691).
104. Hoffmann, N., Gaul, L. (2003). Effects of damping on mode-coupling instability in friction induced oscillations. *ZAMM-Zeitschrift fur Angewandte Mathematik und Mechanik*, 83(8), 524–534.
105. Akay, A., Giannini, O., Massi, F., Sestieri, A. (2009). Disc brake squeal characterisation through simplified test rigs. *Mechanical Systems and Signal Processing*, 23(8), 2590–2607

106. Meehan, P.A. (2020). Prediction of wheel squeal noise under mode coupling. *Journal of Sound and Vibration*, 465, 115025.
107. Ding, B., Squicciarini, G., Thompson, D. (2019). Effect of rail dynamics on curve squeal under constant friction conditions. *Journal of Sound and Vibration*, 442, 183–199.
108. Meehan, P.A., Liu, X. (2018). Modelling and mitigation of wheel squeal noise amplitude. *Journal of Sound and Vibration*, 413, 144–158.
109. Meehan, P.A., Liu, X. (2019). Modelling and mitigation of wheel squeal noise under friction modifiers. *Journal of Sound and Vibration*, 440, 147–160.
110. Liu, X., Meehan, P.A. (2015). Investigation of the effect of crabbing velocity on squeal noise based on a rolling contact two disc test rig. In: *Proceedings of the 10th International Conference of Contact Mechanics and Wear of Rail/wheel Systems CM2015*, Colorado, USA.
111. Thompson, D.J. (2009). *Railway Noise and Vibration: Mechanisms, Modelling and Means of Control*. Elsevier, Oxford.
112. Soobbarayen, K., Besset, S., Sinou, J.-J. (2015). A simplified approach for the calculation of acoustic emission in the case of friction-induced noise and vibration. *Mechanical Systems and Signal Processing*, 50–51, 732–756.
113. Harris, C.M., Piersol, A.G. (2002). *Harris' Shock and Vibration Handbook*, 5th Ed. McGraw Hill, New York.
114. Johnson, K.L. (1985). *Contact Mechanics*. Cambridge University Press, Cambridge.
115. Blevins, R.D. (1979). *Formulas for Natural Frequency and Mode Shape*. Van Nostrand Reinhold Co, New York.
116. Pohrt, R., Popov, V. (2013). Contact stiffness of randomly rough surfaces. *Scientific Reports*, 3, 3293.
117. Meehan, P.A. (2020). Investigation of chaotic instabilities in railway wheel squeal. *Nonlinear Dynamics*, 100, 159–172.
118. Theodorsen, T. (1935). General theory of aerodynamic instability and the mechanism of flutter. NACA Report 496.
119. Theodorsen, T., Garrick, I.E. (1940). Mechanism of flutter: A theoretical and experimental investigation of the flutter problem. NACA Report 685.
120. Pines, S. (1958). An elementary explanation of the flutter mechanism. In: *Proceedings of Dynamics and Aeroelasticity Meeting*, I. A. S., November, New York, 52–59.
121. Dowell, E.H., Scanlan, R.H., Sisto, F., et al. (2021). *A Modern Course in Aeroelasticity,* Solid Mechanics and its Applications, vol 264, 6th Ed. Springer ScienceCham, Heidelberg, New York, Dordrecht, London.
122. Blevins, R.D. (1990). *Flow Induced Vibrations*, 2nd Ed. Van Nostrand Renhold.
123. Venkatramani, J., Sarkar, S., Gupta, S. (2018). Investigations on precursor measures for aeroelastic flutter. *Journal of Sound and Vibration*, 419, 318–336. doi:10.1016/j.jsv.2018.01.009.
124. de Henshaw, M.J.C., Badcock, K.J., Vio, G.A., et al. (2007). Non-linear aeroelastic prediction for aircraft applications. *Progress in Aerospace Sciences*, 43(4–6), 65–137.
125. Sisto, F., Thangam, S., Abdelrahim, A. (1991). Computational prediction of stall flutter in cascaded airfoils. *AIAA Journals*, 29(7), 1161–1167.
126. Zhou, R., Ge, Y., Yang, Y., et al. (2018). Wind-induced nonlinear behaviors of twin-box girder bridges with various aerodynamic shapes. *Nonlinear Dynamics*, 94(2), 1095–1115. doi:10.1007/s11071-018-4411-y.
127. Huang, C., Huang, J., Song, X., et al. (2021). Three dimensional aeroelastic analyses considering free-play nonlinearity using computational fluid dynamics/computational structural dynamics coupling. *Journal of Sound and Vibration*, 494. doi:10.1016/j.jsv.2020.115896.

128. Bouma, A., Yossri, W., Vasconcellos, R., Abdelkefi, A. (2022). Investigations on the interactions between structural and aerodynamic nonlinearities and unsteadiness for aeroelastic systems. *Nonlinear Dynamics*, 107(1), 331–355. doi:10.1007/s11071-021-07011-z.
129. Robinson, B., da Costa, L., Poirel, D., et al. (2020). Aeroelastic oscillations of a pitching flexible wing with structural geometric nonlinearities: Theory and numerical simulation. *Journal of Sound and Vibration*, 484. doi:10.1016/j.jsv.2020.115389.
130. Guo, M., Zheng, G. (2021). Stigma as two degrees of freedom energy sink for flutter suppression. *Journal of Sound and Vibration*, 515. doi:10.1016/j.jsv.2021.116441.
131. Tian, H., Shan, X., Cao, H., et al. (2021). A method for investigating aerodynamic load models of piezoaeroelastic energy harvester. *Journal of Sound and Vibration*, 502 doi:10.1016/j.jsv.2021.116084.
132. Ghommem, M., Hajj, M.R., Nayfeh, A.H. (2010). Uncertainty analysis near bifurcation of an aeroelastic system. *Journal of Sound and Vibration*, 329(16), 3335–3347.
133. Meehan, P.A. (2020). Prediction of wheel squeal noise under mode coupling. *Journal of Sound and Vibration*, 465, 115025.
134. Meehan, P.A., Leslie, A.C. (2021). On the mechanisms, growth, amplitude and mitigation of brake squeal noise. *Mechanical Systems and Signal Processing*, 152, 107469.
135. Meehan, P.A. (2022). Prediction and suppression of chaotic instability in brake squeal. *Nonlinear Dynamics*, 107, 205–225.
136. Thompson, J.M.T., Bokaian, A.R., Ghaffari, R. (1983). Subharmonic resonances and chaotic motions of a bilinear oscillator. *IMA Journal of Applied Mathematics*, 31(3), 207–234.
137. Bolotin, V.V. (1995). Dynamic stability of structures. In: Kounadis, A.N., Krätzig, W.B. (eds) *Nonlinear Stability of Structures. International Centre for Mechanical Sciences*, vol 342. Springer, Vienna.
138. Bolotin, V.V. (1964). *The Dynamic Stability of Elastic Systems*. Translated from the Russian edition (Moscow, 1965) by Weingarten, V.I., Greszcuzuk, L.B., Trirogoff, K.N., Gallegos, K.D., Holden-Day, San Francisco, CA.
139. Elvey, J.S.N. (1983). On the elimination of destabilising motions of articulated mooring towers under steady sea conditions. *IMA Journal of Applied Mathematic*, 31, 235–251.
140. Burton, T., Jenkins, N., Sharpe, D., Bossanyi, E. (2011). Wind *Energy Handbook*, 2nd Ed. John Wiley & Sons, Ltd., Hoboken, NJ.
141. Chopra, I., Dugundji, J. (1979). Non-linear dynamic response of a wind turbine blade. *Journal of Sound and Vibration*, 63(2), 265–286.
142. Li, S., Caracoglia, L. (2020). Experimental error examination and its effects on the aerodynamic properties of wind turbine blades. *Journal of Wind Engineering and Industrial Aerodynamics*, 206, 104357.
143. Torregrosa, A.J., Gil, A., Quintero, P., Cremades, A. (2022). On the effects of orthotropic materials in flutter protection of wind turbine flexible blades. *Journal of Wind Engineering and Industrial Aerodynamics*, 227, 105055.
144. Chen B., Zhang Z., Hua X., et al. (2018). Enhancement of flutter stability in wind turbines with a new type of passive damper of torsional rotation of blades. Journal of Wind Engineering and Industrial Aerodynamics, 173, 171–179.
145. Lu, M.-M., Ke S.-T., Wu H.-X., et al. (2022). A novel forecasting method of flutter critical wind speed for the 15 MW wind turbine blade based on aeroelastic wind tunnel test. *Journal of Wind Engineering and Industrial Aerodynamics*, 230, 105195.
146. Blevins, R.D. (2015). *Formulas for Dynamics, Acoustics and Vibration*. John Wiley & Sons, Ltd., Hoboken, NJ.
147. Carnegie, W. (1959). Vibrations of pre-twisted cantilever blading. *Proceedings of the Institution of Mechanical Engineers*, 173(1), 343–374.
148. Mansfield, E. (1982). On the torsional modes of a uniformly tapered solid wing. *Aeronautical Quarterly*, 33(2), 154–173.

149. Farsadi T., Kayran A. (2021). Classical flutter analysis of composite wind turbine blades including compressibility. *Wind Energy*, 24(1), 69–91.
150. Hayat, K., De Lecea, A.G.M., Moriones, C.D., Ha, S.K. (2016). Flutter performance of bend-twist coupled large-scale wind turbine blades. *Journal of Sound and Vibration*, 370, 149–162.
151. Rao, S.S., Gupta, R.S. (2001). Finite element vibration analysis of rotating timoshenko beams. *Journal of Sound and Vibration*, 242(1), 103–124.
152. Miley, S.J. (1982). *Catalog of Low-Reynolds-Number Airfoil Data for Wind-Turbine Applications*. Texas A and M Univ., College Station (USA). Dept. of Aerospace Engineering, Technical Report RFP-3387 ON: DE82021712, 1982.
153. Nusse, H., Yorke, J. (1994). *Dynamics: Numerical Explorations*. Springer-Verlag, New York.
154. Balcerzak, M., Dabrowski, A., Blazejczyk-Okolewska, B., Stefanski, A. (2020). Determining Lyapunov exponents of non-smooth systems: Perturbation vectors approach. *Mechanical Systems and Signal Processing*, 141, 106734.

Dragline Bucket Swing Chaos

155. Moon, F.C. (1992). *Chaotic and Fractal Dynamics*. John Wiley & Sons, New York.
156. Kapitaniak, T. (1991). *Chaotic UAOscillations in Mechanical Systems*. Manchester University Press Oxford Road, Manchester, UK.
157. Fradkov, A.L., Evans, R.J. (2005). Control of chaos: Methods and applications in engineering. *Annual Reviews in Control*, 29(1), 33–56.
158. Meehan, P.A., Asokanthan, S.F. (2002). Control of chaotic instabilities in a spinning spacecraft with dissipation using Lyapunov's method. *Chaos, Solitons and Fractals*, 13, 1857–1869.
159. Meehan, P.A., Asokanthan, S.F. (2006). Analysis of chaotic instabilities in a rotating body with internal energy dissipation. *International Journal of Bifurcation and Chaos*, 16(1), 1–19.
160. Meehan, P.A., Austin, K.J. (2006). Prediction of chaotic instabilities in a dragline bucket swing. *International Journal of Nonlinear Mechanics*, 41(1), 304–312.
161. Holmes, P.J., Marsden, J.E. (1983). Horseshoes and Arnold diffusion for Hamiltonian systems on lie groups. *Indiana University Mathematics Journal*, 32(2), 273–309.
162. Koiller, J. (1984). A mechanical system with a "wild" horseshoe. *Journal of Mathematical Physics*, 25(5), 1599–1604.
163. Åström, K.J., Furuta, K. (2000). Swinging up a pendulum by energy control. *Automatica*, 36, 278–285.
164. Munoz-Al, F.J. (2007). Bifurcation behaviour of the furuta pendulum. *International Journal of Bifurcation and Chaos*, 17(8), 2571–2578.
165. Ibáñez, C., Azuela, J. (2007). Stabilisation of the furuta pendulum based on a lyapunov function nonlinear dynamics. *Nonlinear Dynamics*, 49(1–2), 1–8.
166. Pagano, D., Pisarro, L., Aracil, J. (2000). Local bifurcation analysis in the furuta pendulum via normal forms. International *Journal of Bifurcation and Chaos*, 10(5), 981–995.
167. Corke, P.I., Winstanley, G.J., Roberts, J.M. (1997). Autonomous control of underground mining vehicles using reactive navigation. *IEEE International Conference on Robotics and Automation*, Albuquerque, New Mexico, 24–28 April 2000, 1657–1662.
168. Meehan, P.A. (2016). Predicting and simulating chaotic instabilities in an inclined furuta pendulum. *Chaotic Modeling and Simulation (CMSIM)*, 1, 99–112.
169. Meehan, P.A. (2025). Prediction and control of nonlinear instabilities in slewing boom lifting machines. *Nonlinear Dynamicss*, 113, 28675–28698.

Spacecraft Chaos

170. Caprara, G. (1986). *The Complete Encyclopedia of Space Satellites.* Bay Books, Sydney.
171. Tonkin, S.W. (1980). A basic attitude instability of spacecraft with imperfect momentum wheels. *Automatica*, 16, 415–418.
172. McIntyre, J.E., Gianelli, M.J. (1971). Bearing axis wobble for a dual-spin vehicle. *Journal of Spacecraft and Rockets,* 16, 945–951.
173. Cronin, D.L., (1976). Influence of nonlinear rotor-platform interface on dual-spin attitude dynamics. *AIAA Journal,* 14(10), 1395–1401.
174. Scher, M.P. (1971). Effects of flexibility in the bearing assemblies of dual-spin spacecraft. *AIAA Journal,* 12, 1721–1725.
175. Tseng, G.T. (1976). Nutational stability of an asymmetric dual-spin spacecraft. *Journal of Spacecraft and Rockets,* 13, 488–493.
176. Tsuchiya, K. (1979). Attitude behaviour of a dual-spin spacecraft composed of asymmetric bodies. *Journal of Guidance and Control*, 2(4), 328–333.
177. Lukich, M.S., Mingori, D.L. (1985). Attitude stability of dual-spin spacecraft with unsymmetrical bodies. *Journal of Guidance, Control and Dynamics*, 8(1), 110–117.
178. Agrawal, B.N. (1976). Effects of flexibilities in an asymmetric dual-spin spacecraft. In: *Proceedings of a Symposium on Dynamics and Control of Non-Rigid Spacecraft,* ESA SP-117, Frascati, Italy, May 1976.
179. Agrawal, B.N. (1983). *Attitude Stability of a Flexible Asymmetric Dual-Spin Spacecraft.* INTELSAT, Washington, DC.
180. Konstantopoulos, T. (1990). Attitude stability of flexible asymmetric dual-spin spacecraft with physical energy dissipation effects. Master of Engineering Science Thesis, Department of Mechanical Engineering, University of Queensland, Brisbane.
181. Van Doorn, E., Asokanthan, S.F. (1993). Attitude stability of an asymmetric spacecraft. In: *Proceedings of the IEAust. Eigth National Space Engineering Symposium*, NCP no. 93/7, Brisbane, 20–21 September 1993.
182. Chang, C.O., Liu, L.Z., Alfriend, K.T. (1996). Dynamics and stability of a freely precessing spacecraft containing a nutation damper. *Journal of Guidance, Control, and Dynamics*, 19(2), 297–304.
183. Ross, I.M. (1996). Formulation of stability conditions for systems containing driven rotors. *Journal of Guidance, Control, and Dynamics*, 19(2), 305–308.
184. Zaslavsky, G.M. (1978). The simplest case of a strange attractor. *Physical Letters* A, 69(3), 145–147.
185. Grebogi, C., Kosterlich, E.J., Ott, E., Yorke, J.A. (1987). Multi-dimensioned intertwined basin boundaries: Basin structure of the kicked double rotor. *Physica*, 25D, 347–360.
186. Holmes, P.J., Marsden, J.E. (1983). Horseshoes and Arnold diffusion for Hamiltonian systems on lie groups. *Indiana University Mathematics Journal*, 32(2), 273–309.
187. Koiller, J. (1984). A mechanical system with a 'wild' horseshoe. *Journal of Mathematical Physics*, 25(5), 1599–1604.
188. Or, A.C. (1998). Chaotic motions of a dual-spin body. *ASME Journal of Applied Mechanics*, 65, 150–156.
189. Peng, J.H., Liu, Y.Z. (2000). Chaotic motion of a gyrostat with asymmetric rotor. *International Journal of Non-Linear Mechanics*, 35, 431–437.
190. Miller, A.J., Gray, G.L., Mazzoleni, A.P. (2001). Nonlinear spacecraft dynamics with a flexible appendage, damping, and moving internal submasses. *Journal of Guidance, Control, and Dynamics*, 24(3), 605–615.
191. Kuang, J., Leung, A.Y.T., Tan, S. (2004), Chaotic attitude oscillations of a satellite filled with a rotating ellipsoidal mass of liquid subject to gravity-gradient torques. *Chaos*, 14(1), 111–117.

192. Kuang, J., Tan, S., Arichandran, K. (2001). Chaotic attitude motion of gyrostat satellite via melnikov method. *International Journal of Bifurcation and Chaos*, 11(5), 1233–1260.
193. Cochran, J.E., Jr., Thompson, J.A. (1980). Nutation dampers vs precession dampers for asymmetric spacecraft. *Journal of Guidance Control*, 3, 22–28.
194. Meehan, P.A., Asokanthan, S.F. (1997). Chaotic motion in a spinning spacecraft with circumferential nutational damper. *Nonlinear Dynamics*, 12, 69–87.
195. Meehan, P.A., Asokanthan, S.F. (1996). Chaotic motion in a rotating body with internal energy dissipation. In: Kliemann, W.H., Langford, W.F., Namachchivaya, N.S. (eds) *Nonlinear Dynamics and Stochastic Mechanics*. Fields Institute Communications, vol 9. AMS, Providence, RI, 175–202.
196. Meehan, P.A., Asokanthan, S.F. (2002). Control of chaotic instabilities in a spinning spacecraft with dissipation using Lyapunov's method. *Chaos, Solitons & Fractals*, 13(9), 1857–1869.
197. Meehan, P.A., Asokanthan, S.F. (2006). Analysis of chaotic instabilities in a rotating body with internal energy dissipation. *International Journal of Bifurcation and Chaos*, 16(1), 1–19.
198. Wolf, A., Swift, J.B., Swinney, H.L., Vasano, J.A. (1985). Determining lyapunov exponents from a time series. *Physica*, 16D, 285–317.

Other Instability Phenomena

Transmission Driveline Chaos

199. Asokanthan, S.F., Meehan, P.A. (2000). Non-linear vibration of a torsional system driven by a Hooke's joint. *Journal of Sound and Vibration*, 233(2), 297–310.
200. Hilborn, R.C. (1994). *Chaos and Nonlinear Dynamics: An Introduction for Scientists and Engineers*. Oxford University Press, New York.

Chaos in Nanosystems

201. Defoort, M., Rufer, L., Fesquet, L., et al. (2021). A dynamical approach to generate chaos in a micromechanical resonator. *Microsystems & Nanoengineering*, 7, 17.
202. Madiot, G., Correia, F., Barbay, S., Braive, R. (2022). Random number generation with a chaotic electromechanical resonator. *Nanotechnology*, 33(47), 475204.
203. Ebrahimi, R. (2022). Chaotic vibrations of carbon nanotubes subjected to a traversing force considering nonlocal elasticity theory. *Proceedings of the Institution of Mechanical Engineers, Part N: Journal of Nanomaterials, Nanoengineering and Nanosystems*, 236(1–2), 31–40.
204. Sobamowo, G., Akanmu, J., Adeleye, O., et al. (2021). Nonlinear vibration of multi-walled carbon nanotubes with initial curvature resting on elastic foundations in a nonlinear thermo-magnetic environment. *Engineering Transactions*, 69(2), 87–165.

Chaos in the Human Body

205. Pascolo, P.B., Marini, A., Carniel, R., Barazza, F. (2004). Posture as a chaotic system and an application to the Parkinson's disease. *Chaos, Solitons and Fractals*, 24, 1343.
206. Di Ieva, A., Grizzi, F., Jelinek, H., et al. (2014). Fractals in the neurosciences, Part I. *The Neuroscientist*, 20, 403–417.

207. Beckers, F., Verheyden, B., Aubert, A. (2006). Aging and nonlinear heart rate control in a healthy population. *American Journal of Physiology-Heart and Circulatory Physiology*, 290(6), H2560.
208. Bassler, K., Gunaratne, G., McCauley, J. (2006). Hurst exponents, Markov processes, and fractional Brownian motion. *Physica A-Statistical Mechanicsand its Applications*, 369(2), 343.
209. Costa, M., Goldberger, A.L., Peng, C.-K. (2002). Multiscale entropy analysis of complex physiologic time series. *Physical Review Letters*, 89, 068102.
210. Darbin, O., Soares, J., Wichmann, T. (2006). Nonlinear analysis of discharge patterns in monkey basal ganglia. *Brain Research*, 1118, 84–93.
211. Ponomarev, A., Zavyalov, A., Smirnov, V., Lockner, D. (1997). Physical modeling of the formation and evolution of seismically active fault zones. *Tectonophysics*, 277(1–3), 57.
212. Knezevic, A., Martinis, M. (2006). Traveling waves for an integrable higher order KDV type wave equations. *International Journal of Bifurcation and Chaos*, 16(7), 2103.
213. Ivanov, P., Amaral, L., Goldberger, A., et al. (1999). Multifractality in human heartbeat dynamics. *Nature*, 399, 461–465. doi:10.1038/20924.
214. Yulmetyev, R., Demin, S., Panischev, O.Y., et al. (2006). Regular and stochastic behavior of Parkinsonian pathological tremor signals, *Physica A*, 369, 655–678.
215. Yulmetyeva, R., Demin, S., Emelyanova, N., et al. (2003). Stratification of the phase clouds and statistical effects of the non-Markovity in chaotic time series of human gait for healthy people and Parkinson patients. *Physica A*, 319, 432–446.
216. Yulmetyeva, R., Hanggi, P., Yulmetyeva, D., et al. (2007). Anomalous diffusion in the dynamics of complex processes. *Physica* A, 383, 443–454.
217. Meehan, P.A., Bellette, P.A. (2009). Chaotic signal analysis of Parkinson's disease STN brain signals, topics on chaotic systems. *World Scientific*, 6, 232–241.
218. Madalena Costa, Ary L. Goldberger, and C.-K. Peng. (2002) Multiscale Entropy Analysis of Complex Physiologic Time Series. *Physical Review Letters*. 89, 068102.

5 Tools for Control of Instabilities

Teaching my daughter to ride a bicycle is an example of training her brain's innate nonlinear instability controller. The problem of riding a bike is an inherently unstable one because the centre of mass of the system is above the hinge points of the contact between the tyres and the ground. This is analogous to balancing a broom on your hand. In both cases, you need to sense the motion of the centre of mass relative to the hinge point and then finely actuate the muscles to stabilise the inverted position. In fact, this is a challenging fundamental technical problem for control engineers that the human brain and body accomplishes with superior instinctual ease, speed and effectiveness. This chapter looks at generalised, fundamental tools for automatically performing similar tasks in engineering (without requiring a human body!).

The predictive modelling and identification of the instability mechanisms exemplified in Chapter 4 can now be used to identify appropriate tools for control of the phenomenon. For instance, how fast the instability grows was shown to be a function of critical model parameters or conditions, which can then be manipulated to reduce or eliminate the phenomenon. Understanding the mechanism of instability can also lead to a means of controlling the phenomenon. For instance, banks manipulate conditions they control to maximise interest rates, which we know determine the unstable growth of investments. Alternatively, governments schedule and mandate close contact isolation rules to minimise the rate of pandemic exponential spread via transmission. Vaccines are tuned specifically to enhance the human body's defences and to interrupt virus mechanisms of unstable proliferation.

This chapter explores cutting-edge modelling tools for effectively controlling instabilities in engineering systems, delving into both classical and nonlinear realms. Diverse control methods employed in mitigating instability types are first elucidated. Subsequently, the intricate process of crafting controllers, tailored to predictive modelling and mechanisms identified in Chapter 3, provide a comprehensive guide for engineers and researchers to control instability.

5.1 TYPES OF CONTROL

Control of instability essentially boils down to changing the system dynamic behaviour to stabilise it to a desired condition. Passive and active types of control are fundamental approaches used to stabilise unstable systems across a wide range of applications in engineering. Both passive and active control strategies have their advantages and limitations, and making a choice between them depends on factors such as system dynamics, performance requirements and resource constraints.

DOI: 10.1201/9781003496533-5

Often, a combination of passive and active control techniques is employed to achieve optimal stability and efficiency in unstable systems. In the following section, some more details are provided.

Passive Control

Passive control involves modifying the system's inherent properties or introducing passive elements to dampen instability without the need for external energy input. This can include adding damping elements, such as shock absorbers in vehicles or resistors in electrical circuits, to dissipate energy and stabilise the system. Passive control methods are often simpler and more cost-effective but may have limited effectiveness in highly dynamic or unpredictable environments. In this book, we show passive control applied to a number of unstable systems in engineering, typically based on a sensitivity analysis of critical system parameters to achieve stability. Hence, in practice, the control engineer need only design/modify the system parameters to avoid instability. We also show the use of a passive device to stabilise rolling mill chatter. In particular, a tuned mass damper is used to increase the critical mill speed for instability, where it is difficult to change system parameters or the parameter configuration may be required for other reasons. An alternative to passive control is to actively use sensors and actuators to automatically measure and apply control signals to the system, as described below.

Active Control

Active control employs external energy sources and feedback mechanisms to actively intervene and stabilise the system in real-time. This can include using sensors to detect deviations from the desired state and actuators to apply corrective actions, such as adjusting control inputs or manipulating system parameters. Active control allows for precise and adaptive stabilisation, making it particularly suitable for complex, nonlinear systems or situations where passive methods are insufficient. For example, active suspension systems in vehicles use sensors to monitor road conditions and vehicle dynamics. Based on this feedback, the control adjusts individual shock absorbers or suspension components in real-time to improve ride comfort, handling and stability. These systems can compensate for uneven road surfaces, cornering forces and other factors affecting vehicle stability. In this book, we show examples of actively designed control techniques for stabilising a range of nonlinear engineering systems using methods described in the following sections. However, we refer the reader to many text sources of modern methods of control of nonlinear systems such as Slotine and Li [1]

5.2 DESIGNING AN INSTABILITY CONTROLLER

If the tasks of a control system involve instability with typically large range behaviour from stable equilibrium, nonlinear effects will be significant to the dynamics of the system. For instance, in rotating multibody systems like a spinning spacecraft, nonlinearities arise due to relative accelerations and Coriolis forces, causing large excursions from equilibrium in the form of limit cycle or chaotic oscillations. In this case, nonlinear control may be necessary to achieve the desired performance for the system. For the purpose of eliminating any unstable and/or nonlinear behaviour that may arise in the system, a nonlinear controller must be employed to ensure global dynamical stability.

In this section, various nonlinear control methods are described. The methods are considered to be of two separate classes. Firstly, more classical nonlinear methods of control will be reviewed before we outline the recently developed nonlinear techniques devised to eliminate chaotic instability. The methods detailed are for understanding of the application cases detailed in later chapters, but they are a sample of the plethora of techniques for nonlinear control e.g. [1]. However, the methods are all based on the fundamental control aim to break the fundamental instability mechanism or "vicious cycle" that is causing energy input and exponential growth.

Classical Instability Control

For nonlinear and/or unstable systems, there is no general method for designing nonlinear controllers. However, a rich diversity of techniques for particular classes of nonlinear control problems have been formulated over the past century, such as feedback linearisation, adaptive control and gain-scheduling. Slotine and Li [1], amongst others, describe a wide variety of methods available for nonlinear control design. The first control method for instability, we detail, is an obvious passive control method based on sensitivity analysis, which can be used effectively for engineers to avoid instability by parameter design. Many of these passive and active controller design methods are based upon Lyapunov stability concepts. This is because for most control systems, the desire is to stabilise the system around an equilibrium point or trajectory. Some examples of simple control stabilisation tasks are altitude control of aircraft and positional control of robot arms. Nonlinear control systems may also be designed directly from the use of Lyapunov's direct method. Due to the general applicability of the method on mechanical systems, Lyapunov's direct control method will be reviewed. Lyapunov control has been directly implemented on a wide variety of mechanical systems. These are included for control of flexural and torsional vibrations in cantilevered beams (see Bailey and Hubbard [2] and Spearritt and Asokanthan, [3]) and nonlinear manoeuvres of a rigid body, as described in Junkins and Kim [4]. In the following section, a general discussion of the use of stability sensitivity analysis and Lyapunov's direct method for designing stable control systems will be provided.

Stability Sensitivity Analysis

Stability sensitivity analysis is a passive technique used to assess how sensitive a system's stability is to changes in parameters or operating conditions. It is highlighted here as it is often an obvious but crucial first (and sometimes last) step we use in controlling instabilities. Essentially, stability sensitivity analysis involves evaluating how small variations in system parameters or inputs affect the stability margins of the system. This analysis helps control system designers identify critical parameters that significantly influence stability and prioritise them for active or passive control design. In control theory, root locus analysis determines stability sensitivity. Specifically, the effects of system parameter changes on the roots (i.e., modes/poles) are plotted graphically, and their real axis positions (i.e., signs of their real components) determine system stability. Stability methods/theory are powerful and efficient predictors as they typically do not require a full solution to the equations of motion and are valid for a range of initial conditions. By quantifying the sensitivity of system stability to various factors, engineers can develop robust control strategies that effectively mitigate instabilities and enhance system resilience. For instance, in aircraft flight control systems, sensitivity analysis helps identify key aerodynamic parameters or environmental conditions that may compromise stability. This enables the design of aircraft parameters, i.e., geometry or control algorithms, that actively compensate for these factors in real-time. The sensitivity analysis may be performed with an efficient numerical or analytical model to investigate a wide range of parameters very quickly. For instance, an analytical model for brake squeal based on the stability analysis techniques discussed in Chapter 2 (an analytical eigenvalue analysis) can be used to perform a stability sensitivity analysis over all the system parameters instantaneously. This avoids prohibitive computational expense required from a numerical finite-element analysis of many case studies of the brake system. Similarly, an analytical prediction of limit cycle amplitude can be used for efficient squeal noise amplitude sensitivity analysis. These methods can reveal simple changes in the geometry of the brake parts to modify dynamic characteristics to avoid brake squeal. This avoids the need for restrictive, expensive and physically intrusive, active control systems on the brake. A similar approach is used for tool chatter in which operators avoid certain machining parameters revealed by stability sensitivity analysis to ensure smooth cutting operation. For nonlinear systems, Lyapunov stability theory is most useful and a general approach for performing stability sensitivity analysis, as detailed in Chapter 2. This method is also extended directly to control system design, as discussed subsequently.

Lyapunov's Direct Method

Given a control system, the first and most important characteristic is whether it is stable because an unstable control system is typically useless and potentially dangerous in practice. In Section 2.3, Lyapunov's direct method was described for the determination and classification of system stability. The method may also be incorporated into the design of globally stable nonlinear control systems. There are basically two methods of using Lyapunov's method for control design, and both have a trial and error impression. The first technique involves suggestion of a suitable control law for the system and then finding a suitable Lyapunov function to justify the choice. Conversely, the second technique requires suggestion of a suitable Lyapunov function and then finding a suitable control law to justify its applicability to the controlled system. The control is then designed to minimise the time derivative of the characteristic Lyapunov function at every point in time. In both cases, Lyapunov's direct method outlined in Section 2.3 guarantees stability of the control system.

The first approach is to consider a suitable control law $\mathbf{u}(t,x)$ usually based on physical intuition. Then, a suitable Lyapunov function $U(\mathbf{x})$ must be found to ensure stability of this choice. The second approach is to consider a Lyapunov function candidate $U(\mathbf{x})$ for the uncontrolled system, which is usually an approximate energy integral for the system. When a general forcing control $\mathbf{u}(t)$ is then considered, $\dot{U}(\mathbf{x})$ is usually found to depend on the control $\mathbf{u}(t)$ through the equations of motion for the system. A suitable range of control laws $\mathbf{u}(t)$ can then be designed to ensure $\dot{U}(\mathbf{x})$ is negative and thus stabilising. One strategy is then to select a specific control function $\mathbf{u}(t,x)$ to make $\dot{U}(\mathbf{x})$ as negative as possible so that $U(\mathbf{x})$ approaches 0 as fast as possible. This optimal strategy will ensure the system converges on the equilibrium state $\mathbf{x}_e$ as fast as possible. In many cases, it is also advantageous to consider both the Lyapunov function $U(\mathbf{x})$ and the control law $\mathbf{u}(t)$ as design variables, which are both adjusted to achieve optimum performance of the control system.

Finally, it is noted that the control methods reviewed in this section are model-dependent. Typically, these are more efficient and effective techniques (if the model is accurate) as they are designed with knowledge of the open-loop system dynamics. Also, both methods are general, in that they may be used to control chaotic instability. In the following section, we review other methods designed specifically for eliminating chaos following instability.

Chaotic Instability Control

It is not surprising that chaotic motion has been shown to occur widely in engineering systems, particularly since many are inherently nonlinear and subjected to some form of perturbations/forcing in operation. Usually, the presence of these unpredictable vibrations is considered undesirable in engineering systems as they cause fatigue failures in components and cause inaccuracies in smooth operation. With the knowledge of its presence, chaotic vibration may be eliminated with suitable design or parameter changes of a system. However, it is advantageous if methods of feedback control can be employed to do this automatically.

The recent research on dynamics and control has seen the emergence of new breakthroughs. Probably, the most significant work has been achieved by Ott, Grebogi and York [5]. They proposed the OGY method, by which chaos can always be suppressed by shadowing one of the infinitely many unstable periodic orbits entrenched in a chaotic attractor. Due to the realisation that typically only very few control perturbations were required to stabilise these periodic orbits using this method, it became clear that chaotic behaviour could be exploited. For example, chaotic behaviour may be deliberately designed into a system so that a wide range of periodic motions could be attained by employing only minute perturbations. Therefore, although the extreme sensitivity of a chaotic system to initial conditions and parameter variations makes long-term prediction impractical, it also allows for precise, weak control adjustments to effectively control system behaviour. This design philosophy recognises that nonlinear systems exhibit multiple basins of dynamic behaviour that can lead to more creative and efficient solutions to those based on linear dynamic systems. Practical implementations of the usefulness of chaos in this sense have yet to be created in mechanical systems. For most of the cases detailed in this book, chaotic vibrations are detrimental to the normal operations of the systems considered, although this is not always the case, i.e., human body and mixing applications.

The methods of control considered in the following section have been designed using the same exploitive goal as the OGY method; to stabilise one of the unstable periodic orbits using small perturbations. Owing to the universality of their design, they are "black-box" or model-independent techniques. The applications of chaotic instability control in mechanics is growing ie [6–15]. As an introduction to these methods, a brief overview of the OGY control strategy will be given.

OGY Method

When chaotic instabilities occur in a system, all the stable periodic orbits break up in phase space into an infinite number of unstable periodic orbits (UPOs). Thus, chaotic motion may be considered to be the result of the system moving haphazardly between these UPOs, never staying on any particular one for a full orbit or more. An illustration of an UPO in the phase space and Poincaré map is depicted in Figure 5.1. The frequencies of these UPOs correspond to a broad range of frequencies obtained in the frequency spectrum of a chaotic attractor. The aim of the OGY technique is to stabilise one of the many UPOs entrenched in a chaotic attractor with the use of a small perturbation to the system.

For implementation of the OGY method, the phase space flow

$$\dot{\mathbf{X}} = \mathbf{F}(\mathbf{X}, p), \quad \mathbf{X} \in R^N,$$

is first reduced to a map on a Poincaré section

$$\mathbf{X}_{n+1} = \mathbf{F}_M(\mathbf{X}_n, p), \quad \mathbf{X}_n \in R^{N-1}.$$

A point on the Poincaré section denotes an UPO of the flow. This fixed point as follows:

$$\mathbf{X}_f = \mathbf{F}(\mathbf{X}_f, p_0),$$

where p_0 denotes the nominal unperturbed parameter value. The aim of the OGY method is to provide an appropriate perturbation to this parameter $\delta p_n \equiv p_n - p_0$. By also considering the relative displacement of the present Poincaré map point

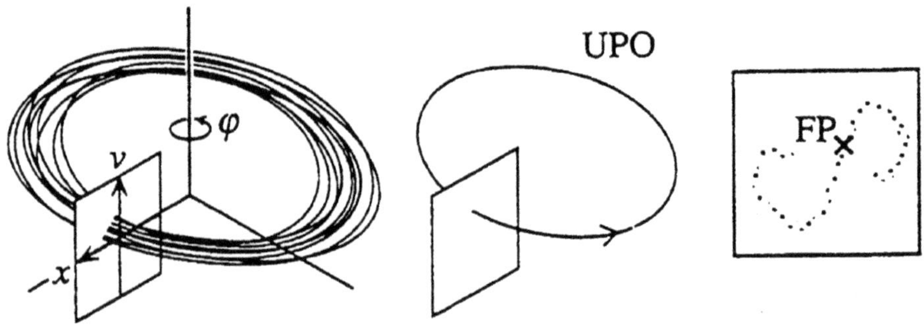

FIGURE 5.1 Unstable periodic orbit in phase space and Poincaré section [16, p. 797].

to the desired fixed point $\delta x_n \equiv x_n - x_{f,}$ the Poincaré map may be linearised about the fixed point as

$$\delta x_{n+1} = \mathbf{M}\delta x_n + \mathbf{s}\delta p_n, \tag{5.1}$$

where $\mathbf{s} = \partial \mathbf{F}/\partial p$ is a shift vector and $\mathbf{M} = \partial \mathbf{F}/\partial \mathbf{X}$ is the Poincare section matrix of partial derivatives, both of which are evaluated at the fixed point. If the displacement δx_n is small enough, the most appropriate choice of relative perturbation is a linear control law of the form

$$\delta p_n = -\mathbf{K}^T \delta x_n, \tag{5.2}$$

where $\mathbf{K}$ is the proportionality vector. Thus, by combining equations (5.1) and (5.2), the linearised map near the fixed point becomes

$$\delta x_{n+1} = \mathbf{C}\delta x_n, \tag{5.3}$$

where $\mathbf{C}$ is the control matrix defined as $\mathbf{M} - \mathbf{s}\mathbf{K}^T$. Designing a suitable proportionality vector $\mathbf{K}$ is a typical control problem. It is a straight-forward result to show that if $\mathbf{K}$ is chosen such that the eigenvalues of the control matrix $\mathbf{C}$ do not cause growth (i.e., the moduli less than unity), then iterates of the Poincaré section will converge toward the fixed point, ($\lim_{n\to\infty} \delta x_{n+1} = 0$) and the UPO will be stabilised. The OGY technique further refines the choice of $\mathbf{K}$ for a saddle fixed point. A brief account of the underlying concepts involved with this method is summarised in Figure 5.2.

In the Poincaré map depicted in Figure 5.2a, a nearby iterate will miss the fixed point. However, with a suitable perturbation, the Poincaré map is shifted, as depicted in Figure 5.2b, so that the next iterate falls on the stable (attracting) manifold of the saddle. Subsequent iterates will then converge onto the fixed point along the stable manifold, as depicted in Figure 5.2c.

In practice, the map $\mathbf{M}$ and shift $\mathbf{s}$ may be detected by performing a linear fit to close returns in the section. Due to the assumption of a linear mapping, the control perturbations are only activated when the map iterates "ergodically" wander close to the desired fixed point. Notably, it is typical in chaotic systems for the iterates to

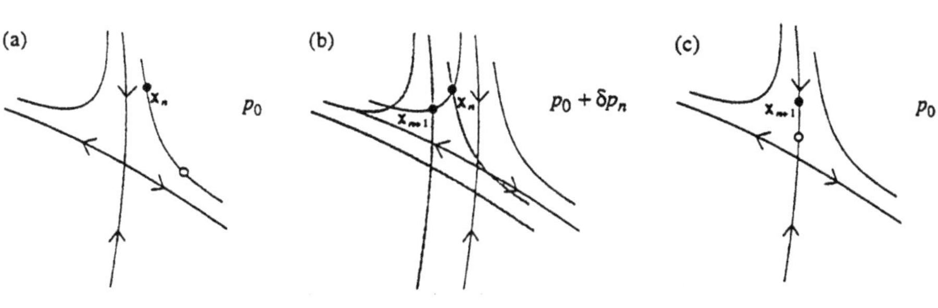

FIGURE 5.2 OGY control method [16, p. 800].

wander close to any chosen point of the strange attractor. By introducing a perturbation limitation $\delta p_{\max}$, the condition for the activation of control may be given as

$$\delta p_n = -\mathbf{K}^T \delta x_n, \tag{5.4}$$

and the general form of the OGY method may summarised as

$$\delta p_n = \begin{Bmatrix} -\mathbf{K}\delta x_n, & |\mathbf{K}\delta x_n| \le \delta p_{\max} \\ 0, & |\mathbf{K}\delta x_n| > \delta p_{\max} \end{Bmatrix}. \tag{5.5}$$

Various control strategies have been devised based on equation (5.5), as described in Lindner and Ditto [16] such as the occasional proportional feedback (OPF) technique and the recursive proportional feedback (RPF) technique. Other techniques have also subsequently been devised based on the goal to stabilise an UPO embedded in a chaotic attractor such as continuous delayed feedback method. These methods will be described subsequently.

Continuous Delayed Feedback

The OGY method is based upon the Poincaré map of a dynamical system, and hence the small control perturbations are applied intermittently. Since the corrections of the parameter are rare and small, under the presence of noise, the controlled system may suffer from occasional bursts of loss of control, as described in Blazejczyk [17]. Furthermore, the OGY method can only stabilise unstable periodic orbits (UPOs) with sufficiently small expanding directions. Specifically, the interval between perturbations must be shorter than the reciprocal of the maximum Lyapunov exponent to ensure effective control. A method of time continuous control was proposed to overcome these limitations [18]. This method is based on the construction of a special form of a time continuous perturbation that does not change the form of a desired unstable periodic orbit or fixed point. Two control methods of this form have been proposed. The first involves the design of an external periodic forcing function that is directly proportional to the difference between the actual orbit and the desired periodic orbit. In this case, prior knowledge of the complete form of the desired periodic orbit is required. The second method overcomes this limitation by using a delayed output signal from the system instead of the desired periodic orbit. Due to the physical applicability of this method, it was considered for investigation in this book and is described in more detail.

The control can be applied to any one or more of the system equations of motion in the form,

$$\dot{\mathbf{X}} = \mathbf{F}(\mathbf{X}, p) + \hat{\mathbf{u}}_i C(\tau)$$

where $\dot{\mathbf{X}} = \mathbf{F}(\mathbf{X}, p)$ are the state space equations of motion, $\hat{\mathbf{u}}_i$ is a unit direction vector for the ith variable and the control signal each ith variable is given by,

$$C(\tau) = \mathrm{K}\left[x_i(\tau - \Delta) - x_i(\tau)\right], \tag{5.6}$$

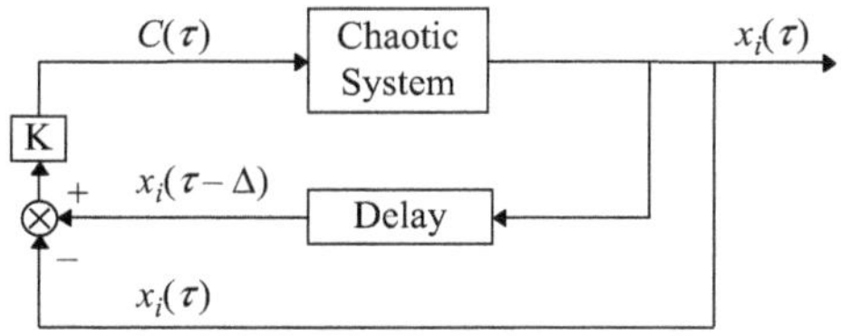

FIGURE 5.3 Block diagram of continuous delayed feedback method [8].

where Δ is the delay time and x_i is the ith variable. Note that $C(\tau)=0$ when $x_i(\tau-\Delta)=x_i(\tau)$. The block diagram of this system is seen in Figure 5.3.

This occurs when the system has settled on an UPO of period, Δ. This means that the perturbation in the form of equation (5.6) does not change the solution of the system corresponding to an embedded UPO of period Δ. Control design involves tuning of the delay, Δ, to the period of the desired unstable periodic orbit and adjustment weight, K, of the feedback to achieve convergence. Barrett [19] has presented analytical results describing a method of determining the required feedback gain, K. It may be noted that only sufficient conditions for stability are employed using this method, and in practice, adjustment of the feedback K is required for more effective controller performance. Typically, K is set low and gradually increased until desired stabilisation occurs (see Pyragas [18] and Pyragas and Tamasevicius [20]). One advantage of this strategy is the simplicity of design for practical implementation.

Recursive Proportional Feedback

The RPF method extends the OPF technique, originally introduced by Hunt [21] and derived from the OGY method [5]. In highly dissipative systems, the phase space contraction is so significant that the attractor is compressed into a low-dimensional structure. When this contraction is sufficient to allow the attractor to be well described by a one-dimensional Poincaré map, the OGY technique simplifies into a geometrical algorithm known as OPF. In this case, the feedback is applied in proportion to the distance from the fixed point. However, OPF can fail if the attractor changes when a control parameter is modified. To address this limitation, Rollins et al. [22] refined the method, developing the RPF technique. Similar to OPF, RPF is particularly effective in highly dissipative systems. However, RPF includes an additional derivative-like term that accounts for attractor dynamics, enhancing stability and control effectiveness. Note this method applies to discrete mappings (i.e., Poincaré sections) of a dynamical system. The control strategy introduces small perturbations to a system parameter, which are proportional to the deviation from the desired output of a sensor variable. For a discrete mapping, the control law for the nth iteration is given by

$$\delta p_n = \left\{ \begin{array}{ll} K_1\delta x_n + K_2\delta p_{n-1} & |K_1\delta x_n| \le \delta p_{\max} \\ 0 & |K_1\delta x_n| > \delta p_{\max} \end{array} \right\}, \tag{5.7}$$

where δp_n defined as $p_n - p_0$ is the relative control parameter perturbation while p_n and p_0 are the control parameters for the nth iterate and uncontrolled value,

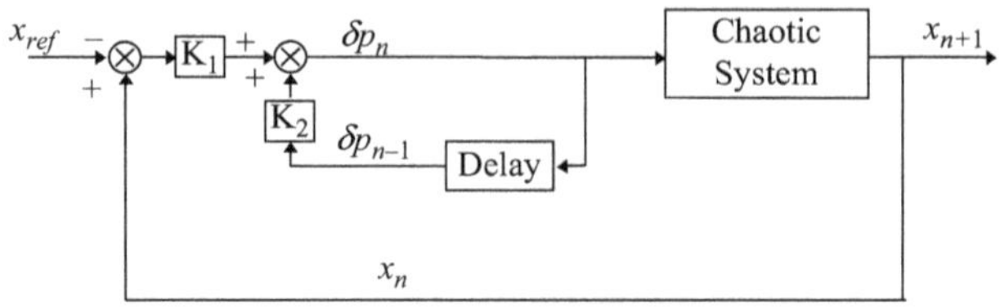

FIGURE 5.4 Block diagram of the RPF method [8].

respectively. The block diagram of this system is seen in Figure 5.4. The relative sensor displacement δx_n from desired fixed point x_f is defined as $x_n - x_f$ where x_n signifies the sensor value at the nth iterate. Also, $|K_1 \delta x_n| \leq \delta p_{\max}$ defines a region or "window" in phase space around the desired fixed point where linear approximation is valid. $K_2 \delta p_{n-1}$ is a "derivative-like" term which can add stability and optimise the control method.

Typically, the mapping frequency is selected as per the Poincaré section to match either the external forcing frequency or a natural frequency of the system. Rollins et al. [22] provides a method of determination of feedback gains K_1 and K_2 via experimentation and measurement from one-dimensional return maps of the controlled system. This method assumes linear behaviour of the system about the fixed point and is valid for small control perturbations. In practice, adjustment of the two feedback gains K_1 and K_2 as well as adjustment of the "window" size $\delta p_{\max}$ is required for more effective controller performance. Typically, the procedure involves tuning gains K_1 and K_2 with a small "window" $\delta p_{\max}$ then gradually increasing the size and adjusting the gains for optimal controller performance.

The RPF and continuous delayed feedback methods described above are both model-independent or "blackbox" techniques for the control of chaotic motion in any dynamical system.

5.3 CONCLUSION

In this chapter, a brief review of modern control techniques used for controlling instabilities was described. A nonlinear controller must be employed to eliminate any large amplitude, instability behaviour that may arise in nonlinear systems so that global dynamical stability is assured. In the subsequent chapter, the control methods are implemented to stabilise a range of engineering systems with instabilities.

REFERENCES

1. Slotine, J.-J.E., Li, W. (1991). *Applied Nonlinear Control*. Prentice-Hall, Englewood Cliffs, NJ.
2. Bailey, T., Hubbard, J. (1985). Distributed piezoelectric polymers active vibration control of a cantilever beam. *Journal of Guidance, Control and Dynamics*, 8, 605–611.
3. Spearritt, D.J., Asokanthan, S.F. (1996). Torsional vibration control of a flexible beam using laminated PVDF actuators. *Journal of Sound and Vibration*, 193(5), 941–956.
4. Junkins, J.L., Kim, Y. (1993). *Introduction to Dynamics and Control of Flexible Structures*. American Institute of Aeronautics and Astronautics, Washington, DC, (ISBN 1-56347-054-3).

5. Ott, E., Grebogi, C., Yorke, J.A. (1990). Controlling chaos. *Physical Review Letters*, 64, 1196–1199.
6. Fradkov Alexander, L., Evans Robin, J., Andrievsky Boris, R. (2006). Control of chaos: Methods and applications in mechanics. *Philosophical Transactions of the Royal Society A*, 364, 2279–2307.
7. Roy, R., Murphy, T.W., Jr., Maier, T.D., et al. (1992). Dynamical control of a chaotic laser: Experimental stabilisation of a globally coupled system. *Physical Review Letters*, 68(9), 1259–1262.
8. Meehan, P.A. (1996). Nonlinear instabilities in rotating multibody systems. Doctor of Philosophy Thesis, Department of Mechanical Engineering, University of Queensland, Australia.
9. Meehan, P.A., Asokanthan, S.F. (2002). Control of chaotic insabilities in a spinning spacecraft with dissipation using Lyapunov's method. *Chaos, Solitons & Fractals*, 13(9), 1857–1869.
10. Meehan, P.A., Asokanthan, S.F. (2002). Control of chaotic instability in a dual-spin spacecraft with dissipation using energy methods. *Multibody System Dynamics*, 7(2), 171–188.
11. Meehan, P.A., Asokanthan, S.F. (2006). Analysis of chaotic instabilities in a rotating body with internal energy dissipation. *International Journal of Bifurcation and Chaos*, 16(1), 1–19.
12. Meehan, P.A., Asokanthan, S.F. (2002). Control of chaotic instabilities in a spinning spacecraft with dissipation using Lyapunov's method. *Chaos, Solitons and Fractals*, 13(9), 1857–1869.
13. Meehan, P.A., Asokanthan, S.F. (2002). Control of chaotic motion in a dual-spin spacecraft with nutational damping. *Journal of Guidance, Control, and Dynamics*, 25(2), 209–214.
14. Meehan, P.A., Asokanthan, S.F. (1998). Control of chaotic motion in a spinning spacecraft with a circumferential nutational damper. *Nonlinear Dynamics*, 17(3), 269–284.
15. Meehan, P.A. (2025). Prediction and control of nonlinear instabilities in slewing boom lifting machines. *Nonlinear Dynamics,* 113, pp. 28675–28698.
16. Lindner, J.F., Ditto, W.L. (1995). Removal, suppression and control of chaos by nonlinear design. *Applied Mechanics Reviews*, 48, 795–780.
17. Blazejczyk, B., Kapitaniak, T., Wojewoda, J., Brindley, J. (1993). Controlling chaos in mechanical systems. *Applied Mechanics Reviews*, 46(7), 385–391.
18. Pyragas, K. (1992). Continuous control of chaos by self-controlling feedback. *Physics Letters A*, 170, 421–428.
19. Barrett, M.D. (1996). Continuous control of chaos. *Physica D*, 91, 340–348.
20. Pyragas, K., Tamasevicius, A. (1993). Experimental control of chaos by delayed self-controlling feedback. *Physics Letters A*, 180, 92–102.
21. Hunt, E.R. (1991). Stabilising high periodic orbits in a chaotic system: The diode resonator. *Physical Review Letters*, 67(15), 1953–1955.
22. Rollins, R.W., Parmananda, P., Sherard, P. (1993). Controlling chaos in highly dissipative systems: A simple recursive algorithm. *Physical Review E*, 47(2), R780.

6 Controlling Instabilities in Engineering Systems

After a recent dinner party, I was showing my son how to make a wineglass sing. This is a frictional instability like railway wheel squeal, whereby under the right frictional conditions between my sliding finger (controlled by dipping my finger into the wine) and the rim, you can excite a vibration mode of the crystal wineglass. These vibrations propagate through the air as sound, which we hear as a typically high-pitched tone. The pitch is dependent on the structural properties of the glass (and how much wine is in it!). Controlling or eliminating this instability (after it has reached its steady limit cycle amplitude) is rather easy once we understand the mechanism of its occurrence. In particular, we can simply change one of the critical parameters such as the friction via drying my finger or changing the structural properties by holding the bowl of the wineglass with my other hand. Although controlling instabilities in engineering systems is invariably not as easy to implement, the fundamental approach is typically the same; understand the process and then passively or actively control the critical parameter conditions of the mechanism of instability.

One of the most important global challenges the world faces at present is climate change due to excessive global warming. An unstable process of global warming occurs due to excessive levels of carbon dioxide. Carbon dioxide in the atmosphere acts as an insulator of heat, which raises the Earth's temperature. This causes a viscous cycle when ocean temperatures rise, which in turn releases more stored carbon dioxide to the atmosphere, like a warm soft drink. Predictive models have been developed and tuned to be used to gain insights into this instability mechanism. Subsequently, this is used to design climate change mitigation methods, based on reducing a critical parameter: atmospheric level of carbon dioxide. This chapter explores a similar design for controlling classical and chaotic instabilities across a spectrum of engineering applications. Controlling classical and chaotic instabilities in engineering systems is a paramount challenge faced by researchers and engineers as they manifest in diverse systems ranging from mechanical and civil engineering to aerospace and beyond. Understanding and mitigating these instabilities are crucial for ensuring the safety, reliability and performance of engineering structures and machinery. In this chapter, we will delve into controlling several prominent examples of classical and chaotic instabilities encountered in engineering systems, including railway wheel squeal, brake squeal, aerodynamic flutter, rolling mill chatter and spacecraft attitude instabilities. For instance, from our understanding and ability to predict wheel squeal vibrations and noise, mitigation methods can be used to minimise discomfort to passengers and residents and potential wear and tear on the wheels and tracks. Similarly, controlling brake squeal is essential for improving braking performance, reducing noise pollution and enhancing overall vehicle safety.

DOI: 10.1201/9781003496533-6

Aerodynamic flutter can compromise the structural integrity of aerospace vehicles and may result in catastrophic failures if left uncontrolled. Effective flutter suppression techniques are essential for ensuring the safety and reliability of aerospace systems. Controlling rolling mill chatter allows increases in mill speed and therefore not only improves product quality but also increases mill throughput and efficiency. In space engineering, controlling spacecraft attitude instabilities improves spacecraft manoeuvrability, communication systems, scientific observations and hence enables the success of space missions.

The control of these instabilities will be detailed in the following. Firstly, the suppression of mode-coupled instabilities in various contexts is explored, including railway wheel, brake squeal and aerodynamic flutter encompassing both classical and chaotic phenomena. Then, innovative approaches for mitigating instabilities are demonstrated, on rolling mill chatter, rail corrugation, attitude instabilities in spacecraft and other chaotic instability. This concise exploration aims to provide practical insights for engineers and researchers grappling with instability challenges in diverse engineering domains. Addressing these challenges often requires interdisciplinary knowledge encompassing advanced modelling, analysis and control techniques tailored to specific engineering applications. By advancing our understanding and control of instabilities, scientists and engineers can pave the way for safer, more efficient and technologically advanced engineering systems in the future.

6.1 RAILWAY WHEEL SQUEAL

Controlling classical and chaotic railway wheel squeal is important for successful railway engineering operations. Wheel squeal, characterised by high-pitched noise and vibration, arises from intricate interactions in the tiny contact patch between the wheel and rail surfaces. This phenomenon not only causes discomfort to passengers and nearby residents but also poses significant safety and maintenance challenges. Classical approaches to mitigate wheel squeal often involve modifying track and wheel profiles, adjusting frictional properties and employing damping mechanisms ie [1,2]. Friction modifiers have been employed to control squeal as they can alter a negatively sloped creep curve into a positive one [3,4]. While top-of-rail friction modifiers are effective in reducing squeal noise and lateral forces, studies have shown that they do not always eliminate squeal entirely [3,5,6]. A possible reason for this limitation is the mode-coupling mechanism [7–9], which can induce squeal regardless of the slope of the creep curve. However, friction modifiers were shown to have varying performance under test rig falling friction [6]. To obtain insights into this, an efficient analytical model was developed to model and mitigate wheel squeal under friction modifiers [10,11]. In any case, these methods may not always provide effective and sustainable solutions, especially in complex railway environments with varying operating conditions.

In contrast, understanding and controlling the chaos in wheel squeal has only just begun and involves exploring the nonlinear dynamics and instability growth mechanisms underlying the squealing behaviour. It appears that chaotic squeal occurs at a higher amplitude than periodic limit cycle squeal [12] and therefore may be important to be addressed practically. By integrating classical and chaotic control approaches, railway engineers aim to develop more robust and adaptive solutions for mitigating wheel squeal phenomena and ensuring safer, quieter and more reliable railway operations.

Controlling Wheel Squeal

In Chapter 4, wheel squeal predictions were made using an analytical model that validated and identified the underlying instability mechanisms of falling friction and mode coupling. The strong correlation between predictions based on falling friction and field results, i.e. [13], establishes a reliable foundation for using analytical models to efficiently explore squeal noise mitigation strategies. Given this, the model was utilised to conduct a parametric investigation aimed at significantly reducing falling friction-induced squeal noise through sensitivity analysis. Specifically, the analytical predictions were employed to assess changes in squeal noise sound pressure levels (SPL) resulting from factor variations in critical parameters, as illustrated in Figure 6.1. For clarity, the creep parameters were separately presented in Figure 6.1b.

Figure 6.1a indicates that squeal vibration amplitude is most strongly dependent upon crabbing velocity (i.e., angle of attack) and then rolling speed and critical creep, as indicated in the experimental and field results. Reductions in crabbing velocity and rolling speed below a factor of 1, as well as increases in critical creep above this factor, result in significant reductions in squeal sound pressure levels, typically around 10 dB. This behaviour occurs because the squeal vibration amplitude is directly proportional to how much the angle of attack exceeds its critical creep value. To confirm this relationship, simplified solutions assuming vibration amplitude as follows, $A\omega \propto V_c - \zeta_c V_o$, were plotted at marked points (+, ×, and o) for comparison. The strong agreement between these simplified solutions and the full analysis confirms that this linear relationship dominates changes in squeal noise levels for these parameters.

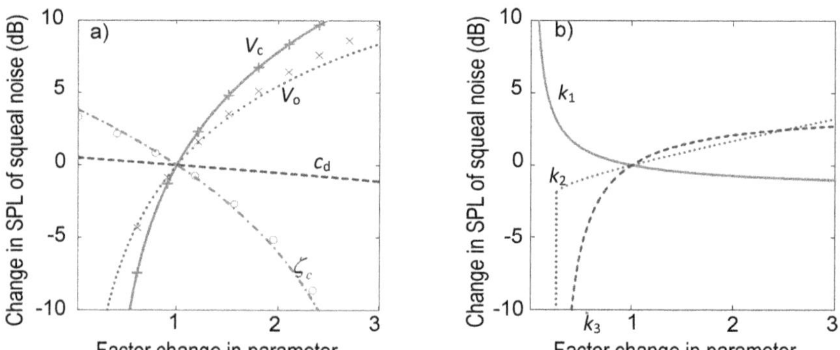

FIGURE 6.1 Sensitivity of sound pressure level of squeal noise using analytical prediction to (a) crabbing velocity (-) rolling velocity (··) modal damping (- -) critical creep (-.-) and creep parameters. (b) slip region slope (-), full sliding region slope (··) and contact parameter (- -). (+, × and o) denote simplified solution proportional to the difference between the crabbing velocity and critical creep [25].

Figure 6.1b illustrates that substantial reductions in squeal vibration amplitude can be achieved by modifying the creep curve, particularly for parameter values below 1. This can be accomplished through top-of-rail lubrication using friction modifiers. Specifically, reducing the slope, k_2, of the full sliding region of the creep curve, below approximately 0.27, stabilises the squeal instability by eliminating the system's negative damping term, $c_d+k_{1,2}k_3/V_0$, thereby suppressing squeal vibrations. However, for values of k_2 above this threshold, squeal noise levels remain relatively insensitive to changes. This suggests that while steady-state limit cycle amplitude is largely unaffected by this parameter, the critical squeal onset and growth rate may still be influenced. A similar trend is observed for creep constants k_1 and k_3, though k_1 exhibits an opposite effect due to its positive slope in the slip region, in contrast to the negative slope of k_2 in the full sliding region. Conversely, Figure 6.1a indicates that modal damping changes have a relatively minor effect on squeal vibration amplitude. This effect is further examined by plotting squeal growth time to steady-state amplitude in Figure 6.2, providing a more detailed analysis of how damping influences the system's response time.

Figure 6.2 shows that squeal noise levels remain largely unaffected by variations in modal damping, c_d, up to a threshold of approximately 3.8. At this point, the modal damping stabilises the squeal instability by eliminating the system's negative damping term, $(c_d+k_{1,2}k_3/V_0)$, thereby suppressing squeal vibrations. Conversely, the squeal growth time (denoted by the o symbol) increases significantly with higher modal damping (i.e., exponentially). The maximum growth time plotted is 2.12 seconds, which remains within the typical duration for a wheel to traverse a curve and reach full squeal noise levels in field conditions. This analysis suggests that increasing structural damping, as a squeal noise control measure, is only effective if it can completely counteract the system's negative damping. However, even if it does not fully suppress squeal, higher damping can significantly extend the time required for

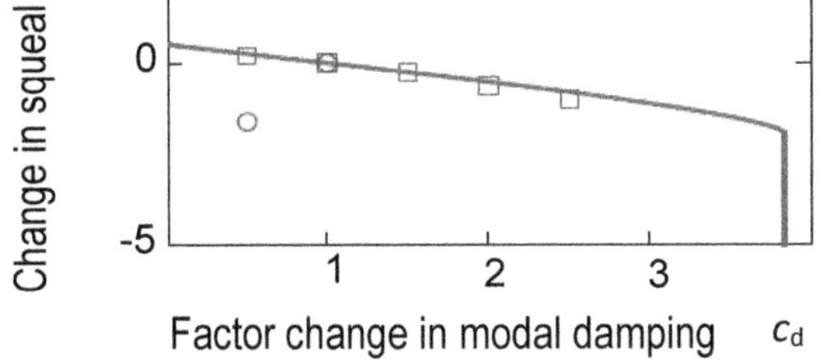

FIGURE 6.2 Sensitivity of sound pressure level of squeal noise to modal damping; using analytical prediction (-) numerical model (□) and time to grow to steady-state limit cycle amplitude (o) [25].

TABLE 6.1
Parameter change required to decrease squeal noise by 6dB[a] [25]

Parameter Description	Change Required[a]
Crabbing velocity (& angle of attack, V_c/V_o) (V_c)	−34%
Rolling velocity (for angle of attack, V_c/V_o constant) (V_o)	−51%
Squeal modal damping constant (c_d)	> +280%
Critical creep (ζ_c)	+100%
Creep curve slip region slope (k_1)	Not Possible
Creep curve full sliding region slope (k_2)	<−73%
Contact parameter (patch size/friction coef) (k_3)	<−50%

Squeal nominal conditions:

V_c	V_o	c_d	W	ζ_c	k_1	k_2
(m/s)	(m/s)	(Ns/m)	(kN)	–	–	–
0.39	17.8	42	1	0.007	0.27	−0.02

[a] *Note:* Parameter change required for Squeal Noise reduction of 6 dB or vibration amplitude reduction of 50%.

squeal to develop, thereby reducing squeal exposure time. To summarise, the percentage changes in key squeal model parameters required to achieve a 50% reduction in squeal vibration amplitude (equivalent to a 6 dB decrease in squeal noise) were calculated and are presented in Table 6.1.

The sensitivity results in Table 6.1 support both theoretical and experimental findings, demonstrating that squeal vibration amplitude is highly dependent on crabbing velocity (or angle of attack). Specifically, a 34% reduction in crabbing velocity (or angle of attack) is sufficient to achieve a 50% decrease in squeal vibration amplitude. This relationship is illustrated in Figure 6.3, where the crabbing velocity moves closer to the critical slip velocity. As a result, the required amplitude of creep oscillations, to balance the energy flow using the positively sloped slip region, is reduced.

Figure 6.3 illustrates how the range of creep oscillations (indicated by arrows) is significantly reduced as the quasistatic lateral creep (crabbing velocity relative to rolling velocity) approaches the critical creep value. Achieving the same 6-dB reduction in squeal noise requires much larger percentage changes in other parameters. Specifically, a 280% increase in modal damping is needed to fully stabilise the system, as predicted by the stability analysis in Chapter 3 and as shown in Figure 6.3. The limited effects of moderate increases in modal damping on squeal noise are consistent with the success of practical attempts to slightly increase wheel damping, though they could still reduce squeal exposure time. Similar trends are observed for the slope of the creep curve in the full sliding region, k_2, and the contact parameter, k_3. According to the squeal stability criteria in Chapter 3, squeal is fully eliminated when these parameters are substantially reduced (by 73% and 50%, respectively). However smaller reductions of these parameters cannot reduce the squeal amplitude by 50%.

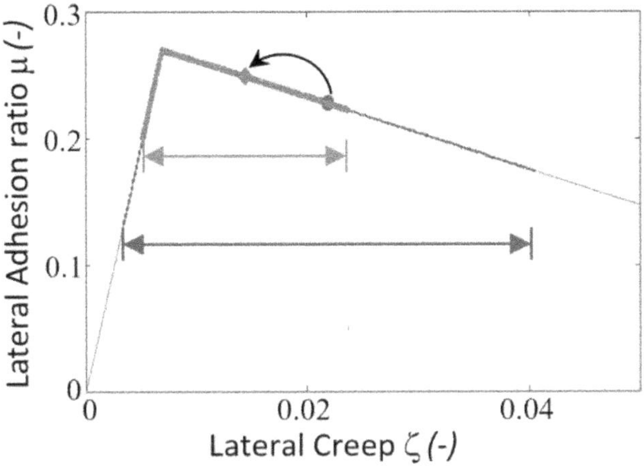

FIGURE 6.3 The approximated creep curve (-), analytical solution for creep oscillations range for nominal conditions (bolder) for crabbing velocity (o) and quasistatic lateral creepage (-) for crabbing velocity reduced by 34% (♦) [25].

In practice, such changes in the creep slope k_2 are achieved by the use of friction modifiers and other lubricants including water under certain conditions. The results of Table 6.1 are consistent with what has been found with the use of such lubricants, in that when successful, squeal is eliminated entirely rather than reduced in amplitude. However, there are complications in using a quasistatic creep curve for squeal prediction under lubricated conditions, as discussed in [6].

The critical creep is another parameter that can be controlled by specialised lubricants [14] or other third-body conditions (i.e., wet track). Large increases of the order of 100% may substantially reduce the amplitude of squeal noise. This is consistent with the squeal creep curve behaviour discussed previously. As the critical creep moves closer to the angle of attack (crabbing velocity on the rolling velocity), the amplitude of creep oscillations required to encounter the stabilising effect of the positive sloped slip region is lower. Hence, the steady-state squeal vibration velocity is also lowered.

The effects of friction modifiers were investigated in more detail in [10,11] including comparisons with experimental results. A rolling contact two-disc test rig was used to measure and control lateral traction/creep curves, angle of attack and sound pressure levels under various friction-modified conditions. The results revealed that instantaneous creep behaviour can deviate from measured quasistatic curves, which may explain why conventional (falling friction) wheel squeal can still persist despite the application of friction modifiers. This finding aligns with previous observations of friction coefficient variations under friction modification at higher temperatures, as reported in [15]. Building on this, [11] investigated the prediction of vibration amplitude and sound pressure levels of wheel squeal noise when friction modifiers were applied. For this study, measured quasistatic traction–creep curves were used to infer critical parameters of the instantaneous creep curve. The resulting

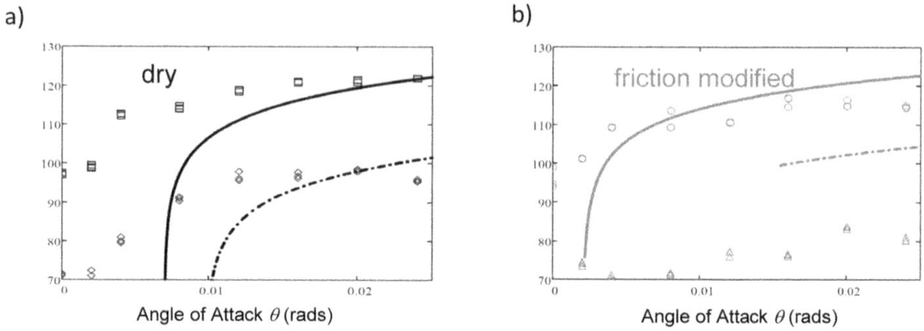

FIGURE 6.4 (a) Analytical at a rolling speed of 17.8 m/s (-) and 2.225 m/s (.-.) and experimental at a rolling speed of 17.8 m/s (□) and 2.225 m/s (◊) sound pressure level vs angle of attack under dry conditions; (b) analytical at a rolling speed of 17.8 m/s (-) and 2.225 m/s (.-.) and experimental at a rolling speed of 17.8 m/s (○) and 2.225 m/s (Δ) sound pressure level vs angle of attack under (oil-based) friction modified conditions. Note the analytical prediction for friction modifiers at a rolling speed of 2.225 m/s (.-.) completely vanishes if the modal damping is increased by 65% [11].

mathematical predictions were then compared with previous measurements [6] of squeal sound pressure levels obtained from a two-disc test rig, across a range of angles of attack and rolling speeds, as illustrated in Figure 6.4.

Reasonably strong correlations are seen between the analytical predictions and the experimentally measured changes in squeal noise levels in Figure 6.4a, due to a substantial speed reduction from 800 to 100 rpm for all angles of attack above the critical creep threshold. At higher angles of attack, the lower speed results in a sound pressure level (SPL) reduction of approximately 20 dB, consistent with the analytical prediction of an approximately proportional reduction with speed. Near the critical creep threshold, the experimental noise levels exceed the predicted values, likely due to the presence of additional noise sources such as rolling noise and operational equipment noise. These extraneous noises become more significant at lower speeds, where the overall sound pressure levels are lower, leading to a greater impact on the measurements. Figure 6.4b further supports the analytical prediction, showing that squeal noise disappears at angles of attack below 15 mrad. This aligns with the low noise levels (<80 dB) observed from other sources. The disappearance of squeal occurs because the system damping becomes positive as the positive slip region and structural modal damping, surpasses the diminishing (with angle of attack) negative damping effect from the full sliding region of the instantaneous creep curve. This result is consistent with the significant reduction in the negative slope magnitude of the creep curve at lower speeds [11]. Above 15 mrad, the analytical model predicts squeal occurring around 100 dB, converse to the experimental measurements. It is noted that by increasing the modal damping by 65% (from 42 to 69 Nms), the squeal vanished for the higher angles of attack as well. Modal damping is difficult to experimentally measure accurately, and the increase of 65% is likely possible due to the effects of rotation of the disc. In particular, the impact hammer measurements of modal damping were taken under stationary conditions. However, specialised non-contact devices have measured increases in modal damping in discs up

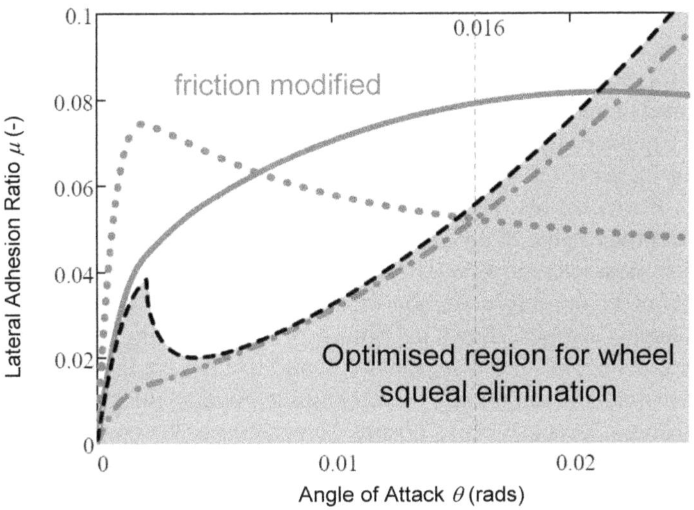

FIGURE 6.5 Optimal friction-modified creep curve critical value (--) and region (greyed out below) to eliminate wheel squeal at all angles of attack; the simulated quasistatic creep curve for (oil-based) friction modifier (tuned to measured creep curve) at a rolling speed of 17.8 m/s (-) that still results in squeal noise (see Figure 6.7b); a solution for a new friction-modified quasistatic curve that eliminates squeal at 17.8 m/s for $K_L = 0.01$, $K_{S1=}0.075$ and $K_{S2=}0.539$ (.-.) and the corresponding instantaneous creep curve at 16 mrad (..) [11].

to 70% due to rotation [16] (explained by aerodynamic damping forces). The presence of oil-based friction modifier on disc surfaces may also have increased this effect. Also, since this change does not affect the other results noticeably (<1 dB), this could explain the observed differences between the predicted and experimental results for friction-modified conditions at low speeds and high angles of attack. This finding also provides indirect confirmation that structural damping control can, in some cases, completely eliminate squeal, rather than merely reducing its limit cycle amplitude, as suggested in [13,17].

With this understanding, the experimental occurrence and non-occurrence of wheel squeal under friction modifiers appear to have been successfully predicted analytically using a newly proposed friction modifier model. The efficient modelling framework was then applied to determine the optimal friction-modified traction–creep curve required to completely mitigate wheel squeal noise under friction-modified conditions. Using the validated predictive model, the friction coefficient was appropriately constrained at a given angle of attack. The analysis revealed that the maximum allowable friction modification, below which squeal is eliminated, is directly proportional to the modal damping. As a result, lower modal damping necessitates a greater reduction in friction to fully suppress squeal. The optimal friction-modified curve, capable of eliminating wheel squeal at all angles of attack based on test rig parameters, is presented in Figure 6.5. In this case, k_2 is derived from the local slope of the instantaneous creep curve, rather than an estimate over a range of creep. This distinction is crucial because squeal onset is dictated by the local slope, whereas the squeal limit cycle amplitude occurs over a range of creep.

Figure 6.5 indicates that although squeal still occurred under the measured friction-modified creep curve at 17.8 m/s, it is possible to eliminate squeal for all angles of attack at this speed by modifying the friction further to the optimised grey region. This no-squeal region envelopes the existing friction modified curve at high angles of attacks and explains why the simulation results showed no squeal when they exceeded 23 mrad (see Figure 6.4b). An example friction-modified curve that eliminated squeal at all angles of attack is also shown along with the corresponding instantaneous creep curve at a nominal value of 16 mrads. This was achieved by modifying the quasi-static friction parameters to $K_L = 0.01$, $K_{S1} = 0.075$ and $K_{S2} = 0.539$. Simulations using this example of an optimised friction modified curve confirm that squeal is eliminated at all angles of attack shown in Figure 6.5. It is interesting to note that the slope in the instantaneous full sliding region is still negative but at a lower magnitude, such that the positive modal damping, c_d, overcomes it. Further, the grey region of squeal elimination does not specify that a positive sloped quasistatic creep curve is required to eliminate wheel squeal. In particular, a negatively sloped quasistatic curve within the grey region could also eliminate wheel squeal. These results highlight that the effect of friction modification on squeal appears to be more dependent upon the overall size (height) of the measured quasistatic creep curve rather than the slope in the full sliding region. Ultimately, this is primarily because the instantaneous creep curve is only quasistatically affected by the friction modification due to the slower time constant (than the squeal vibration period) associated with temperature effects in the contact. This needs to be confirmed with more focussed experiments, although the verification of squeal amplitude, occurrence and non-occurrence with the experimental measurements under friction modification presented in this paper appears somewhat compelling. In addition, the insights gained by the development and use of a closed-form analytical solution have enabled the efficient prediction and mitigation of wheel squeal using friction modifiers. Note a further investigation of water-based friction modifiers has been performed in [10]. This additionally showed that increasing the critical creep achieved substantial reductions (although not complete elimination) in squeal noise levels across a reasonable range of angles of attack while still maintaining moderate friction levels. However, the squeal noise reductions decreased with increased angle of attack. This may explain why water-based friction modifiers, that maintain a moderate friction level for traction and steering, are generally effective for squeal mitigation unless there are high angles of attack [18]. In any case, the realisation of such optimised friction modifiers, perhaps using nanoparticle technology, could be an important avenue of further research. Also, control of mode-coupled wheel squeal could be researched further in the future, although at present, its manifestations in the field appear to be rare and limited to the outer wheelset [19].

Finally, we note the control of wheel squeal chaos can be performed in a similar manner using an efficient analytical model to perform a wide sensitivity analysis of chaotic amplitude to the identified critical parameters. However, the elimination of wheel squeal limit cycles, demonstrated in this section, will also eradicate any chaos which follows as it is driven by the same instability mechanisms. There may however be cases where chaos can be eliminated, leaving a lower amplitude limit cycle squeal which can be investigated using the predictive algorithms developed in

Chapter 4. We note this seems to be more likely with the high growth mechanism of mode coupling [12]. This is investigated in more detail subsequently in controlling brake squeal.

6.2 BRAKE SQUEAL

Controlling classical and chaotic brake squeal represents a significant challenge in automotive and industrial engineering. We have shown in the previous chapters that brake squeal is characterised by high-frequency vibrations and noise emissions during braking and arises from complex interactions between brake components, frictional forces and environmental conditions. Classical approaches to mitigating brake squeal typically involve altering brake pad materials, optimising rotor designs and adjusting operating parameters. In the following section, we will show why and how these control measures work based on the underlying efficient predictive models developed previously. While these methods can effectively reduce squeal under certain conditions, they may not address the chaotic dynamics inherent in the braking system.

Understanding and controlling brake squeal chaos requires an understanding of the nonlinear dynamics and instabilities governing the braking process. Chaotic vibrations in brake squeal essentially emerge from high-growth-coupled self-excited oscillations and nonlinear frictional effects within the brake system. Hence, control can be focussed on mitigating any one or more of these two fundamental characteristics. Therefore, the efficient closed-form solutions detailed in Chapters 3 and 4 can be used to investigate a wide range of parameters to mitigate its occurrence. By integrating classical and chaotic control techniques, we can develop more robust and adaptive solutions for mitigating brake squeal, enhancing braking performance and ensuring quieter and safer operation of automotive and industrial braking systems. This can be achieved using the closed-form modelling and solutions to provide important insights into the occurrence and mitigation of the underlying mechanisms of brake squeal.

Controlling Classical Brake Squeal

The control of brake squeal in a real brake assembly was demonstrated on a full brake system with detailed experimental measurements [20] using the 2-degree-of-freedom sprag model shown in Chapters 3 and 4. Note that many cases of squeal involving multiple rotor and/or calliper modes can be represented by this modelling. The comprehensive set of brake squeal experiments are detailed in [20] and are summarised here for convenience. A quarter vehicle dynamometer test was first performed to identify the squeal noise frequency, brake pressures, speed and braking conditions under which brake squeal occurred on a floating calliper brake assembly. A static assembly modal test was then performed to determine the dominant modes of vibration under a range of pressures for the brake assembly. Further operational deflection shape measurements were performed using accelerometers and a laser scanner under dynamic (rotating) conditions to confirm the modal analysis. Subsequent analysis and design to reduce brake squeal were verified on the brake dynamometer. The nominal case for squeal on this real brake system was shown to be caused by stiffness mode coupling.

This was due to the closeness of coupled in-plane radial (IPR) and out-of-plane diametral (OPD) modes of the rotor which was dependent on the brake pressure, i.e., contact stiffness [20,21]. This nominal case was investigated for control.

The effect of all critical parameters on the occurrence of squeal in this case is explored further in the following. In particular, the % change in nominal value required to eliminate or reduce squeal amplitude by 6dB is determined using the verified analytical solution provided in Chapter 3, as shown in Table 6.2.

The sensitivity results in Table 6.2 confirm the previous theoretical and experimental observations that the occurrence of the stiffness mode-coupled squeal is strongly dependent on the closeness of the uncoupled modes. In particular, a number of important points are identified as follows:

- Relatively small changes in modal mass or stiffness (>15%) of both modes eliminate stiffness mode coupling by causing the determinant Δ to change its sign to positive. This confirms the successful investigations in Park et al. [20], in which squeal was eliminated by making small structural changes to the rotor.

TABLE 6.2
Parameter change required to eliminate or decrease squeal noise by 6dB [21]

Parameter Description	Change Required for (−6 dB/Elimination)	Example Practical Mitigation Method (Control/Change in)
	Structural Vibration Parameters	
Modal mass (for the sprag angle, θ constant) (m_a, m_b)	+15%, −10% (elim)	Rotor/calliper structural design
Modal stiffness (for the sprag angle, θ constant) (k_a, k_b)	−10%, +15% (elim)	Rotor/calliper structural design
Modal damping (for the sprag angle, θ constant) (c_a, c_b)	+151%, +163% (elim)	Structural/material/shim damping
Contact stiffness (for the sprag angle, θ constant) (k_H)	+119% (elim)	Material/geometry/surface roughness
Sprag angle (θ)	−86%, +39% (elim)	Structural geometry/dynamic parameter
	Friction Parameters	
Static friction coefficient constant (μ_s)	−24% (elim)	Pad material/friction modifier
Critical sliding ratio (ζ_c)	Not possible	–
Stick slope ($k_{\mu 1}$)	Not possible	–
Sliding slope ($k_{\mu 2}$)	−31,000% (elim)	Pad material/friction modifier
Brake pad angle of attack (θ_A)	−50% (−6 dB)	Structural geometry/pad end condition
Nominal sliding velocity (for angle of attack, θ_A constant) (V)	−49% (−6 dB)	Braking functional requirement (difficult)
Normal loading (N)	Not possible	–

- Larger changes in contact stiffness are required to detune the modes due to less direct and combined effects on the coupled modal frequencies.
- Larger changes in modal damping are also required (>+163%) to damp out the negative damping due to the high complex stiffness due to stiffness mode coupling.
- Another key parameter to eliminate squeal is the sprag angle [22]. Substantial increases (>+39%) or decreases (<–86%) in the sprag angle, θ, can eliminate squeal by violating the necessary sprag condition defined in Chapter 3. Physically, this could be achieved by changing the structural geometry and or dynamic parameters. The same effect can be achieved by reversing the sliding direction.

The friction parameters generally have a lesser effect on squeal apart from the friction coefficient:

- As found in the previous results of Chapter 3, reductions in the friction coefficient (<–24%) can eliminate squeal by firstly reducing the negative damping from stiffness mode coupling. For larger reductions, it eliminates stiffness mode coupling and spragging altogether (although this may be undesirable for brake performance).
- Changes in the critical sliding ratio (where friction is maximised), the stick slope and normal contact load are not able to reduce or eliminate squeal. This is because falling friction has only a small role to play compared to stiffness mode coupling, in this case.
- In particular, the sliding slope needs to change its sign to positive and increase in magnitude by over 309 times in order to dampen out the negative damping from stiffness mode coupling. This magnitude is not physically achievable using brake pad friction modifiers or other third-body conditions (i.e., wet brakes) in this case, but could be under other parameter conditions where the magnitude of the complex stiffness is relatively low.
- In contrast, the squeal vibration amplitude can be reduced by approximately 50% with 50% reductions in sliding velocity, V, or brake pad angle of attack, θ_A, [23,24] in accordance with the predictive algorithm described in Chapter 3. This is because the sliding amplitude reduces as the angle of attack or nominal sliding velocity reduces as it moves closer to the stabilising effect of the positive-sloped stick region [25]. This could be achieved by careful control of the brake pad centre of pressure.

The results highlight that the analytical model can be used to very quickly investigate a wide range of mitigation methods for brake squeal.

Controlling Brake Squeal Chaos

In Chapter 4, brake squeal chaos was shown to be firstly determined by the onset of stiffness mode coupling and secondly large nonlinearities in the friction curve due to negative sliding [26–28]. Suppression of chaotic instability can therefore be achieved using the quantitative understanding of the parameter effects on verified criteria developed (or alternative techniques ie [29–31]). Hence, here the suppression of the onset of chaotic instability in the real brake squeal model was explored further in the following Table 6.3 with nominal damping at 10% of the values of Table 6.1 in [26]. In particular, the % change in nominal parameter value required to avoid chaotic instability by ensuring the critical static friction coefficient exceeds realistic values of $\mu_{crit} > 1$, was determined, using the verified analytical criterion for negative sliding developed in Chapter 4. Note, in this investigation of suppression, the friction sliding slope, $k_{\mu 2}$, is held constant as the friction coefficient changes, to isolate its effects.

Table 6.3 highlights multiple means of eliminating chaotic instability in brake squeal by parameter control. In general, the results appear to reflect that chaotic instability is dependent upon the stiffness mode coupling occurrence, which in turn is dependent upon how close the uncoupled mode frequencies are. In particular, the following observations can be made from the results of Table 6.3:

TABLE 6.3
Parameter change required to suppress brake squeal chaotic instability to excessive friction levels, $\mu_{crit} > 1$ [26]

Parameter Description	Change Required for Chaos Elimination ($\mu_{crit} > 1$)	Example Practical Suppression Method (Control/Change in)
	Structural Vibration Parameters	
Modal mass (for the sprag angle, θ constant) (m_a, m_b)	−62% or +29%, −23% or +166%	Rotor/calliper structural design
Modal stiffness (for the sprag angle, θ constant) (k_a, k_b)	−24% or +88%, +31%	Rotor/calliper structural design
Modal damping (for the sprag angle, θ constant) (c_a, c_b)	+6000%, +7160%	Structural/material/shim damping
Contact stiffness (for the sprag angle, θ constant) (k_H)	−77% or +2300%	Material/geometry/surface roughness
Brake pressure (p)	−72% or +346%	Braking functional requirement
Sprag angle (θ)	−98% or +158%	Structural geometry/dynamic parameter
	Friction Parameters	
Critical sliding ratio (ζ_c)	Not possible	–
Sliding slope ($k_{\mu 2}$)	−789,700%	Pad Material/friction modifier
Brake pad angle of attack (θ_A)	−95%	Structural geometry/pad end condition
Nominal sliding velocity (for the angle of attack, θ_A constant) (V)	−99%	Braking functional requirement (difficult)
Normal loading (N)	+6100%	–

- Brake squeal chaos can be suppressed by small changes in modal mass or stiffness (>±31%) by greatly reducing the squeal growth rate due to stiffness mode coupling. These results are similar to means of avoiding brake squeal entirely, as found in [21].
- Very large changes in modal damping are needed (>+73x) to override the negative damping due to the very high complex stiffness from stiffness mode coupling [32].
- Large changes in contact stiffness are required (>–77%) to eliminate chaos as it has a less direct effect of detuning the modes to lessen stiffness mode coupling.
- Brake pressure has a stronger effect than contact stiffness as it affects this and the sprag modal properties. As the brake pressure necessarily changes as the brakes are applied over a full range, it is difficult to avoid chaos by controlling this parameter.
- Brake squeal can be eliminated by avoiding the sprag condition for mode coupling [21] by large increases (>+158%) or decreases (<–98%) in the sprag angle, θ. The sprag angle is determined by the mode shapes and brake structural geometry.

Brake squeal chaos suppression via friction parameter control is generally harder to achieve according to the following:

- Brake squeal chaos cannot be eliminated by controlling the critical sliding ratio (at maximum friction) because the falling friction mechanism is secondary in this case compared to stiffness mode coupling.
- An extremely large increase in magnitude (>7898x) and change in sign of the sliding slope is required for positive friction to dampen the negative damping from stiffness mode coupling. In this case, this change is not realistic due to the very high complex stiffness.
- Chaotic instability can be avoided by large decreases in sliding velocity, V, or brake pad angle of attack, θ_A, to reduce squeal amplitudes by moving the initial sliding oscillations closer to the positive damped stick region [25].
- Very large increases (>62x) in normal loading are required to avoid brake squeal chaos, in this case. Although normal loading may affect the stability characteristics in a way not modelled by the criteria.

These results show the conservative analytical criteria can very efficiently evaluate and quantify several brake squeal chaos suppression techniques. This is without the need for lengthy and extensive multiple numerical integrations of the nonlinear equations of motion. Note that Table 6.3 summarises results for directly eliminating chaotic instability, not necessarily the brake squeal limit cycles which can still occur under positive sliding conditions. By inspection of Figure 9 in [26], it may be deduced that suppressing chaotic squeal to limit cycles can substantially reduce vibration velocity amplitudes by the order of 6 times, leading to large reductions in sound pressure levels of approximately 15 dB (using the squeal sound radiation model in [21]). The frequency content of the squeal limit cycle will be necessarily more tonal than the squeal chaos (as found in [33]).

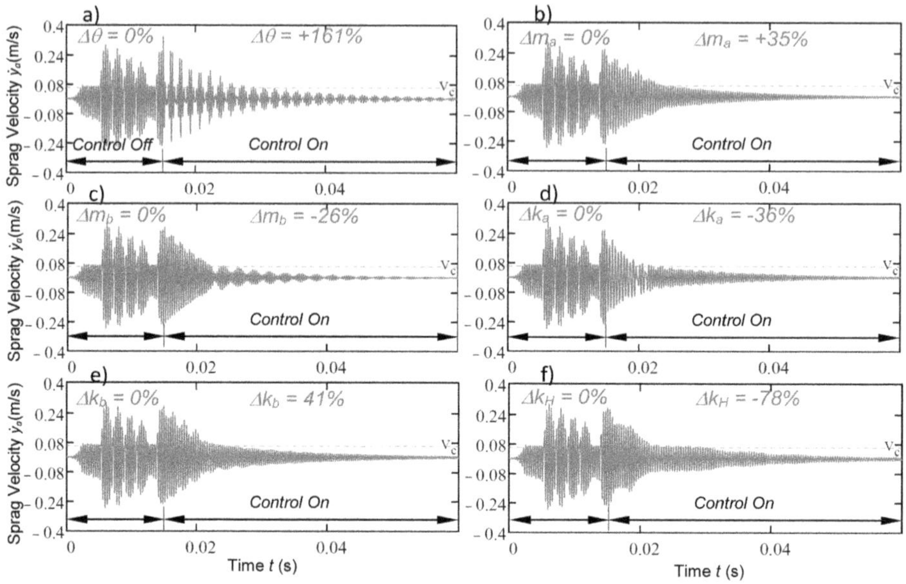

FIGURE 6.6 Simulated time history of squeal sprag mode vibration velocity (-) for a real brake system showing control of chaotic oscillations starting at 0.015 seconds by changing (a) sprag angle; (b) modal mass *a*; (c) modal mass *b*; (d) modal stiffness *a*; (e) modal stiffness *b*; (f) contact stiffness under modal damping $c_a = 360$ Ns/m and $c_b = 3.613c_{a.}$

To confirm these predictions, the smallest (most feasible) control parameter changes were performed on the full nonlinear equations of motion, after large chaotic oscillations occurred at 0.015 seconds, as shown in Figure 6.6. The results show the chaotic squeal is eliminated in all cases. The magnitude of the parameter changes required were found to be slightly higher than the predictions (up to 12% for the modal stiffness), although both chaotic and limit cycle squeal were mitigated. These results highlight the efficiency (almost instantaneous) and accuracy of the analytical model-based predictions for control of brake squeal.

6.3 AERODYNAMIC FLUTTER

Flutter, an aeroelastic phenomenon, poses a significant threat to the structural integrity and stability of aerospace vehicles and structures. Its occurrence arises from the interaction between aerodynamic forces and structural dynamics, leading to self-excited oscillations. Controlling flutter is imperative to ensure the safety and performance of aircraft and other fluid loaded structures.

Typical methods for controlling flutter primarily involve structural modifications such as stiffening or adding damping to alter the natural frequencies of the system [34]. These techniques have been effective but can add weight and complexity to the structure, impacting fuel efficiency and manoeuvrability. Composite materials offer a promising solution due to their high strength-to-weight ratio and flexibility in design. By tailoring material properties, designers can mitigate flutter while minimising weight penalties. Passive control techniques, such as the use of linear and nonlinear dynamic vibration absorbers, have been used to change the stability behaviour and increase the flutter speed and reduce the amplitude of the oscillations [34,35,36–39]. Other approaches integrate active components [40], advanced materials and adaptive control systems [41]. Active flutter control [42] enables real-time adjustments to active aerodynamic surfaces and distributed control actuators to counteract flutter-inducing forces. These systems rely on sophisticated algorithms and sensors to detect and respond to flutter conditions swiftly [43]. Recent research also explores the integration of machine learning algorithms for predictive flutter control. By analysing vast amounts of data from flight tests and simulations, these algorithms can anticipate flutter onset and optimise control strategies pre-emptively.

Here, we apply the insight and predictive modelling obtained on flutter to perform passive control of both classical and chaotic flutter instabilities in a wind turbine airfoil example.

Controlling Classical and Chaotic Flutter Instability

Efficient closed-form conservative analytical conditions for classical and chaos following flutter were identified and discussed in Chapters 3 and 4. These were tuned to a case of a wind turbine section considering the blade geometry and optimal design of the wind turbine. These predictions were verified for a range of conditions including stall slope and damping. The results confirm that flutter arises due to a mode coupling instability, and chaos following flutter can occur due to nonlinearities in the aerodynamics, i.e., due to a bilinear lift law [44,45] or other means ie [46–51]. In particular, the results in Chapter 4 confirm that chaos following flutter is characterised by conventional (stiffness mode coupled) flutter combined with parametric excitation nonlinearities associated with lower damping and sharper stall slope. These results also highlight that the analytical predictions can remarkably be generalised to any parameter set. They also provide almost instantaneous calculations representing many thousands of numerical simulations from many bifurcation diagrams (computational acceleration factor of 10^7 times). Here, we investigate suppression of classical flutter and chaos flowing flutter to unrealistically high wind speeds and/or elimination by quantified variation of system parameters using the predictive model, as detailed in [45].

Methods to avoid or suppress the chaos following blade flutter onset, in a wind turbine, were investigated quantitatively in Table 6.4 with the chaos following flutter conditions of Figure 4.10b, i.e., sharper stall slope (smoothing factor $\beta=20$, $k_{\text{stall}}=1.5$). Specifically, the % change in nominal parameter value required to eliminate or suppress chaos following flutter to wind speeds above the very unlikely value of 30 m/s was calculated using the verified analytical criterion for no overlapping parametric resonances (4.22)–(4.25). Note, in this suppression investigation, the damping ratio is assumed to be constant (unless directly changed) irrespective of the changes in natural frequency. Also, the values may differ slightly from [45] as the slight effect of wind speed via reduced frequency K has been taken into account.

Table 6.4 illuminates various ways of controlling wind turbine parameters to eliminate flutter instability. Elimination of chaos following flutter also means elimination

TABLE 6.4
Parameter change required to suppress chaos following flutter of a wind turbine blade to excessive wind speeds, $U_\infty > 30\,\text{m/s}$ [45]

Parameter Description	Change Required for Chaos Elimination ($U_\infty > 30\,\text{m/s}$)	Example Practical Suppression Method (Control/Change in)
	General Wind Turbine Blade Parameters	
Tip speed ratio (λ)	−15%	Turbine speed control setting
Blade length (R)	NP	Wind turbine size.
	Structural Vibration Parameters	
Mass, Mass moment of inertia: (m, J_α)	−81% or +191%, −81% or +2843%	Blade section design, i.e., member thickness, weighting or material.
Stiffness: (k_h, k_α)	+2429%, −97% or +107%	Blade section design, i.e., member thickness, stiffening or material.
Damping: (c_h, c_α)	NP, +100%[a]	Structural/material damping. Added vibration damping.
Coupled inertia term (S_x)	−80% or +131%	Blade section design to shift elastic centre rel. to centre of gravity.
Chord length ($B = 2b$)	+55%	Wind turbine size/blade section design.
Bending mode shape amplitude (ϕ_{b1})	−50%	Design (mode) coupling higher modes.
Torsional mode shape amplitude (ϕ_{t1})	NP	Design (mode) coupling higher modes.
	Aerodynamic Parameters	
Normalised position of the elastic centre (a)	+65%	Blade section design to shift the elastic centre.
Nominal stall slope factor of $H_3^a(k_{\text{stall}})$	−17%[a]	Blade section outer shape design.

[a] Denotes a case where chaos is eliminated but not flutter.

of flutter in most cases, except those indicated with *. Overall, the findings suggest that the occurrence of chaotic instability is influenced mainly by stiffness mode coupling or classical flutter, which in turn depends on the proximity of the uncoupled mode frequencies. However, the avoidance of overlapping parametric resonances is more important for chaos control using damping and stall conditions. The following discussion of the Table 6.4 predictions, for elimination or suppression of flutter and chaos following blade flutter, is provided:

- The tip speed ratio has a strong effect on both forms of flutter via its effect on the onset of stiffness mode coupling. In particular, lowering the tip speed ratio directly lowers the local wind speed at the tip to below the onset of flutter.

Lowering the tip speed ratio also increases the tip chord via optimal design, which in turn increases the critical flutter speed.

- The blade length has no effect on the critical flutter speed under optimal design conditions. This is because increases in blade length cause increases in tip modal masses and corresponding decreases in the fundamental natural frequencies while the modal stiffnesses remain constant. These effects on the critical flutter speed cancel each other, resulting in the surprising prediction that under optimal wind turbine design, the blade length has no effect on the critical flutter speed. Hence, it also has no effect on the onset of chaos following flutter (if the damping ratios remain unaffected). Note, alternatively, chaos following flutter is more likely to occur if the damping is held constant as blade length increases since the damping ratio will be lowered.
- Wind turbine blade flutter and chaos can be suppressed by changes in modal mass or stiffness by separating (detuning) the plunging and pitching natural frequencies or lowering the modal mass according to [45]. This increases the critical flutter speed due to stiffness mode coupling.
- A reasonable increase in the pitch modal damping (>+1x) damps out any chaotic overlapping harmonic resonances from parametric excitation, caused by the bilinear lift curve. Note unrealistically large increases in damping would be needed to eliminate wind turbine flutter due to stiffness mode coupling.
- Changes in the coupled inertia term dictate the position of the centre of gravity with respect to the elastic centre. This in turn directly affects the dynamic divergence, which is a necessary driving mechanism for wind turbine flutter [44].
- Increases in chord length changes the airfoil section size, which changes several dynamic parameters. If the natural frequencies are held constant, the modal mass will increase. This, in turn, will increase the critical stiffness mode-coupled flutter speed and hence also suppress chaos following flutter.
- The bending mode shape amplitude determines how much unstable aerodynamic force is transmitted to the mode. Hence, decreases in this amplitude via the coupling of a higher mode may be possible to increase the flutter critical velocity.

Wind turbine chaos suppression via aerodynamic parameter control is generally easier to achieve if the blade airfoil section can be redesigned according to the following:

- The position of the elastic centre is another critical parameter associated with dynamic divergence mechanism of airfoil flutter [44]. Moving the elastic centre further away from the midchord reduces the dynamic divergence criteria via the flutter derivatives, which will increase the critical stiffness mode-coupled flutter speed.
- Only small decreases (−17%) in the stall slope factor are required to eliminate harmonics overlapping due to parametric excitation via the criterion

for chaos following flutter. This is well within measured experimental variations at high angles of attack near or in the stall region [52], so larger decreases may be required to reduce the experimental uncertainty.

The results show that small changes in the tip speed ratio (−15%) and stall slope factor (−17%) can eliminate or suppress chaos following flutter. In general, larger magnitude changes in dynamic parameters (i.e., mass, inertia >81%, stiffness >97%, damping >100%) are required to achieve the same, by detuning the coupled plunge and pitch natural frequencies or damping out overlapping parametric resonances. To confirm these predictions, the smallest (most feasible) control parameter changes were performed on the full nonlinear equations of motion, after large chaotic oscillations occurred at 0.015 seconds, as shown in Figure 6.7. The results confirm the chaotic flutter is eliminated to no vibrations in all cases except with control of damping and stall angle. In these cases, the chaotic flutter is eliminated to a limit cycle flutter with phase space shown in Figure 6.7c and f. The magnitude of the parameter changes required was found to be slightly higher than the conservative predictions (up to 12% for the modal mass).

These results show the conservative analytical criteria can very efficiently investigate and evaluate several suppression techniques for flutter and chaos following flutter, without the need for lengthy and extensive multiple numerical integrations of the nonlinear equations of motion. By inspection of Figures 4.10b, 6.7c and f, it may be deduced that suppressing chaos following flutter to limit cycles only slightly reduced vibration velocity amplitudes. Therefore, although the * cases of increasing damping and decreasing

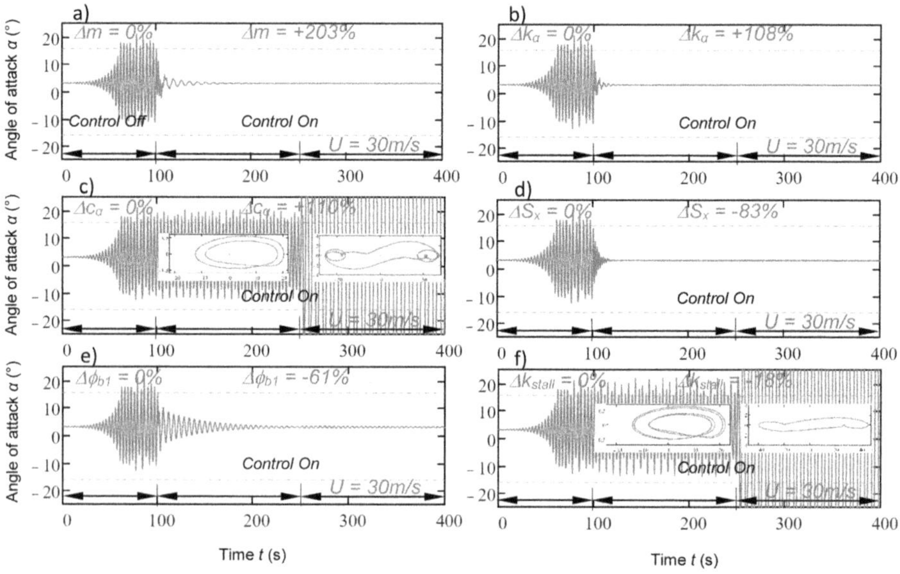

FIGURE 6.7 Simulated time history of flutter (-) for a wind turbine blade tip showing control of chaotic oscillations starting at 0.015 seconds by changing (a) modal mass; (b) torsional stiffness; (c) torsional damping; (d) coupled inertia term; (e) fundamental bending mode shape at the tip; (f) stall slope.

stall slope required generally smaller changes, it may be more desirable to eliminate flutter and chaos using the other means in Table 6.4. Although if annoyance from flutter noise is an issue, the frequency content of the flutter limit cycle will be necessarily more tonal than the chaos following flutter (as found in [33]). These results also highlight that the analytical predictions can remarkably be generalised to any parameter set. Also, they provide almost instantaneous calculations representing many thousands of numerical simulations from many bifurcation diagrams almost instantaneously (computational acceleration factor of 10^7 times). General insight is also provided into the occurrence and suppression of airfoil chaos following flutter in aeroelastic structures like wind turbines.

6.4 ROLLING MILL CHATTER

Controlling rolling mill chatter, a vibration instability that causes strip thickness variations and/or mill and strip corrugations, is essential for maintaining product quality and operational efficiency. The avoidance of chatter restricts mill speeds and causes excessive mill maintenance to regrind surface defects also leading to reduced equipment lifespan. Researchers have investigated both passive and active control methods to mitigate rolling mill chatter.

Passive control methods focus on structural enhancements, damping techniques and lubrication. Adjusting roll configurations, modifying mill stiffness and employing damping materials have been somewhat effective in reducing chatter [53]. For example, increasing the damping capacity of the rolling mill housing or using tuned mass dampers can absorb vibrational energy, thereby decreasing the speed at which chatter first occurs. Tlusty et al. [54] and Critchley and Paton [55] have investigated tuned mass dampers for this purpose, although the optimal tuning of the damper parameters on multiple stands was not investigated. Also, these methods often have limited adaptability and may not be sufficient for all operating conditions.

Active control methods utilise sensors and actuators to dynamically counteract or actively damp vibrations [56]. Advanced control algorithms, such as adaptive and feedback control systems, have been developed to respond to real-time vibration data. These systems can adjust rolling parameters, like roll gap or force, to suppress chatter. For instance, hydraulic actuators can apply counteracting forces to stabilise the rolling process, although there are limitations on accurate actuator frequency response [57]. In [58], an intelligent lubrication control system for preventing chatter was proposed and tested successfully. This was achieved by feedback control of the plate-out oil film to balance the friction coefficients between the final stand and preceding stand.

Perhaps, combining structural modifications with sophisticated real-time control technologies appears to be the most effective approach in the future for mitigating rolling mill chatter. In any case, the success of the control method depends on a deep and efficient model-based understanding of the instability mechanism of rolling mill chatter. This, in turn, ensures higher quality output and prolonged equipment life.

Controlling Rolling Mill Chatter

In Chapter 3, third-octave mode chatter was shown to be due to a vicious cycle between the roll force, vertical roll vibrations and interstand tension dynamics, as depicted in the block diagram shown in Figure 3.24 [53,59]. Rolling mill chatter can therefore be suppressed by breaking or retarding the energy of this cycle, i.e., stabilising it. The chatter predictive criterion was based on a stability analysis, so suppression can be achieved by suitable control of the parameters involved.

Hence, using the chatter criterion (Chapter 3) and simulation model for typical parameter values in a steel rolling environment, the percentage changes in key rolling parameters required to produce a 10% increase in the critical third octave rolling speed were determined. These are presented in Table 6.5. The rolling conditions chosen are provided in Table 6.5 as Item 2 and represent a schedule for typical five-stand tinplate mill rolling thin hard product, known to have a third-octave chatter problem. The penultimate stand of the mill was found to determine the critical speed at which chatter first occurred and is denoted as stand i.

In general, the sensitivity results in Table 6.5 confirm the observations made directly from the chatter stability criterion discussed in the previous section. For instance, the results confirm that in general, an increase in the inherent damping in the rollstack causes an increase in critical speed. In particular, it is interesting to note that increasing the rollgap damping, c_3, appears to be more effective than increasing the damping at the top of the rollstack, c_1. The effects of changes in the

TABLE 6.5
Third Octave Critical Speed Sensitivity to Process Parameters [59]

Parameter Description	Parameter Change[a]
Damping top-BUR (c_1)	+30%
Rollgap damping (c_3)	+15%
Strip Young's modulus (E)	−10%
Sensitivity of roll force to entry tension ($\partial P/\partial T_f$)	−10%
Thickness reduction $\left(H_i/h_i\right)$	−10%
Interstand length (L)	+10%
Stiffness top BUR (k_1)	+15%
Stiffness BUR-WR (k_2)	−35%
Workroll flattening stiffness (k_w)	−50%
Backup roll mass (m_1)	−15%
Strip stiffness ($\partial P/\partial h$)	+15%
Upstream nominal tension force (T_f)	−10%

[a] *Note:*

1. Parameter change required to increase the chatter speed by 10%.
2. Rolling conditions:

	Parameter					
Stand	**h (mm)**	**T_f (kN)**	**t_f (kN)**	**P (kN/mm)**	**μ (-)**	**W (mm)**
$i-1$	0.43	90.2	62.8	8.62	0.03	1000
i	0.27	62.8	31.4	7.55	0.022	1000

strip modulus, E, and interstand length, L, are due to the resultant changes they cause on the interstand tension response time constant, which in turn directly affects parameter a (see equation (6.2)). Table 6.5 also highlights a high sensitivity of critical millspeed to the sensitivity of rollforce to entry tension and exit thickness. These parameters have a highly significant effect on the onset of chatter, primarily as they proportionally determine the amount of feedback between the interstand tension and rollstack dynamics associated with parameter K_1 in (Chapter 3). It may also be noted that changing the stiffness and mass in different components of the rollstack has variant effects on the critical speed, depending upon the location. This would appear to be associated with the different impact each component has on the modal characteristics of the rollstack dynamics for third-octave chatter, in particular, the chatter mode frequency. For instance, a change in the mill frame stiffness, associated with k_1 , results in a larger change in the chatter frequency than a change in the roll contact stiffness, k_2, between the backup roll and work rolls. Therefore, this has a more significant effect on the chatter critical speed. In fact, it may be shown that the chatter critical speed is approximately proportional to the cube of the chatter frequency. The derivation of this result is provided in [53].

Using the key insights provided, a passive control device was developed specifically to suppress the occurrence of chatter on a rolling mill [53]. The effective increase in roll stack damping provided by the device suppresses the occurrence of third-octave chatter (or shudder) to higher rolling speeds. The specialised pistons replace the conventional backup roll balance pistons but do not affect normal roll balance operation. The device is tuned such that it only dissipates energy in a frequency range centred on the chatter frequency. As such, it does not adversely affect AGC/HGC replacing the existing pistons. The performance is only limited by the amount of space available for the pistons.

Figure 6.8 shows the performance based on a simulation of the tinplate tandem mill, which was tuned using the chatter measurements collected at that facility, as described in the previous section. The indicated range of measured chatter frequency is shown as the shaded region on the graph, which provides the estimated level of performance across the product range. The graph shows the performance of two basic designs which have different mass ratios, μ_D. This is defined as the percentage cumulative mass of a set of four vibration-inhibiting pistons (or VI Pistons) with respect to the third-octave modal mass of the rollstack. The first design has a mass ratio of 1% representing the ideal design for a typical rolling mill. For this design, two different tuning approaches were investigated: labelled fine and broadband tuning. The aim of broadband tuning is to broaden the frequency range over which the performance remains above an acceptable high level. The tuning method involves deliberately tuning each VI piston separately to frequencies within a bandwidth centred about the optimal value. Figure 6.8 displays the effect of bandwidth tuning as compared to fine tuning to the optimal frequency.

The performance is based on the conditions described previously in Table 6.5 for a five-stand tandem mill rolling thin, hard product (grade a)). The performance of broadband tuning has shown a range of improvements from 25% to 31% increase in chatter free rolling speed across the critical product range as compared to a range of 20%–50% for fine tuning. It may be deduced from these results that the tighter the

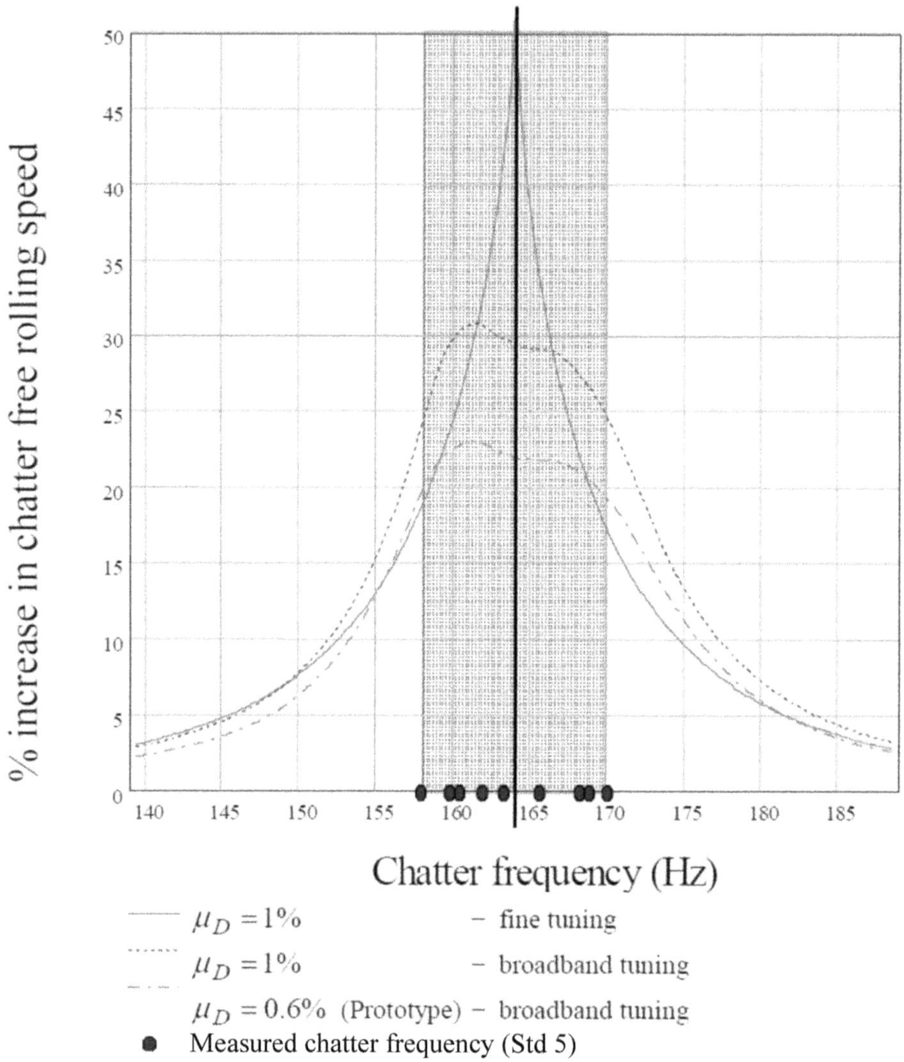

FIGURE 6.8 Performance of a set of vibration-inhibiting pistons (tuned mass dampers) for different frequency tuning methods [53].

variance in mill chatter frequency and tolerances of tuning parameters, the smaller the tuning bandwidth can be. Therefore, the higher is the performance that can be gained. The second, more economical design has a mass ratio of 0.6% and has been built as a prototype for the tinplate tandem mill described previously. The predicted performance of broadband tuning has shown a range of improvements from 20% to 23% increase in chatter free rolling speed across the critical product range.

In order to verify these results, a set of four VI piston prototypes were installed on the five-stand, steel tinplate tandem mill described in Chapter 3 and trialled over a period of 2 months. Installation simply involved replacing the existing roll balance cylinders in the bottom backup roll chocks of the last stand with the specialised VI pistons in the roll shop. Subsequently, the backup roll was installed in the last stand of the five-stand tandem mill during a normal scheduled backup roll change. The vibration-suppressing effect of the VI Piston prototypes is designed only to function when the roll balance is pressurised. Thus, the trials involved pressurising or depressurising the roll balance during the rolling of chatter critical product and adjusting the rolling speed to maintain the vibration levels at the same highest, chatter free level. The performance of the VI piston prototypes was then given by the resultant percentage change in millspeed required to avoid chatter. Figure 6.9 displays the results obtained for two coils rolled for the most critical chatter product rolled during the trial period.

The vibration amplitude shown is for the acceleration of the bottom backup chock in stand 5. Typically, the millspeed was previously restricted to 600–800 m/min during rolling of this product. For the first coil depicted in Figure 6.9, the roll balance is initially depressurised. A severe chatter problem is encountered initially as the millspeed is increased from 630 m/min, as indicated by a sharp increase in the vibration level, resulting ultimately in the millspeed being restricted to 670 m/min. Shortly after, the roll balance is pressurised, allowing the millspeed to be increased to a much higher level between 850 and 900 m/min. Thus, a 25%–35% increase in millspeed was obtained. Similarly, for the second coil in Figure 6.9, a 25%–32% increase in mill speed was obtained. In general, over a range of the chatter critical products, the performance of the VI Piston prototypes varied from 5% to 35%. The variation in performance was attributed in part to the fact that in some cases, chatter occurred primarily on stand 4, in which no VI pistons had been installed. Typically, it would be desirable to install VI pistons in at least two stands for a tandem mill to obtain a more consistent performance. Also, for some products, a high level of vibration at about 200 Hz was encountered, which was outside the tuning range of the prototype pistons. This vibration appeared to emanate from a different source as it did not result in any significant thickness variations or audible noise levels, typically characteristic of third-octave chatter.

The third-octave chatter model and stability criterion has proven critical in providing key insights into chatter phenomena and has successfully predicted rolling speeds for the onset of third-octave chatter across a wide range of products encountered on

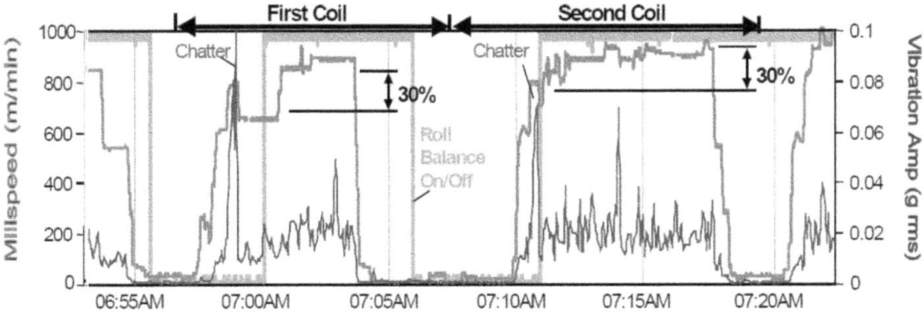

FIGURE 6.9 Performance of a set of vibration inhibiting pistons (tuned mass dampers) on a five-stand tinplate mill [53].

a tandem tinplate mill. The model has then also enabled the development of chatter suppression methods. In particular, the design and implementation of a novel chatter passive control device, a VIP, has been described. The results of rolling mill trials on prototype VI Pistons have shown a maximum performance of a 35% increase in millspeed. The performance is primarily limited, in practice, by the amount of space available for the device and the level of inherent mechanical damping in the rollstack.

6.5 RAIL CORRUGATION

Rail corrugation is a periodic irregularity that is difficult to control as the formation and growth process is analogous to corrugations developing on unsealed roads and rolling mills [60–63]. As predicted in Chapter 3, this irregularity grows in amplitude as a function of the number of wheelset passes. In practice, once an excessive level is reached, typically, removal by grinding is required to ameliorate the excessive noise and vibration caused by the corrugated rail. We have modelled the most prolific type of corrugation, characterised by wear as the damage mechanism. Control of rail corrugation is highly desirable for the railway industry as no reliable alternative exists apart from regrinding the affected surfaces. Due to the persistent nature of this issue, many rail networks enforce periodic preventative rail grinding schedules, placing significant costs on the industry. With increasing rail traffic volume and speed, these maintenance costs will continue to rise unless an alternative method for rail corrugation control is implemented.

The most effective and efficient approach to addressing acoustic and vibration issues is to control their source. This makes it essential to develop a comprehensive theoretical understanding of the mechanisms driving corrugation growth. Over the past few decades, international research has made significant progress in this area. Studies [64–84], among many others, have contributed to the development of numerical models capable of simulating the characteristic behaviour of corrugation formation. Additionally, simpler analytical models have provided valuable insights while avoiding the computational expense of numerical simulations. These models, have been instrumental in identifying fundamental growth mechanisms and analysing parametric trends that could reduce corrugation formation. For instance, in [85,69], analytical predictions revealed that short-pitched wavelength ranges of corrugation are amplified due to a wavelength-based contact filtering effect; however, dynamic wheel-rail contact forces were not considered. In contrast, [71] and [72] explored system stability in the interaction between structural dynamics and contact mechanics over multiple wheelset passages, demonstrating the exponential growth of corrugations and deriving a closed-form expression for their growth rate.

The high costs of rail grinding have spurred research into corrugation mitigation methods, but many have yielded unreliable results (see reviews of [86,87]). Among the most widely recognised approaches are railhead hardening, lubrication using friction modifiers and vibration damping of the vehicle and/or track. Hardening the rail contact surface has long been proposed as a potential solution as it provides increased resistance to wear, theoretically reducing variations in wear patterns and, consequently, corrugation formation. Field tests of HSH rails [73] versus conventional rails support this method, although its validity for all conditions and types of corrugations is debatable.

Arguably, the most promising solution for rail corrugation, at present, is friction modifiers. These are specialised lubricants applied to the wheel-rail contact that effectively control the coefficient of friction to approximately 0.35 [74] and lower traction and wear. Friction modifiers also produce a positive sloped traction–creep curve under full sliding conditions that should eliminate stick–slip instability and squeal (assuming quasistatic behaviour). In practice, lubricants have been shown to reduce the occurrence of certain types of corrugations, but their effectiveness, along with the feasibility of controlled field application, remains under investigation. Additionally, tuned vibration dampers and absorbers have been proposed and tested for railway tracks and wheels to mitigate wayside noise [75]. However, these solutions can be costly, and their impact on corrugation growth appears to be limited.

Controlling Rail Corrugation

An alternative approach proposed and investigated in [72,82] involves controlling train pass speeds using an efficient, mechanics-based predictive model for corrugation growth. A key advantage of such validated analytical models is their extremely fast computational speed, enabling a comprehensive analysis of parameter variations, such as train speed, on corrugation growth rates. To gain further insights, the analytical model developed by [72,82] was employed to predict corrugation growth. This model also provided a mechanics-based analysis of field data, incorporating the influence of environmental conditions, contact patch widening, deformation variations and train speed fluctuations. The conceptual block diagram of the corrugation formation model, shown in Figure 6.10, illustrates the feedback process occurring over multiple wheelset passes. This process arises from the interaction between vehicle and track vibration dynamics, contact mechanics and a damage/wear mechanism, which is triggered by an initial rail profile irregularity.

The model has been used for considering either vertical or lateral vibration modes, interacting with longitudinal or lateral contact forces. In such a way, both straight

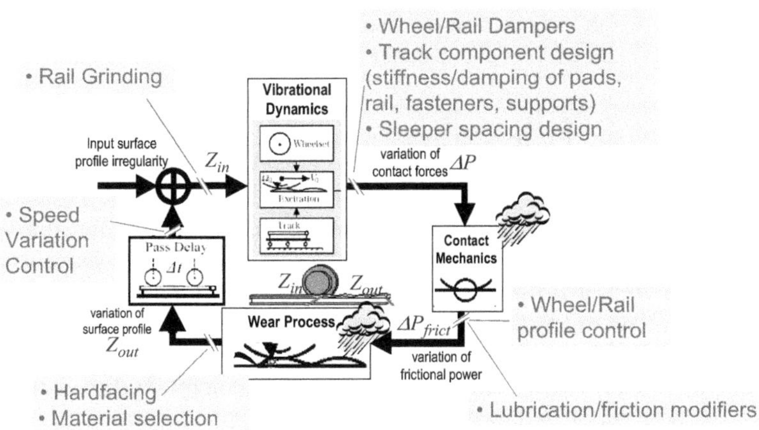

FIGURE 6.10 Corrugation control methods based on interrupting the feedback mechanism of growth. (Based on [90].)

track [71] or curving/cornering conditions [82] may be modelled. In Australia, the model was used to simulate corrugations in curves dominated by vertical vibrations interacting with lateral traction, slip and wear variations [82]. The contact mechanics model is based on quasistatic microslip and considers small linear variations about nominal nonlinear operating conditions using sensitivity coefficients derived from creep theory (i.e., [88,89]). The wear model operates under the assumption that the rate of wear is proportional to the frictional power dissipated. Due to its computational efficiency, this model allows for an extensive analysis of how train pass speed distributions affect corrugation growth rates over multiple vehicle passages. Additionally, it enables simulations across a range of initial track profiles and frictional/wear conditions. Field condition variations, such as contact patch geometry, rainfall, and humidity, primarily influence parameters related to wheel-rail rolling contact mechanics and rail surface wear, as depicted in Figure 6.10. The detailed derivation of these results is summarised in [82]. A key assumption in the model is that a single dominant vibration mode, characterised by a damping constant ζ_i, governs the response of the combined wheel/rail system (and hence the corrugation wavelength). This leads to the formulation of a powerful predictive algorithm for corrugation growth, given by

$$\left|\frac{Z_{n+1}}{Z_n}\right| = 1 + G_{ri}, G_{r_i} \approx K_b\left(1 + \frac{K_{\text{ci}}}{4\zeta_i(1+\zeta_i)}\right), \tag{6.1}$$

where G_r represents the approximate exponential growth rate parameter, analogous to an interest rate per vehicle passage, and Z_n denotes the nth profile. The parameter K_b groups factors related to the sensitivity of wear variations to lateral wheel–rail contact deflection variations, while K_{ci} accounts for the modal sensitivity of wheel–rail relative displacement due to changes in the longitudinal track profile input. For realistic railway conditions, K_{ci} and K_b are always positive-valued. The exponential growth rate described by (6.1) is valid only under stable environmental conditions, where changes in vehicle speed, contact patch geometry, wear and deformation remain small. However, when train pass speeds follow a random distribution, $p(x)$, the expected amplitude ratio of the transfer function in (6.1) evolves into

$$\frac{Z_{n+1}}{Z_n} = \exp\int_{-\infty}^{\infty} \ln\left|K_b \frac{1}{1 + k_c\left(R_r\left(V/\lambda\right) + R_w\left(V/\lambda\right)\right)} + 1\right| p(x)\,dx, \tag{6.2}$$

where multiple wheel passes at varying speeds are accounted for by adjusting the frequency as a ratio of speed, V, relative to a fixed wavelength, λ. The corrugation growth model in (6.2) was first tuned and validated against field measurements collected from suburban rail tracks in curved sections, incorporating non-uniform train pass speeds over multiple passages. The predicted effects of modifying the existing train speed distribution on corrugation growth rates, along with the corresponding grinding cost projections, are illustrated in Figure 6.11.

Figure 6.11 demonstrates a significant reduction in corrugation growth achieved solely by modifying the distribution of train speeds at the affected site. This adjustment leads to a potential reduction in grinding costs by up to 45%, without altering

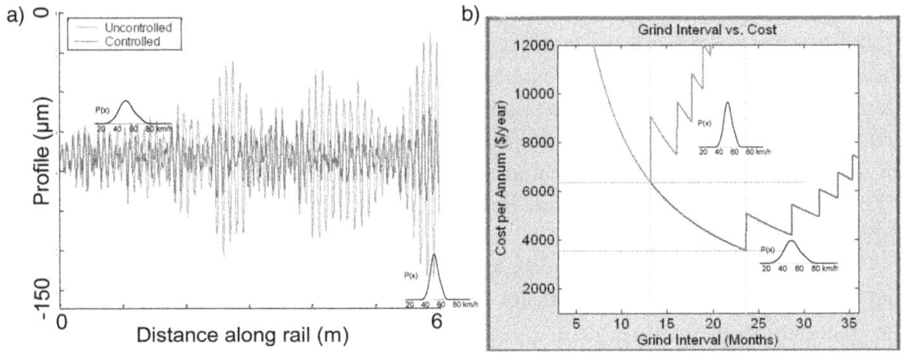

FIGURE 6.11 Results of predictive modelling for corrugation growth showing the effects of controlling successive train pass speed distribution on (a) corrugation profile growth; (b) grinding cost.

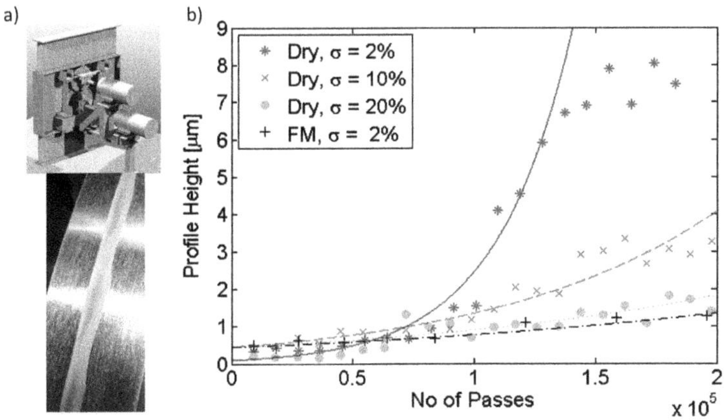

FIGURE 6.12 (a) Experimental apparatus and (b) corrugation growth on a two-disc test rig during speed variation and friction modifier tests. The solid line indicates theoretical prediction [83].

the average train speed. The lowest speed adjustments would result in minimal timetable impact, with delays of less than 20 seconds. This approach highlights the effectiveness of efficient, validated mechanics-based modelling in providing insights and evaluating a proposed control method for corrugation mitigation before its implementation in laboratory and field trials, as detailed in [90]. These results were also confirmed using a two-disc test rig that is scaled in size but matches the contact stress conditions. The speed variation and friction modifier tests were run for approximately 180,000 passes under the following conditions. The results for these tests on peak profile amplitude are shown in Figure 6.12.

Figure 6.12 shows that for the large profile amplitude in the 2% speed variation test, the growth rate reduces, due to increasing conformity in the contact, as discussed previously. The growth rate during the exponential part of the growth was found by fitting a line of best fit to the logarithm of these data. The main sources of uncertainty arise from the standard error of the measured standard deviation in speed and the uncertainty of the steady-state change in the profile per pass measurements. It can be seen that these model-based growth rate predictions and measured growth rates are reasonably close to each other when sources of error are accounted for. Note that friction modification and speed variation with a standard deviation of 20% provide substantial reduction in corrugation growth and therefore are useful corrugation controllers.

Finally, corrugation control was field-implemented under both environmental and deliberate pass speed variations (see [90] and [82] for details). In particular, the distribution of passing train speed was controlled such that the average speed standard deviation over the section was increased from a nominal 4.4 to 7.6 km/h. This speed-based corrugation control system (CCS) was turned on when the corrugation amplitude exceeded approximately 40 μm, as shown in Figure 6.13. During this control phase (CCS ON), the vehicle passing speed standard deviation was increased by 1.7 times, while maintaining the same average speed as the CCS Off condition. The predicted reduction in corrugation growth rate (without considering environmental variations) due to this increase in pass speed distribution was 38% [82,90]. However, the field-measured reduction obtained was approximately 60%, as shown in Figure 6.13 and Table 6.6.

To account for the difference in predicted and measured performance, measurements of rainfall during this period were performed in order to determine if environmental factors had significantly changed. During the control period (CCS ON), the average rainfall increased from 2.1 ± 0.4 up to 4.1 ± 1.2 mm/day. According to the inverse correlation identified in [90], this higher rainfall will lower the corrugation growth rate. Based on this, the linear trend-line for the correlation between

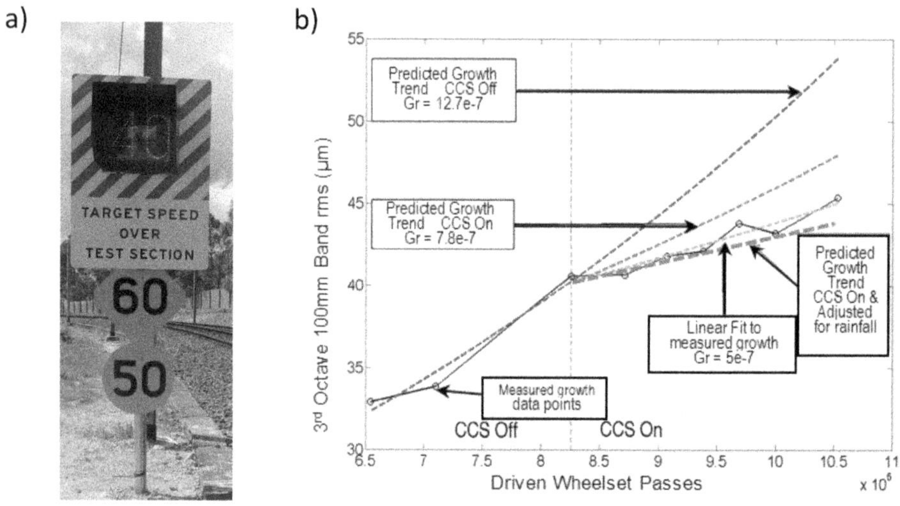

FIGURE 6.13 Corrugation Control System (CCS); (a) field installation and (b) measured, predicted and adjusted corrugation growth rates for the field site test [90].

TABLE 6.6
Corrugation Growth Rate Predictions under Different Conditions [90]

Condition	Growth Rate Parameter Gr
Uncontrolled	1.27×10^{-6}
Uncontrolled, rainfall adjusted (−46%)	0.69×10^{-6}
Controlled (−38%)	0.78×10^{-6}
Controlled & rainfall adjusted (−63%)	0.47×10^{-6}
Measured (−60%)	$\mathbf{0.5 \times 10^{-6}}$

corrugation growth and rainfall was used to predict the influence of the increased rainfall in this period on corrugation growth. Note that any negative predictions of growth rate have been set instead to 0 in these calculations. These growth rate estimates are outlined in Table 6.6 and shown compared to the observed profile growth in Figure 6.13.

The results in Table 6.6 highlight a close match between the predicted (−63%) and field (−60%) results once environmental effects have been considered. Note that the effects of rainfall or vehicle passing speed variation individually provide a substantial drop in corrugation growth. These results highlight the importance of considering environmental variations on corrugation growth. It is noted that in this case study, the effects of contact patch-widening and/or the damage mechanism, based on the results in [90], are assumed to be relatively small and slow after the CCS is activated. They are also considered to be relatively small as compared to the environmental effects over the subsequent period investigated. However, it is recommended to study this more closely in future field investigations.

The rail corrugation research covered in Chapters 3 and 6 summarise a successful approach based on the fundamental stability analysis to predict and control a previously intractable railway defect phenomenon.

6.6 SPACECRAFT INSTABILITY

Controlling spacecraft instabilities, including those stemming from chaotic dynamics, is important in space exploration and satellite operations. As humanity ventures further into space, the complexities of spacecraft dynamics become increasingly apparent and needs to be predicted and addressed prior to launch as most missions are unmanned.

Effective control of spacecraft instabilities typically demands a combination of predictive modelling, advanced control algorithms and real-time monitoring systems to mitigate the effects of instabilities. Adaptive control strategies are often employed to dynamically adjust spacecraft parameters in response to evolving conditions, providing stability in the face of uncertainty. In the context of spacecraft, chaotic instabilities can manifest in various forms, such as erratic movements, irregular trajectories and unstable orbits. Chaotic motion in the attitude dynamics of a communication satellite can severely disrupt the high pointing accuracy required for antennas to maintain precise coverage on the Earth's surface. To prevent such instabilities, designers must either avoid chaotic regions through parameter optimisation or implement active control strategies that allow the use of spacecraft parameters necessary for other mission requirements while mitigating instability. These chaotic phenomena stem from unstable interactions between spacecraft components, gravitational forces and external disturbances, necessitating effective control to prevent mission failure. In recent years, various control techniques have been successfully employed to suppress or manipulate chaotic instabilities. Lindner and Ditto [91] and Blazejczyk et al. [92] provide comprehensive reviews of these methods, while Fradkov [93] presents a dedicated section on spacecraft chaos control. This book explores three primary methods for controlling chaotic vibrations in a simplified dual-spin spacecraft model. The first approach, recursive proportional feedback (RPF), was developed by Rollins et al. [94] as an extension of the occasional proportional feedback (OPF) technique [95,96]. The second method, continuous delayed feedback, was devised by Pyragas [97]. Both techniques are derived from the renowned, model-independent control strategy introduced by Ott, Grebogi, and Yorke (OGY) [98], which does not require prior knowledge of the system dynamics. Meehan and Asokanthan [99–105] have identified chaotic instability in fundamental and more complex dual-spin spacecraft systems. They then demonstrated the effectiveness of these chaotic control techniques.

Hence, predictive modelling and research into instabilities continually inform the development of innovative control methodologies, by providing crucial insights into the underlying mechanisms driving spacecraft instabilities. By harnessing these insights, engineers can design spacecraft control systems that are more resilient to instabilities, enhancing mission reliability and ensuring the safety of space exploration endeavours. Some of these spacecraft control applications will be described and summarised subsequently.

Controlling Spacecraft Instability

The first two control methods are employed on a simple spacecraft: the rotating body with internal energy dissipation, as model-independent methods designed to control any chaotic system. The third method employed is designed using the dynamics of the model and is thus a model-dependent method that stabilises the inherent instability. Each of the employed control methods is capable of detecting unwanted motion, activating control only when necessary, and switching off once the stability is restored. The objective for this system is to stabilise the desired state of a constant spacecraft reference rotation rate with no nutation or precession, ensuring optimal satellite operation. Hence, for simplicity, the spacecraft's angular velocity $\hat{\omega}$ is selected as the sensor variable, while the control actuator is an applied torque $\hat{M}_C$ acting on the spacecraft. This control torque could be implemented using thrusters or an external rotor in a dual-spin satellite. When the system experiences perturbations from a periodically varying excitational torque $\hat{M} = \hat{M}_E + \hat{M}_C$, the satellite is subjected to a combination of applied torques along the spinning axis. Due to practical constraints, the maximum allowable control torque is limited to $M_{\max}$ to ensure feasibility in field application.

Continuous Delayed Feedback Control

Referring to the model and results described in "Prediction of Spacecraft Chaos" section of Chapter 4, the control law may be expressed as a function of the control parameter $x_i \equiv \hat{\omega}$ and torque limitation $M_{\max}$, as

$$\hat{M}_C(\tau) = \begin{Bmatrix} C(\tau), & \left|\hat{M}_C(\tau)\right| < M_{\max} \\ M_{\max}, & \hat{M}_C(\tau) > M_{\max} \\ -M_{\max}, & \hat{M}_C(\tau) < -M_{\max} \end{Bmatrix}, \quad C(\tau) = K\left[\hat{\omega}(\tau - \Delta) - \hat{\omega}(\tau)\right].$$

The block diagram of this system is seen in Figure 6.14. In this case, the delay, Δ, is chosen to be smaller than the forcing period, in accordance with recommendations in Chapter 5 to stabilise a fixed point.

Also, the initial control actuation conditions are described by

$$\left|\hat{\omega}\right| < \frac{\hat{I}\hat{k}}{\hat{h}} \quad \text{and} \quad \hat{y}\hat{y}' < 0,$$

are employed so that the equilibrium state described by $\begin{bmatrix}\hat{y} & \hat{y}' & \hat{\omega}\end{bmatrix}^T = \begin{bmatrix}0 & 0 & \bar{\omega}^*\end{bmatrix}^T$ is stabilised rather than a limit cycle about the equilibrium states $\begin{bmatrix}\hat{y} & \hat{y}' & \hat{\omega}\end{bmatrix}^T = \begin{bmatrix}\pm\bar{y}^* & 0 & \pm\sqrt{\hat{k}}\end{bmatrix}^T$. The first condition corresponds to the second equilibrium of the phase space curves in Section 4.5. The second condition ensures that the control is activated when the movement of the damper mass, $\hat{y}$, is towards the desired fixed point position $\hat{y} = 0$.

RPF Control

Referring to the results described in "Prediction of Spacecraft Chaos" section of Chapter 4, for the present model, a suitable choice for RPF control parameters is

$$p_n \equiv \hat{M}_{C\langle n\rangle}, \quad p_0 \equiv 0, \quad x_n \equiv \hat{\omega}_{\langle n\rangle}, \quad x_f \equiv \omega_{\text{ref}},$$

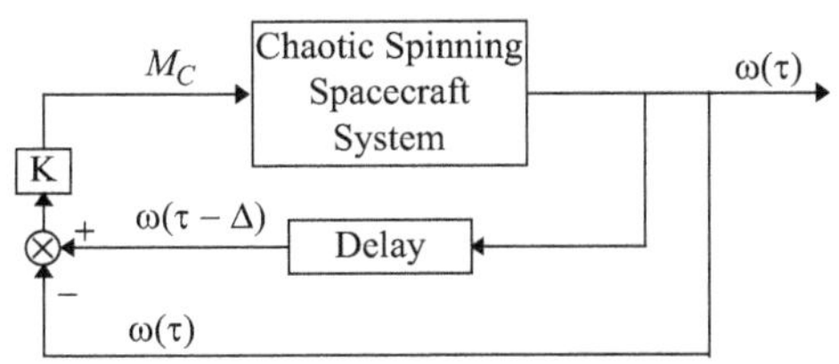

FIGURE 6.14 Block diagram of the continuous delayed feedback method [99].

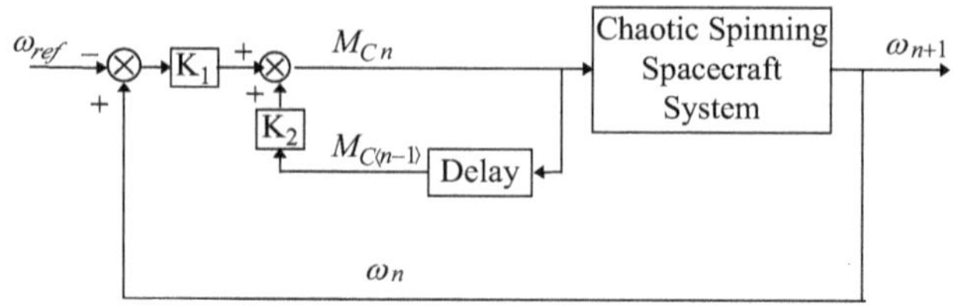

FIGURE 6.15 Block diagram of the recursive proportional feedback (RPF) method [99].

such that

$$\delta p_n \equiv \hat{M}_{C\langle n\rangle}, \quad \delta x_n \equiv \hat{\omega}_{\langle n\rangle} - \omega_{\text{ref}} \quad \text{and} \quad \delta p_{\max} \equiv M_{\max},$$

where $(\ )_{\langle n\rangle}$ is the nth map iterate. Therefore, the RPF control law may be given by

$$\hat{M}_{C\langle n\rangle} = \left\{ \begin{array}{ll} K_1\left(\hat{\omega}_{\langle n\rangle} - \omega_{\text{ref}}\right) + K_2\hat{M}_{C\langle n-1\rangle} & \left|\hat{M}_{C\langle n\rangle}\right| \le M_{\max} \\ 0 & \left|\hat{M}_{C\langle n\rangle}\right| > M_{\max} \end{array} \right\}.$$

The block diagram of this system is seen in Figure 6.15.

In practice, the control design is implemented on a continuous system by holding the control constant for each period of the excitation torque.

Lyapunov Control

A Lyapunov function for the spacecraft system, incorporating a control torque $\hat{M}_C$, can be derived from the total mechanical energy of the system, as outlined in Chapter 5. By minimising the time derivative of this function, the system can be brought to equilibrium as quickly as possible, effectively dissipating energy at each moment. Using the Lyapunov function $\hat{V} = \hat{E} - \hat{E}_C$, where $\hat{E}_C$ is a constant representing the total energy at a specific critical point, it follows that:

$$\hat{V}' = \hat{E}' = -cX_2^2 + \hat{M}X_3.$$

This result is intuitive as it accounts for power input from the external torque $\hat{M}$ and power dissipation due to the damping constant, c. To ensure stability, a controller can be designed to make the second term in the equation as negative as possible, thereby dissipating energy effectively. Since $\hat{M} = \hat{M}_E + \hat{M}_C$, it follows that asymptotic stability is achieved when $\hat{M}_C\omega \le -\hat{M}_E\omega$. Therefore, a suitable control torque is thus designed as

$$\hat{M}_C = \left\{ \begin{array}{ll} C(\tau), & |C(\tau)| < M_{\max} \\ M_{\max}, & C(\tau) > M_{\max} \\ -M_{\max}, & C(\tau) < M_{\max} \end{array} \right\}, \quad C(\tau) = -\left[K_1|\hat{y}'|\text{sgn}(\hat{\omega}) + K_2(\hat{\omega} - \omega_{\text{ref}})\right],$$

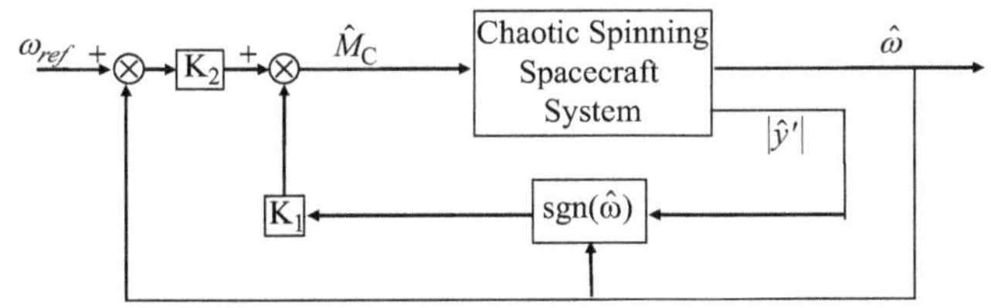

FIGURE 6.16 Block diagram of the Lyapunov control method [99].

where $M_{\max}$ represents the limiting torque magnitude. The block diagram of this control system is depicted in Figure 6.16, illustrating how the control torque is applied to stabilise the spacecraft.

This control law possesses multiple desirable properties. It allows for a specified final angular velocity, ω_{ref}, enabling control over the total angular momentum of the spacecraft, and hence any desired equilibrium point can be reached. Once the system attains the target state without damper vibrations, the control torque naturally reduces to 0, maintaining stability without further intervention. Stability is guaranteed if the control gains K_1 and K_2 are sufficiently large, ensuring $\hat{M}_C\omega \le -\hat{M}_E\omega$. Here, K_1 serves to increase the system's damping, optimising controller performance. However, for a simpler design, stability can still be achieved using only the K_2 term.

Application of Control Methods

To assess the effectiveness of the control methods on the chaotic dynamics of the system, numerical integration of the full nonlinear equations of motion was conducted while the system was subjected to a time-varying excitational torque $\hat{M}_E$ causing chaos without control. Numerical integration was executed using the fourth-order Runge–Kutta method [102–105]. Various analytical tools were employed to examine the impact of control on nonlinear behaviour, including time history analysis, bifurcation diagrams, Poincaré maps and Lyapunov exponent calculations. The Lyapunov spectrum was computed over at least 1000 forcing periods, with renormalisation applied at half the forcing period. The equations of motion were integrated in state-space form, incorporating the combined external torques $\hat{M} = \hat{M}_E + \hat{M}_C$. Other simulation details are provided in [103,105].

Initially, the characteristic nonlinear phenomena of the open-loop system were explored by varying the excitation torque amplitude $\hat{M}_E$. An intermittent route to chaos was identified, where chaotic intervals increased with the amplitude of the excitation torque. At $\hat{M}_E = 1.584$, fully developed chaotic behaviour was observed, as illustrated in Figure 6.17a and b through phase space and Poincaré maps, respectively. Time history and power spectrum analyses further confirmed these findings. The characteristic Lyapunov exponents were calculated, with a positive exponent confirming the presence of chaotic motion. Additionally, the correlation dimension was derived from the Poincaré map, verifying the system's chaotic nature. Given

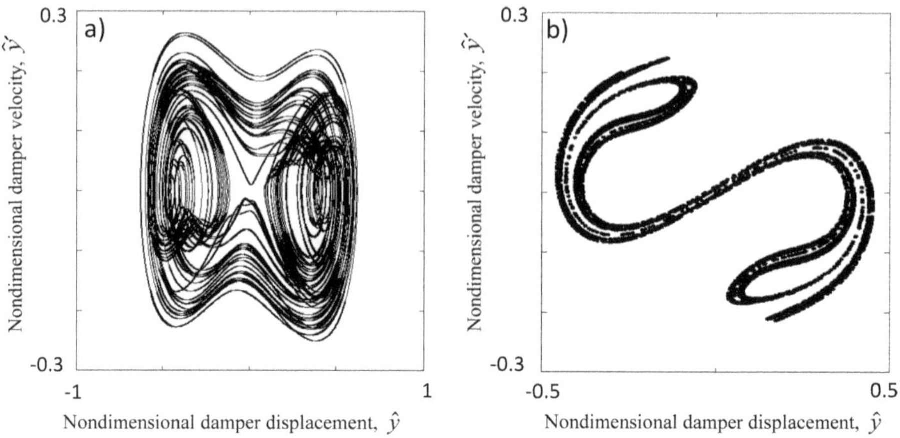

FIGURE 6.17 Numerical simulation for an excitational torque amplitude of $\hat{M}_E = 1.584$ showing chaotic motion via: (a) phase space; (b) Poincaré map [99].

these findings, the effectiveness of two different control methods was evaluated under the same excitation torque to determine their ability to mitigate chaos.

Figures 6.18–6.20 illustrate the effect of each control method by depicting the time histories of the control torque, damper mass position and angular velocity of the system. In each scenario, the control loop is manually activated after a short initial interval, as indicated in the diagrams, allowing for a direct comparison between the open-loop and closed-loop systems. Additionally, a torque limitation ($M_{max} = 10^3$) is imposed in each case, ensuring that the applied control remains within a predefined range.

Figure 6.18a–c shows the effect of applying the delayed feedback method to the system, while it is behaving chaotically. Control is achieved via adjustment of the control parameters to $K = 3.3 \times 10^2$ and $\Delta = 0.25$.

Figure 6.18 shows that once the continuous delayed feedback control loop is closed, a large control torque spike is immediately activated, effectively suppressing the large-amplitude instability, reducing it to a small oscillation around a constant angular velocity.

Figure 6.19 demonstrates the performance of the RPF control method under the same conditions. The control parameters are set to $K_1 = 52.8$ and $K_2 = 0.02$, with the desired reference angular velocity matching that of the continuous delayed feedback method for comparison. While the RPF method requires a control torque amplitude that is an order of magnitude smaller, it results in a larger final oscillation in angular velocity. Although optimisation of control parameters can further reduce this amplitude, the discrete nature of the RPF control method limits how much the oscillation can be minimised.

Figure 6.20 illustrates the effect of the Lyapunov control method through time histories of the control torque, damper mass position and angular velocity of the system. Control is achieved by adjusting the parameters to $K_1 = 1$ and $K_2 = 60$. As depicted in Figure 6.20a, the controller initially applies a large control torque upon

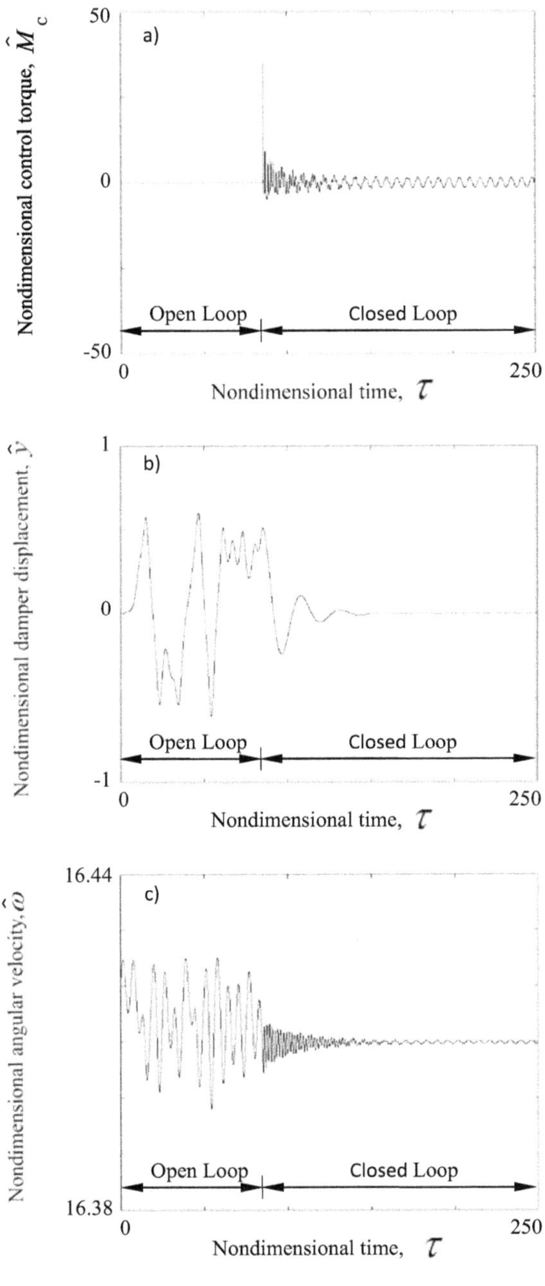

FIGURE 6.18 Numerical simulation for an excitational torque amplitude of $\hat{M}_E = 1.584$ using delayed feedback showing elimination of chaotic motion via time histories of; (a) control torque; (b) damper displacement; (c) angular velocity [99].

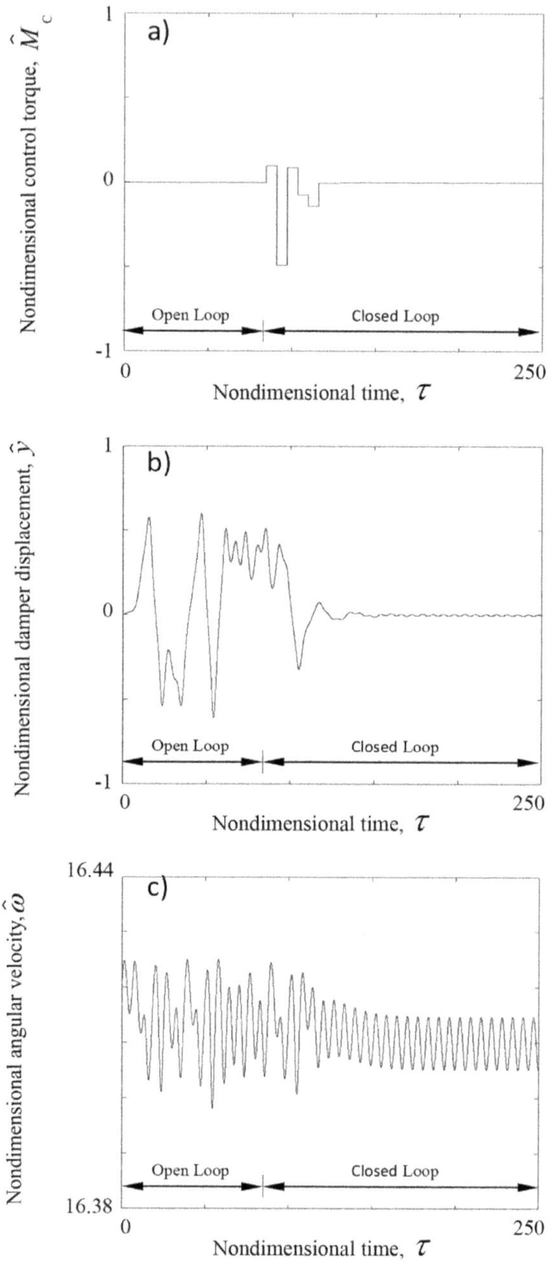

FIGURE 6.19 Numerical simulation for an excitational torque amplitude of $\hat{M}_E = 1.584$ using RPF and $\omega_{ref} = 16.41$ showing elimination of chaotic motion via time histories of (a) control torque; (b) damper displacement; (c) angular velocity [99].

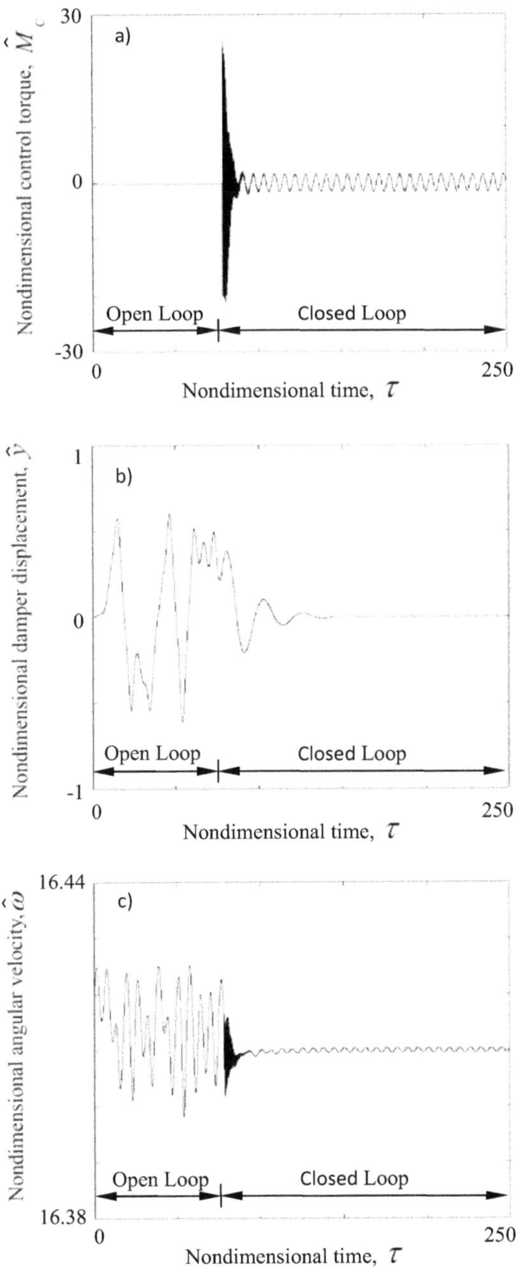

FIGURE 6.20 Numerical simulation for an excitational torque amplitude of $\hat{M}_E = 1.584$ using Lyapunov controller with $\omega_{\text{ref}} = 16.41$ showing elimination of chaotic motion via time histories of (a) control torque; (b) damper displacement; (c) angular velocity [99].

activation, which then stabilises to a small steady value. Figure 6.20c confirms the effectiveness of the control method in eliminating chaotic instability, reducing the system to a similar small periodic oscillation in angular velocity compared to the continuous delayed feedback method but at a faster rate.

To evaluate the robustness of each control technique, the excitational torque amplitude was varied over a range, and a bifurcation diagram of the Lyapunov spectrum was generated for both open-loop and closed-loop cases. Figure 6.21 illustrates these results. In the open-loop case (Figure 6.21a), chaotic motion is present for specific excitational torque amplitudes, $\hat{M}_E > 1.33$, as indicated by a positive maximal Lyapunov exponent. The observed uncertainty in the Lyapunov exponents is attributed to the intermittent nature of the chaotic behaviour. The closed-loop results for the Delayed Feedback, RPF and Lyapunov Controller methods are displayed in Figure 6.21b–d, respectively. In each case, the control parameters were set to the same previously defined values. For all control methods, the maximal Lyapunov exponent is reduced to 0 over the same excitation torque amplitude range as the open-loop case, confirming that the chaotic instability has been successfully suppressed. The zero maximal Lyapunov exponent at the fixed point results from a zero eigenvalue, with an eigenvector aligned along the equilibrium line for angular velocity.

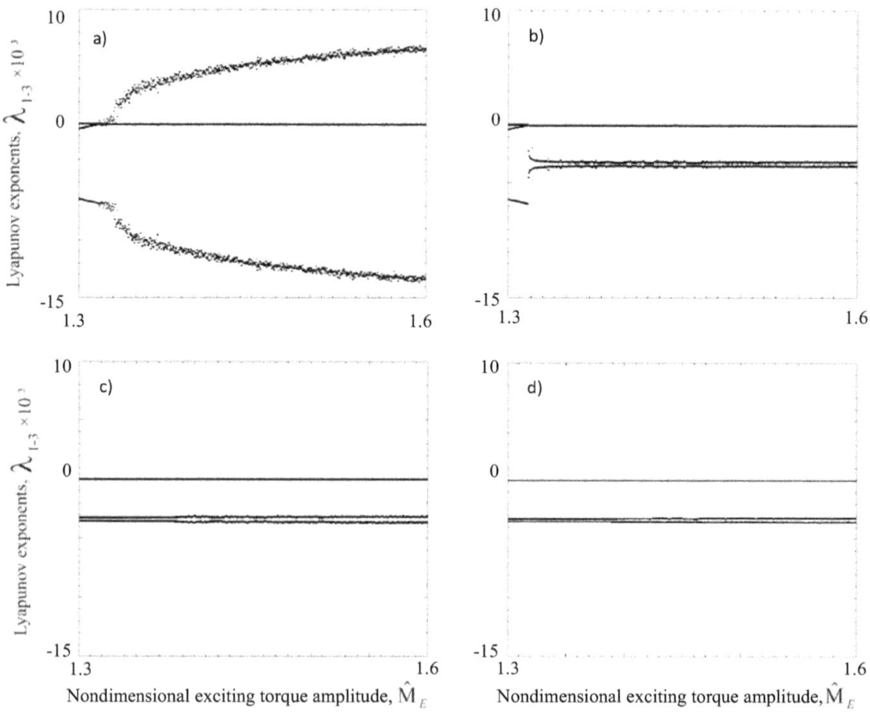

FIGURE 6.21 Bifurcation diagrams of the Lyapunov spectrum of exponents for the (a) open-loop system; (b) closed-loop system with delayed feedback; (c) closed-loop system with recursive proportional feedback (RPF) and (d) closed-loop system with Lyapunov control [99].

Additionally, bifurcation diagrams of the time history confirmed that in all cases, the desired equilibrium point was successfully stabilised under closed-loop control.

Although all control methods successfully eliminate chaotic instabilities across a range of excitation torque amplitudes, each approach has its advantages and limitations. The Delayed Feedback and Lyapunov control methods proved to be more effective in suppressing residual oscillations in angular velocity, primarily due to their continuous nature. In contrast, the RPF method, being discrete, exhibited certain limitations. As noted by Blazejczyk et al. [92], the RPF method can only stabilise periodic orbits where the maximal Lyapunov exponent is small relative to the inverse of the control update interval. Since the control torque adjustments are infrequent and minimal, any external noise fluctuations can occasionally push the system far from the desired periodic orbit. However, the RPF method offers practical advantages, particularly for spacecraft applications. It is simpler in design, requiring only one sensor and low control torques, making it well-suited for spacecraft systems equipped with torque thrusters as actuators. Additionally, unlike the delayed feedback method, the RPF method allows for control at any desired angular velocity, ω_{ref}, providing greater operational flexibility.

In contrast, the delayed feedback method requires sensor input from three state variables due to required initial actuation conditions, and the final angular velocity is determined by the total angular momentum of the system. The continuous nature of the method would require the use of an external rotor or momentum wheel for torque actuation rather than thrusters. In practice, delayed feedback and RPF techniques would be simple to implement on a spacecraft as they require very little or no prior knowledge of the system dynamics.

In comparison, the Lyapunov control method was the most effective control method in eliminating the chaotic instabilities. The control method had the lowest settling times and final steady-state amplitudes for a range of conditions. The method is simple in design, requiring only one or two sensors, low control torques and no prior knowledge of the system parameters. Additionally, it offers the flexibility to control the system at any desired angular velocity. However, due to its continuous nature, implementing this method would necessitate the use of an external rotor or momentum wheel for torque actuation rather than relying on thrusters.

For a more complex practical application, similar control techniques were applied to a dual-spin spacecraft [99,104]. Figure 6.22a–d show the effectiveness of an energy-based control method (based on Lyapunov control) for an external rotor dual-spin satellite and conditions. The control parameters are chosen with a desired reference angular velocity $\omega_{ref} = 50$ to simulate rotor spin-up conditions. The initial control torque amplitude is large to obtain the desired final rotor angular momentum; however, stability is achieved relatively quickly.

To evaluate the robustness of each control method, the rotor excitational torque amplitude was varied, and a bifurcation diagram of the first three Lyapunov exponents was generated for both open-loop and closed-loop cases, as shown in Figure 6.23. In the open-loop case, shown in Figure 6.23a, chaotic motion is indicated by a positive maximal Lyapunov exponent across a range of excitational torque amplitudes. In contrast, the closed-loop results for the delayed feedback, RPF, and energy-based control methods are displayed in Figure 6.23b–d, respectively. In all controlled cases, the

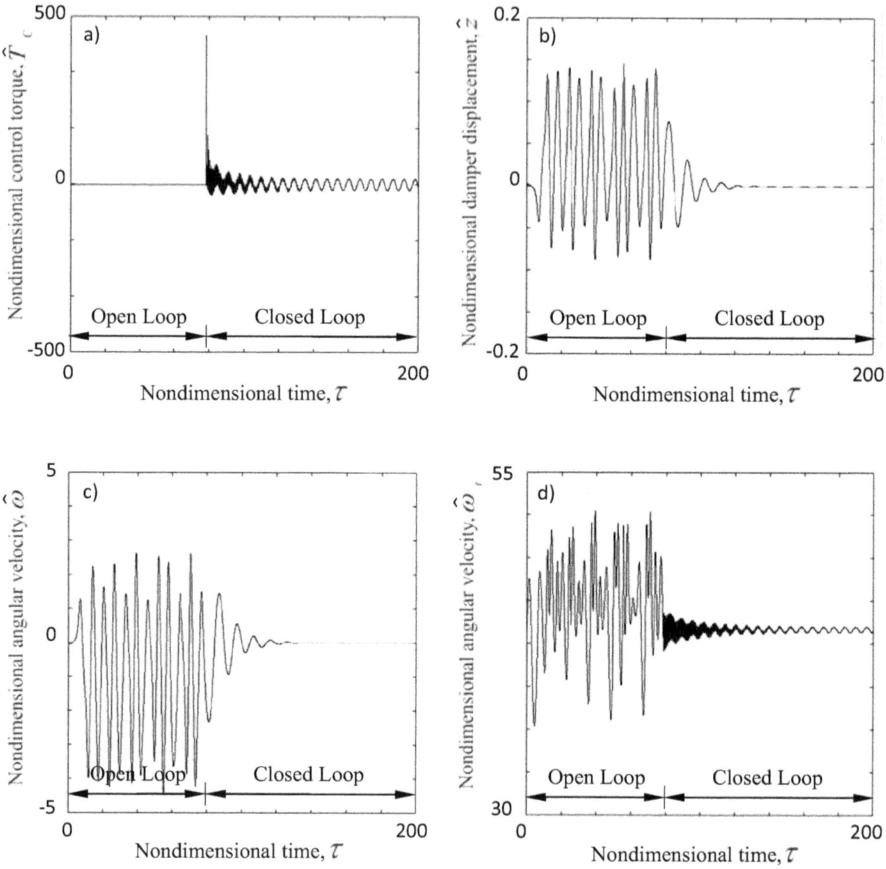

FIGURE 6.22 Numerical simulation for external rotor dual-spin satellite during rotor spin-up ($\omega_{ref} = 50$) showing elimination of chaotic motion via time histories of (a) control torque; (b) damper displacement; (c) angular velocity about body fixed axis x; (d) rotor angular velocity [99].

maximal Lyapunov exponent is reduced to 0, demonstrating that the chaotic instability has been eliminated. Similar to the simpler single spinning spacecraft case of Figure 6.21, the zero maximal Lyapunov exponent at the fixed point is attributed to the presence of a zero eigenvalue due to the corresponding eigenvector aligned along the equilibrium z and rotor angular velocity lines. Also, bifurcation diagrams of time history confirm that in each case, the system successfully stabilised at the desired equilibrium point. Similar results for constant rotor speed and a bias momentum configuration dual spin satellite were also obtained showing stability over the same range of excitational torque amplitudes.

Although all control methods were again successful in eliminating dual-spin spacecraft instability, the energy-based control method (similar to the Lyapunov controller for the simpler spacecraft) was the most effective in eliminating the chaotic instabilities in both external rotor and bias momentum configurations. This is most

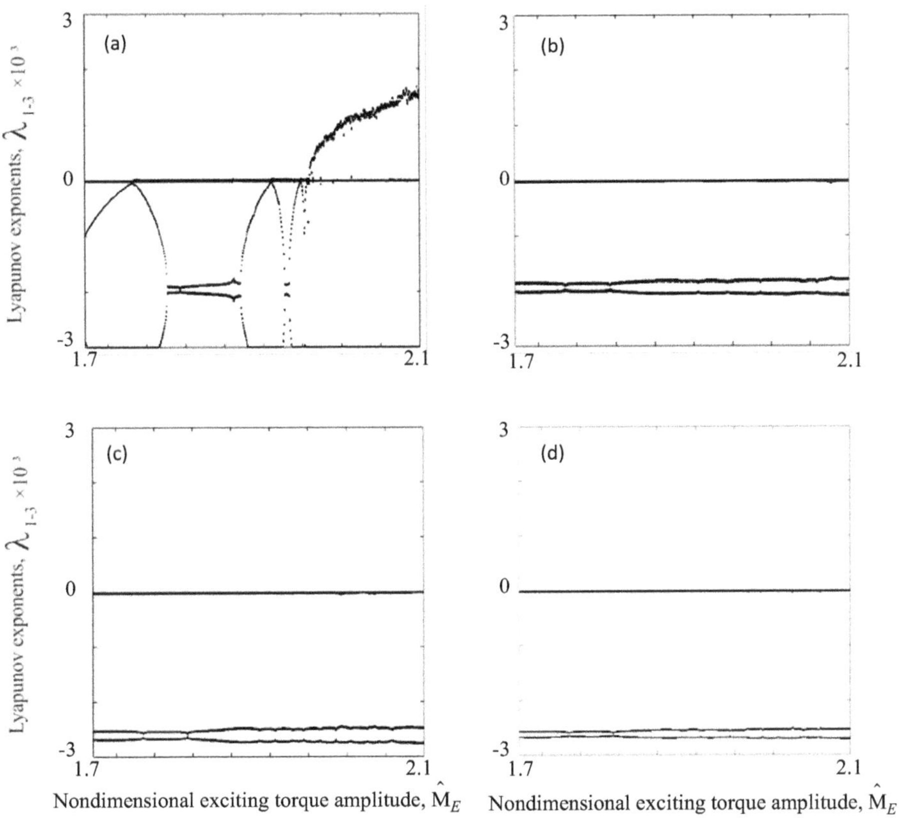

FIGURE 6.23 Bifurcation diagrams of the Lyapunov spectrum of exponents of a chaotic external rotor dual-spin spacecraft (a) open-loop system; (b) closed-loop system with delayed feedback; (c) closed-loop system with RPF; (d) closed-loop system with the energy-based method during spin-up [99].

likely because it is based on maximising the loss of energy associated with chaotic instability. This controller is straightforward in design, requiring only one or two sensors and low control torques, despite its more complex design analysis. Additionally, it offers flexibility in regulating the system to any desired angular velocity but would necessitate the use of a momentum wheel for torque actuation instead of thrusters.

The control of vibrational instabilities in a simplified and dual-spin spacecraft with a nutational damper has been achieved using three techniques. Each technique is outlined, and the effectiveness of the three controllers in eliminating instabilities is then numerically confirmed. The control was shown to be effective on typical simple and complex spacecraft parameter configurations that have previously been shown to exhibit chaotic instability. The robustness of each chaos control technique was evaluated across a range of rotor excitational torque amplitudes for both configurations using Lyapunov exponent analysis. In practical applications, these methods are straightforward in design, requiring only one or two sensors, low control torques and minimal prior knowledge of the system parameters. From an engineering perspective, implementing these control strategies would enable designers to utilise system parameters that are optimal for other requirements while still preventing chaotic instability.

6.7 OTHER CHAOTIC INSTABILITY PHENOMENA

Chaos has been controlled in many engineering systems. Here, we touch on a couple of additional processes to those detailed in the previous sections. However, the reader is directed to other more focussed reviews on control of this phenomenon, such as [91–93,106,107].

Transmission Driveline Chaos

In Chapters 3 and 4, we showed that a transmission driveline incorporating Hooke's (Universal) joint can exhibit twisting instabilities due to parametric excitations at running speed combinations of the natural frequencies of the shafts [108]. The parametric excitation amplitude enlarges as the angle between the shafts enlarge and can lead to larger amplitude chaotic oscillations. These transmission drivelines are commonly used in larger rear wheel-driven vehicles such as trucks, so it is of interest to control these instabilities to avoid excessive vibrations and associated fatigue. In particular, at critical input shaft frequencies, and for sufficiently large Hooke's joint angles, the response of the driveline becomes highly sensitive and unstable. These instability regions occur for ranges about critical input shaft frequencies. For operation close to critical input shaft frequencies, very small variations in input shaft frequency as well as Hooke's joint angle could influence strongly the type of torsional vibrations of the system. It has been shown for the increasing angle of misalignment in shafts that the onset of chaotic motion was characterised by the quasiperiodic route. From a practical point of view, the results obtained have shown that parametric instabilities of the subharmonic type, close to the higher natural frequency, as well as the combination type can be a problem even for low (and high) Hooke's joint angles.

Two model-independent controllers, continuous delayed feedback and the RPF method, have been employed to detect the presence of chaotic motion, switch on the control and then switch it off once it is no longer required. For this system, the aim is to stabilise the motion as close to the desired operation conditions for a Hooke's joint driveline of no torsional vibration. For simplicity in design of the control mechanism, the time differential of the output shaft twist $\dot{y}$ is chosen as the sensor variable and a small change in the angular velocity of input shaft $\Delta\Omega_c$ is used as the control signal. Due to practical constraints, a maximum change in input speed of $\Delta\Omega_{Max}$ is chosen for the control actuator. The feedback control diagrams of the methods are shown in Figures 6.24 and 6.25, with more details available in [99].

The effect of the control methods upon the chaotic dynamics of the system was investigated numerically by integrating the nonlinear state-space equations while the system is subjected to a small control perturbation $\Delta\Omega_c$ to the angular velocity of input shaft. The system parameters chosen for the present study conform to the University of Queensland experimental system, as described previously in the

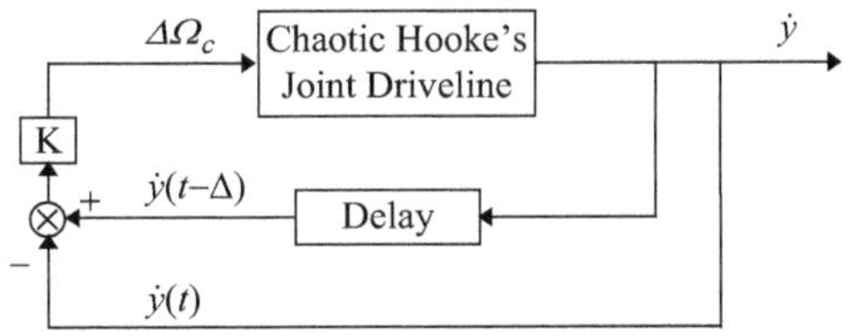

FIGURE 6.24 Block diagram of the continuous delayed feedback method.

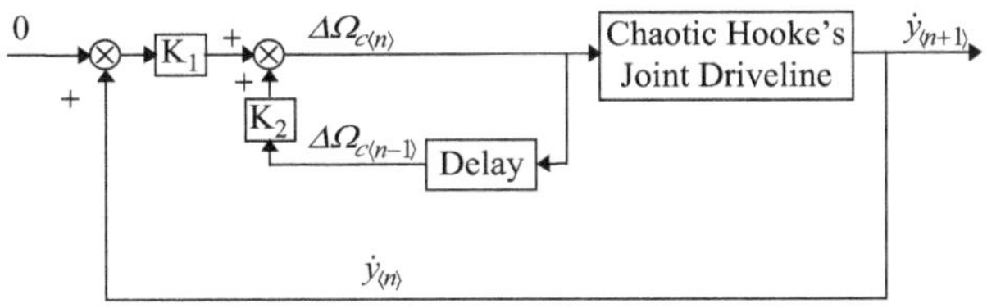

FIGURE 6.25 Block diagram of the RPF method [99].

stability analysis. For these parameters, the natural frequencies ω_1, ω_2 associated with the torsional vibration in the first and second modes take the values 13.78 and 96.61 rad/s, respectively. In this case, the primary combination resonance frequency takes the value of 55.19 rad/s. For all cases, the nominal angular velocity of input shaft Ω was chosen to be 100 rad/s and is subjected to a small control perturbation, $\Delta\Omega_c$, under closed-loop conditions. The open-loop behaviour evolution showed a quasiperiodic route to chaos. A Hopf bifurcation into quasiperiodic motion occurs for $\theta \approx 39°$ as indicated by a 0 maximal Lyapunov exponent. Chaotic motion occurs for Hooke's joint angles greater than $\theta \approx 40.5°$, as indicated by a positive Lyapunov exponent. The amplitude of vibrations of the shafts is found to be greatest when the motion is chaotic. Subsequently, it was of interest to investigate the effectiveness of the two control methods on the system, while the Hooke's joint angle has the value $\theta \approx 42.5°$.

Figure 6.26a–d show the effect of each control method on the Hooke's joint driveline, while it is behaving chaotically via time histories of the control input, $\Delta\Omega_c$ and output shaft twist, y. In each case, the control loop is manually closed after a short interval, as indicated on the diagrams so that a comparison of the open-loop and closed-loop systems can be made. The maximum change in input speed is chosen in each case to be $\Delta\Omega_{\text{Max}} = 15$ rad/s.

Continuous delayed feedback control is achieved via adjustment of the control parameters to $K = 0.028$ and $\Delta = \pi / 200$ seconds. A small control input is activated after the control loop is closed, and a steady-state amplitude of ± 2 rad/s persists. Figure 6.26a shows the control is effective in eliminating the large amplitude chaotic vibrations down to a substantially smaller periodic oscillation in the output shaft twisting motion. Figure 6.26c and d show the effectiveness of the RPF control method for the same conditions. The control parameters are chosen to be $K_1 = 0.46$ and $K_2 = 0.02$. In comparison with the delayed feedback method, the transient control input is large and saturates initially to $\Delta\Omega_{\text{Max}} = 15$ rad/s. However, the steady-state

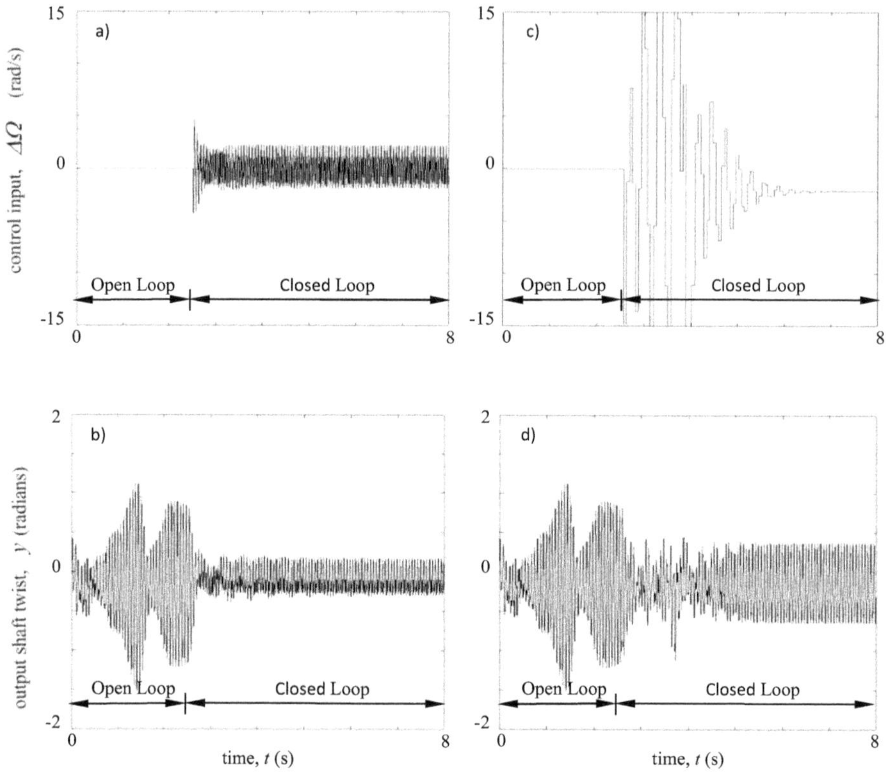

FIGURE 6.26 Numerical simulation showing elimination of chaotic motion via time histories using continuous delayed feedback (a) control input; (b) output shaft twist and RPF; (c) control input; (d) output shaft twist [99].

control input is constant at $\Delta\Omega_c = -2$ rad/s. Also, the amplitude of the final periodic oscillation in the output shaft twisting motion is larger as compared to the delayed feedback method. Optimisation of control parameters can decrease this amplitude; however, the discreteness of the control method is a limitation in the reduction of this final amplitude.

To illustrate the robustness of each control technique, the numerical simulations were performed for the range of Hooke's joint angles, $\theta = 0$–$45°$. Figure 6.27a–c illustrate these results in the form of bifurcation diagrams of time history.

These figures depict the extreme values of the twist of the output shaft, y, for a range of Hooke's joint angles after transient conditions have subsided. The open-loop case is depicted in Figure 6.27a. In this case, the chaotic vibrations, predicted in Chapter 4 for Hooke's joint angles greater than $\theta \approx 40.5°$, are indicated by large amplitudes of the output shaft twisting motion. The closed-loop results for the delayed feedback and RPF method are shown in Figure 6.27b and c, respectively. In each case, the control parameters were chosen as previously. The amplitude of the output shaft twisting motion as compared to the open-loop case is significantly reduced for each method over the range of Hooke's joint angles. Bifurcation diagrams of Poincaré map history

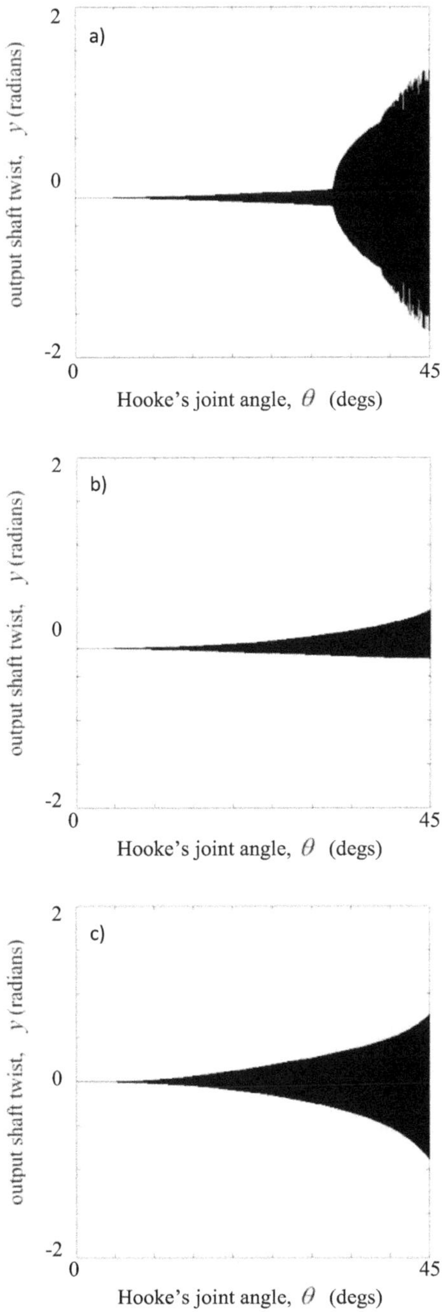

FIGURE 6.27 Bifurcation diagrams of time history showing amplitude of the twist of the output shaft for the (a) open-loop system; (b) closed-loop system with continuous delayed feedback; (c) closed-loop system with RPF [99].

confirmed that the chaotic and quasiperiodic instabilities have been quenched in both cases to a periodic oscillation.

Both control methods effectively eliminated chaotic instabilities across a range of Hooke's joint angles, but each has distinct advantages and limitations. The delayed feedback method was more effective at suppressing the amplitude of the output shaft's twisting motion, likely due to its continuous nature. In contrast, the RPF method exhibited a small residual constant control input, which may be undesirable in certain applications. A key limitation of the RPF method is its discrete nature, which can restrict its ability to stabilise certain periodic orbits, as noted in [92]. This means that rare and small control corrections can allow fluctuation noise to push the system into unwanted states. From a practical perspective, both techniques are straightforward to implement on a driveline system as they do not require prior knowledge of the system's dynamics. However, they are only effective in eliminating chaotic behaviour, while residual periodic limit cycles may still persist after instability is controlled.

Tower Crane Chaos

In Chapter 4 we demonstrated that mining draglines can exhibit undesirable chaotic instabilities in their bucket swing while slewing under normal operating cycles. Tower cranes, another class of slewing horizontal boom lifting machines, have similarly been shown to display chaotic motion during routine slewing operations [109]. In recent years, research on tower cranes has accelerated, driven largely by advances in automation [109] and the increasing need for safe motion control of their inherently complex nonlinear dynamics [110–112]. Building on this progress, [109] extended earlier studies by incorporating more realistic operating conditions, in which the slewing boom oscillates over measured periodic cycles, while rope lengths and load masses vary to represent the full load–unload-return cycle of actual crane operation. That work also introduced a significant step forward in understanding and suppression of chaotic instabilities in slewing boom lifting machines: the development and implementation of a Lyapunov-based control strategy capable of maintaining stable behaviour even at high slewing rates. Although various control approaches have been applied to regulate chaos in simpler rotating multibody systems, the intricate coupling and high dimensionality of slewing boom machines pose substantially greater challenges.

To assess the effect of this control strategy on chaotic behaviour, the full equations of motion were numerically integrated under realistic, chaos-inducing conditions [109]. The controller was then applied in both closed-loop and open-loop configurations—following the methodology developed and demonstrated previously in Chapters 5 and 6—to evaluate its effectiveness. A suite of nonlinear dynamics tools was employed, including phase-space analysis, Poincaré sections, and Lyapunov-based indicators, to quantify the degree of chaos suppression. Representative results are shown in Figure 6.28 and Figure 6.29, illustrating the controller's ability to stabilise the swinging load motion and eliminate the chaotic oscillations.

The closed-loop control strategy was implemented through a continuous adjustment of the control parameters detailed in [109]. Under open-loop conditions, the

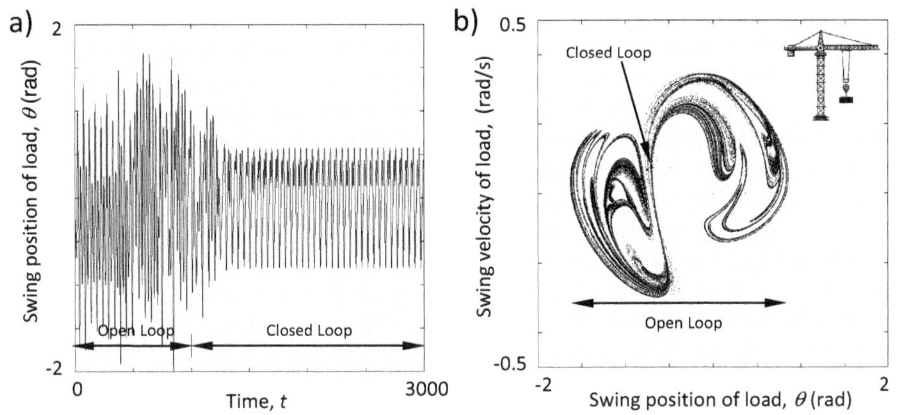

FIGURE 6.28 Numerical simulation of large tower crane case R2 b [109] with an excitational slew torque amplitude of $M_E = 1.44$ MNm and cycle time of 50s using Lyapunov controller showing elimination of chaotic motion via a) time history b) Poincaré map.

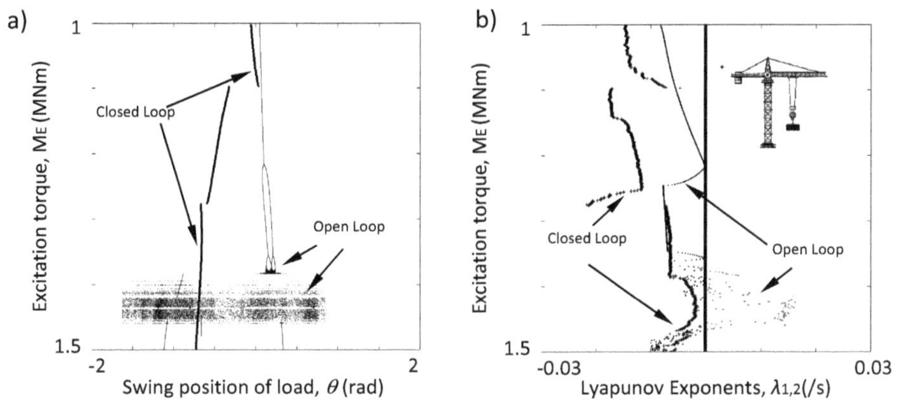

FIGURE 6.29 Control robustness simulations of large tower crane case R2 b [109] over a range of excitational slew torque amplitude ME and cycle time of 50s using Lyapunov controller showing elimination of chaotic motion via bifurcation diagrams of a) swing motion b) first two Lyapunov exponents where one is zero corresponding to the unrestrained slew motion.

tower crane exhibits fully developed chaotic oscillations at M = 1.44 MN, as shown in Figures 6.28a and b. This represents the uncontrolled state—large-amplitude, erratic swing motion that would be unacceptable in real operation. Once the closed-loop controller is activated, however, the chaotic behaviour is suppressed quickly into a much smaller amplitude (3x), periodic oscillation, illustrated in Figure 6.28b. The resulting motion is a stable period-one limit cycle, driven naturally by the slow variations in rope length and suspended mass that occur during actual load–unload–return cycles. Figure 6.29 confirms the robustness of the controller across a wide range of excitation (slew-torque) amplitudes. Not only is stability retained, but the system converges more quickly to the desired limit cycle, as evidenced by the more negative Lyapunov exponent in Figure 6.29b. This confirms that the controller does more than suppress chaos—it actively accelerates the return to stable behaviour. Overall, the method eliminates the chaotic instability with impressive reliability and ensures rapid convergence to the final periodic state.

The Lyapunov-based control method therefore proves highly effective in stabilising the nonlinear dynamics of a tower crane operating under realistic conditions, including time-varying rope lengths and fluctuating loading. The design is intentionally simple and practical; it requires only one or two sensors, uses the existing slew-torque actuator, and demands no prior knowledge of system parameters [109]. Simulations incorporating field-measured values confirm that the corrective control torques remain well within the operating limits of standard commercial slew drives. Moreover, the torque input is energy-efficient—applied only when the crane's motion begins to drift from the stable region. Because the controller is derived analytically from a Lyapunov function, it guarantees global asymptotic stability without relying on excessively high-gain or aggressive inputs, which are commonly impractical for large industrial machinery. The demonstrated robustness across a range of operating conditions suggests that such a control architecture could enable future, larger tower cranes to perform slewing manoeuvres at significantly higher speeds, while maintaining improved safety, precision, and suppression of chaotic instabilities. This tower crane study is another example of effective control and active suppression of chaotic instability based on complete understanding of the nonlinear behavior of the system.

REFERENCES

Wheel Squeal Control

1. Thompson, D. J. (2024). *Railway Noise and Vibration: Mechanisms, Modelling, and Means of Control* (2nd ed.). Elsevier, Oxford/Amsterdam.
2. Thompson, D.J. (2009). Railway noise and vibration: Mechanisms, modelling and means of control. Elsevier, Oxford.
3. Eadie, D.T., Santoro, M. (2006). Top-of-rail friction control for curve noise mitigation and corrugation rate reduction. *Journal of Sound and Vibration*, 293(3), 747–757.
4. Suda, Y., Iwasa, T., Komine, H., et al. (2005). Development of on-board friction control. *Wear*, 258, 1109–1114.
5. Eadie, D.T., Santoro, M., Powell, W. (2003). Local control of noise and vibration with KELTRACK(TM) friction modifier and Protector® trackside application: an integrated solution. *Journal of Sound and Vibration*, 267, 761–772.
6. Liu, X., Meehan, P.A. (2016). Investigation of squeal noise under positive friction characteristics condition provided by friction modifiers. *Journal of Sound and Vibration*, 371, 393–405.
7. Hoffmann, N., Fischer, M., Allgaier, R., Gaul, L. (2002). A minimal model for studying properties of the mode-coupling type instability in friction induced oscillations. *Mechanics Research Communications*, 29, 197–205.

8. Pieringer, A. (2014). A numerical investigation of curve squeal in the case of constant wheel/rail friction. *Journal of Sound and Vibration*, 333(18), 4295–4313.
9. Zenzerovic, I., Kropp, W., Pieringer, A. (2018). Influence of spin creepage and contact angle on curve squeal: A numerical approach. *Journal of Sound and Vibration*, 419, 268–280.
10. Meehan, P.A., Liu, X. (2019). Wheel squeal noise control under water-based friction modifiers based on instantaneous rolling contact mechanics. *Wear*, 440–441, 203052.
11. Meehan, P.A., Liu, X. (2019). Modelling and mitigation of wheel squeal noise under friction modifiers. *Journal of Sound and Vibration*, 440, 147–160.
12. Meehan, P.A. (2020). Investigation of chaotic instabilities in railway wheel squeal. *Nonlinear Dynamics*, 100, 159–172.
13. Meehan, P.A., Liu, X. (2018). Modelling and mitigation of wheel squeal noise amplitude. *Journal of Sound and Vibration*, 413, 144–158.
14. Eadie, D.T., Santoroa, M., Kalousek, J. (2005). Railway noise and the effect of top of rail liquid friction modifiers: Changes in sound and vibration spectral distributions in curves. *Wear*, 258, 1148–1155.
15. Blau, P.J. (2008). Lubrication to control friction. In: Blau, P.J. (ed) *Friction Science and Technology*. CRC Press, Boca Raton, 221–268.
16. Irretier, H., Reuter, F. (1995). Experimental modal analysis of rotating disk systems. In: *Proceedings of the 1995 ASME Design Engineering Technical Conference*, Boston, MA, September 17–20, 1995, 1201–1206, 84(3B).
17. Brunel, J.F., Dufrénoy, P., Naït, M., et al. (2006). Transient models for curve squeal noise. *Journal of Sound and Vibration*, 293(3), 758–765.
18. Anderson, D., Wheatley, N. (2008). Mitigation of wheel squeal and flanging noise on the Australian rail network. In: *Proceedings of the 9th International Workshop on Railway Noise*, Notes on Numerical Fluid Mechanics and Multidisciplinary Design Book Series, vol. 99, NNFM, 2008, 399–405.
19. Liu, S., De Silva, U., Chen, D., et al. (2023). Investigation of wheel squeal noise under mode coupling using two-disk testrig experiments. *Wear*, 530–531, 205035.

Brake Squeal Control

20. Park, J., Lee, J.C., Cho, S., Yoon, K. (2012). Reduction of brake squeal analyzed in terms of coupling between in-plane and out-of-plane modes. *SAE International Journal of Passenger Cars - Mechanical Systems*, 5(4), 1230–1243.
21. Meehan, P.A., Leslie, A.C. (2021). On the mechanisms, growth, amplitude and mitigation of brake squeal noise. *Mechanical Systems and Signal Processing*, 152, 107469.
22. Spurr, R.T. (1961). A theory of brake squeal. *Proceedings of Automotive Division, Institution Mechanical Engineers (AD)*, 1, 33–40.
23. Fieldhouse, J., Ashraf, N., Talbot, C., et al. (2006). Measurement of the dynamic center of pressure of a brake pad during a braking operation. SAE Technical Paper 2006-01-3208.
24. North, M.R. (1976). Disc brake squeal. In: *Braking of Road Vehicles,* Institution of Mechanical Engineers, Automobile Division (Ed.), Mechanical Engineering Publications Limited, London, England, 169–176.
25. Meehan, P.A., Liu, X. (2018). Modelling and mitigation of wheel squeal noise amplitude. *Journal of Sound and Vibration*, 413, 144–158.
26. Meehan, P.A. (2022). Prediction and suppression of chaotic instability in brake squeal. *Nonlinear Dynamics*, 107, 205–225.
27. Oberst, S., Lai, J.C.S. (2011). Chaos in brake squeal noise. *Journal of Sound and Vibration*, 330(5), 955–975.

28. Feeny, B., Moon, F.C. (1994). Chaos in a forced dry-friction oscillator: Experiments and numerical modelling. *Journal of Sound and Vibration*, 170(3), 303–323.
29. Lin, B., Chang, S., Hu, J., Lue, Y. (2015). Controlling chaos for automotive disc brake squeal suppression. *Journal of Mechanical Science and Technology*, 29(6), 2313–2322.
30. Feeny, B.F., Moon, F.C. (2000). Quenching stick–slip chaos with dither. *Journal of Sound and Vibration*, 237(1), 173–180.
31. Denimal, E., Sinou, J.-J., Nacivet, S. (2019). Influence of structural modifications of automotive brake systems for squeal events with kriging meta-modelling method. *Journal of Sound and Vibration*, 463, 114938.
32. Hoffmann, N., Gaul, L. (2003). Effects of damping on mode-coupling instability in friction induced oscillations. *ZAMM-Zeitschrift fur Angewandte Mathematik und Mechanik*, 83(8), 524–534.
33. Meehan, P.A. (2020). Investigation of chaotic instabilities in railway wheel squeal. *Nonlinear Dynamics*, 100, 159–172.

Flutter Control

34. Blevins, R.D. (1990). *Flow Induced Vibrations*, 2nd Ed. Van Nostrand Renhold, New York, USA.
35. Verstraelen, E., Habib, G., Kerschen, G., Dimitriadis, G. (2016). Experimental passive flutter suppression using a linear tuned vibration absorber. *AIAA Journal*, 55(5), 1–16.
36. Malher, A., Touzé, C., Doaré, O., et al. (2017). Flutter control of a two-degrees-of-freedom airfoil using a nonlinear tuned vibration absorber. *Journal of Computational and Nonlinear Dynamics*, 12(5), 051016.
37. Guo, M., Zheng, G. (2021). Stigma as two degrees of freedom energy sink for flutter suppression. *Journal of Sound and Vibration*, 515. doi:10.1016/j.jsv.2021.116441.
38. Zhang, Z., Chen, B., Nielsen, S.R.K. (2017). Coupled-mode flutter of wind turbines and its suppression using torsional viscous damper. *Procedia Engineering*, 199, 3254–3259.
39. Chen, B., Zhang, Z., Hua, X., et al. (2018). Enhancement of flutter stability in wind turbines with a new type of passive damper of torsional rotation of blades. *Journal of Wind Engineering and Industrial Aerodynamics*, 173, 171–179.
40. Kassem, M., Yang, Z., Gu, Y., et al. (2020). Active dynamic vibration absorber for flutter suppression. *Journal of Sound and Vibration*, 469, 115110.
41. Cassaro, M., Battipede, M., Marzocca, P., Behal, A. (2015). Comparison of adaptive control architectures for flutter suppression. *Journal of Guidance, Control, and Dynamics*, 38(2), 346–355.
42. Livne, E. (2018). Aircraft active flutter suppression: State of the art and technology maturation needs. *Journal of Aircraft*, 55(1), 410–452.
43. Venkatramani, J., Sarkar, S., Gupta, S. (2018). Investigations on precursor measures for aeroelastic flutter. *Journal of Sound and Vibration*, 419, 318–336. doi:10.1016/j.jsv.2018.01.009.
44. Meehan, P.A. (2022). Flutter prediction of its occurrence, amplitude and nonlinear behaviour. *Journal of Sound and Vibration*, 535, 117117.
45. Meehan, P.A. (2023). Prediction and suppression of chaos following flutter in wind turbines. *Nonlinear Dynamics*, 111, 22153–22176.
46. Zhao, L.C., Yang, Z.C. (1990). Chaotic motions of an airfoil with non-linear stiffness in incompressible flow. *Journal of Sound and Vibration*, 138(2), 245–254.
47. Dai, H., Yue, X., Xie, D., Atluri, S.N. (2014). Chaos and chaotic transients in an aeroelastic system. *Journal of Sound and Vibration*, 333(26), 7267–7285.

48. Esbati Lavasani, R., Shams, S. (2020). A new dynamic stall approach for investigating bifurcation and chaos in aeroelastic response of a blade section with flap free-play section. *International Journal of Bifurcation and Chaos*, 30(14), 2050200.
49. Moon, F.C. (1992). *Chaotic and Fractal Dynamics: An Introduction for Applied Scientists and Engineers*. Wiley Interscience, New York.
50. Akhavan, H., Ribeiro, P. (2021). Stability and bifurcations in oscillations of composite laminates with curvilinear fibres under a supersonic airflow. *Nonlinear Dynamics*, 103(4), 3037–3058.
51. Bose, C., Gupta, S., & Sarkar, S. 2019. Transition to chaos in the flow-induced vibration of a pitching–plunging airfoil at low Reynolds numbers: Ruelle–Takens–Newhouse scenario. International journal of non-Linear mechanics, 109, 189–203.
52. Li, S., Caracoglia, L. (2020). Experimental error examination and its effects on the aerodynamic properties of wind turbine blades. *Journal of Wind Engineering and Industrial Aerodynamics*, 206, 104357.

Rolling Mill Chatter Control

53. Boulton, G., Domanti, T., Edwards, J., Meehan, P.A., Wallace, G. (2000). *Modelling and Suppression of Rolling Mill Chatter*. Association of Iron and Steel Engineers. Pittsburgh, PA, USA.
54. Tlusty, J., Chandra, G., Critchley, S., Paton, D. (1982). Chatter in cold rolling. *Annals of CIRP*, 31(1), 195–199.
55. Critchley, S., Paton, D.L. (1987). Tandem mill vibration. *Proceedings, 4th International Steel Rolling Conference, Deauville-France*, 2, E.12.1–E.12.6.
56. Krueger, M.M. (2018). Chatter management in cold rolling mills. *Background and Solutions Light Metal Age*, 76(6), 10–15.
57. Keintzel, G., Pröll, C., Krimpelstätter, K., Gerlinde, D. (2015). Chatterblock® Control: Elimination of mill chatter vibration in cold rolling. In: *52º Seminário de Laminação*, Rio de Janeiro, 331–338.
58. Fujita, N., Kimura, Y., Kobayashi, K., et al. (2015). Dynamic control of lubrication characteristics in high speed tandem cold rolling. *Journal of Materials Processing Technology*, 229, art. no. 14577, 407–416.
59. Meehan, P.A. (2002). Vibration instability in rolling mills: Modelling and experimental results. *Journal of Vibration and Acoustics*, 124, 221–227.

Rail Corrugation Control

60. Sato, Y., Matsumoto, A., Knothe, K. (2002). Review on rail corrugation studies, *Wear*, 253, 130–139.
61. Both, J.A., Hong, D.C., Kurtze, D.A. (2001). Corrugation of roads. *Physica A*, 301, 545–559.
62. Yun, I.S., Wilson, W.R.D., Ehmann, K.F. (1998). Review of chatter studies in cold rolling. *International Journal of Machine Tools & Manufacture*, 38, 1499–1530.
63. Shelley, S., Zwart, J., Fournier, A. (1997). New insights into calender barring prevention. In: *Proceedings of the 1997 83rd Annual Meeting of the Technical Section of Canadian Pulp and Paper Association. Part A*. 1997, A251–A258.
64. Hempelmann, K., Knothe, K. (1996). An extended linear model for the prediction of short pitch corrugation. *Wear*, 191, 161–169.

65. Igeland, A., Ilias, H. (1997). Rail head corrugation growth predictions based on non-linear high frequency vehicle/track interaction. *Wear*, 213, 90–97.
66. Matsumoto, A., Sato, Y., Tanimoto, M., Qi, K. (1996). Study on the formation mechanism of rail corrugation on curved track. *Vehicle System Dynamics*, 25, 450–465.
67. Wu, T.W., Thompson, D.J. (2003). An investigation into rail corrugation due to microslip under multiple wheel/rail interactions. In: *Proceedings of the 6th International Conference on Contact Mechanics and Wear of Rail/Wheel Systems(CM2003) in Gothenburg*, Sweden, June 10–13, 2003, 59–67.
68. Bhaskar, A., Johnson, K.L., Wood, G.D., Woodhouse, J. (1997). Wheel–rail dynamics with closely conformal contact. Part 1. Dynamic modelling and stability analysis. *Proceedings of the Institution of Mechanical Engineers*, 211(F), 11–26.
69. Nielsen, J.B. (1999). Evolution of rail corrugation predicted with a non-linear wear model. *Journal of Sound and Vibration*, 227, 915–933.
70. Frederick, C.O. (1987). A rail corrugation theory. In: *Proceedings of the 2nd International Conference on Contact Mechanics and Wear of Rail/Wheel Systems in Kingston*, Rhode Island, July, Waterloo Press, 1987, 181–211.
71. Meehan, P.A., Daniel, W.J.T., Campey, T. (2005). Prediction of growth of wear-type rail corrugation. *Wear*, 258, 1001–1013.
72. Bellette, P.A., Meehan, P.A., Daniel, W., J.T. (2008). Effects of variable pass speed on wear-type corrugation growth. *Journal of Sound and Vibration*, 314, 616–634.
73. Girsch, G., Heyder, R. (2003). Testing of HSH-rails in high speed tracks to minimize rail damage. In: *6th International Conference on Contact Mechanics and Wear of Wheel/Rail Systems, (*CM2003*)*, Sweden, June 10–13, 2003, 265–272.
74. Eadie, D.T., Kalousek, J., Chiddick, K.C. (2002). The role of positive friction (HPF) modifier in the control of short pitch corrugations and related phenomenon. *Wear*, 253, 185–192.
75. Schrey & Veit Gmbh, (2003). Vibration absorber at railway tracks to reduce the wayside noise. Final Report No. R0152003.
76. Meehan, P.A., Daniel, W.J.T., Campey, T. (2003). Wear-type rail corrugation prediction and prevention. In: *Proceedings of the 6th International Conference on Contact Mechanics and Wear of Rail/Wheel Systems (CM2003) in Gothenburg*, Sweden, June 10–13, 445–454.
77. Johnson, K.L., Gray, G.G. (1975). Development of corrugations on surfaces in rolling contact. *Proceedings of the Institution of Electrical Engineers*, 189, 567–580.
78. Johansson, A. (2005). Out-of-round railway wheels – causes and consequences. PhD Thesis, Chalmers University of Technology, Goteborg, Sweden, F15–18.
79. Igeland, A. (1996). Railhead corrugation growth explained by dynamic interaction between track and bogie wheelsets. *Proceedings of the Institution of Mechanical Engineers, Part F: Journal of Rail and Rapid Transit*, 210, 11–20.
80. Bellette, P.A., Meehan, P.A., Daniel, W.J.T. (2010). Contact induced wear filtering and its influence on corrugation growth. *Wear*, 268(11–12), 1320–1328.
81. Meehan, P.A., Daniel, W.J.T. (2008). Effects of wheel passing frequency on wear-type corrugations. *Wear*, 265(9–10), 1202–1211.
82. Meehan, P.A., Bellette, P.A., Batten, R.D., et al. (2009). A case study of wear-type rail corrugation prediction and control using speed variation. *Journal of Sound and Vibration*, 325, 85–105.
83. Bellette, P.A., Meehan, P.A., Daniel, W.J.T. (2011). Validation of a tangent track corrugation model with a two disk test rig. *Wear*, 271(1–2), 268–277.
84. Batten, R.D., Bellette, P.A., Meehan, P.A., et al. (2011). Field and theoretical investigation of the mechanism of corrugation wavelength fixation under speed variation. *Wear*, 271(1–2), 278–286.

85. Muller, S. (2011). A linear wheel-rail model to investigate stability and corrugation on straight track. *Wear*, 249, 1117–1127.
86. Grassie, S.L., Kalousek, J. (1993). Rail corrugation: Characteristics, causes and treatments. In: *Proceedings of the Institution of Mechanical Engineers, Part F*, 207, 57–68.
87. Nelson, J.T. (1997). *Transit Cooperative Research Program Report 23, "Wheel/Rail Noise Control Manual"*. ch. 10. National Academy Press, Washington, DC.
88. Shen, Z.Y., Hendrick, J.K., Elkins, J.A. (1993). A comparison of alternative creep force models for rail vehicle dynamic analysis. In: *Proceedings of the 8th IAVSD Symposium*, Cambridge, MA, 1993.
89. Polach, O. (1999). A fast wheel-rail forces calculation computer code. *Vehicle Systems Dynamics Supplement*, 33, 728–739.
90. Meehan, P.A., Bellette, P.A., Horwood, R.J. (2014). "Does god play dice with corrugations?": Environmental effects on growth. *Wear*, 314(1–2), 254–260.

Spacecraft Instability Control

91. Lindner, J.F., Ditto, W.L. (1995). Removal, suppression and control of chaos by nonlinear design. *Applied Mechanics Reviews*, 48, 795–780.
92. Blazejczyk, B., Kapitaniak, T., Wojewoda, J., Brindley, J. (1993). Controlling chaos in mechanical systems. *Applied Mechanics Reviews*, 46(7), 385–391.
93. Fradkov Alexander, L., Evans Robin, J., Andrievsky Boris, R. (2006). Control of chaos: Methods and applications in mechanics. *Philosophical Transactions of the Royal Society A*, 364, 2279–2307.
94. Rollins, R.W., Parmananda, P., Sherard, P. (1993). Controlling chaos in highly dissipative systems: A simple recursive algorithm. *Physical Review E*, 47(2), R780–R783.
95. Hunt, E.R. (1991). Stabilising high periodic orbits in a chaotic system: The diode resonator. *Physical Review Letters*, 67(15), 1953–1955.
96. Roy, R., Murphy, T.W., Jr., Maier, T.D., et al. (1992). Dynamical control of a chaotic laser: Experimental stabilisation of a globally coupled system. *Physical Review Letters*, 68(9), 1259–1262.
97. Pyragas, K. (1992). Continuous control of chaos by self-controlling feedback. *Physics Letters A*, 170, 421–428.
98. Ott, E., Grebogi, C., Yorke, J.A. (1990). Controlling chaos. *Physical Review Letters*, 64, 1196–1199.
99. Meehan, P.A. (1996). Nonlinear instabilities in rotating multibody systems. Doctor of Philosophy Thesis, Department of Mechanical Engineering, University of Queensland, Australia.
100. Meehan, P.A., Asokanthan, S.F. (2002). Control of chaotic insabilities in a spinning spacecraft with dissipation using Lyapunov's method. *Chaos, Solitons & Fractals*, 13(9), 1857–1869.
101. Meehan, P.A., Asokanthan, S.F. (2002). Control of chaotic instability in a dual-spin spacecraft with dissipation using energy methods. *Multibody System Dynamics*, 7(2), 171–188.
102. Meehan, P.A., Asokanthan, S.F. (2006). Analysis of chaotic instabilities in a rotating body with internal energy dissipation. *International Journal of Bifurcation and Chaos*, 16(1), 1–19.
103. Meehan, P.A., Asokanthan, S.F. (2002). Control of chaotic instabilities in a spinning spacecraft with dissipation using Lyapunov's method. *Chaos, Solitons and Fractals*, 13(9), 1857–1869.

104. Meehan, P.A., Asokanthan, S.F. (2002). Control of chaotic motion in a dual-spin spacecraft with nutational damping. *Journal of Guidance, Control, and Dynamics*, 25(2), 209–214.
105. Meehan, P.A., Asokanthan, S.F. (1998). Control of chaotic motion in a spinning spacecraft with a circumferential nutational damper. *Nonlinear Dynamics*, 17(3), 269–284.

Other Instability Control

106. Basit, A., Khan, N.B., Ali, S., et al. (2024). Chatter detection and suppression in machining processes: A comprehensive analysis. *International Journal of Interactive Design and Manufacturing*, 18, 3751–3771.
107. Beckers, F., Verheyden, B., Aubert, A. (2006). Aging and nonlinear heart rate control in a healthy population. *American Journal of Physiology-Heart and Circulatory Physiology*, 290(6), H2560.
108. Asokanthan, S.F., Meehan, P.A. (2000). Non-linear vibration of a torsional system driven by a Hooke's joint. *Journal of Sound and Vibration*, 233(2), 297–310.
109. Meehan, P.A. (2025). Prediction and control of nonlinear instabilities in slewing boom lifting machines. *Nonlinear Dynamics,* 113, 28675–28698.
110. Guo, H., Peng, W., Zhang, M. et al. (2025). Adaptive control for 5-DOF varying-cable-length tower cranes with multivariable state constraints. *Nonlinear Dynamics,* 113, 2239–2257.
111. Schatz, J., Caverly, R.J. (2023). Payload trajectory tracking of a 5-DOF tower crane with a varying-length hoist cable: A passivity-based adaptive control approach. *Mechatronics,* 94, 103027.
112. Xia, J., Ouyang, H., Li, S. (2023). Fixed-time observer-based back-stepping controller design for tower cranes with mismatched disturbance. *Nonlinear Dynamics,* 111, 355–367.

Index

Note: **Bold** page numbers refer to tables and *italic* page numbers refer to figures.

S

T

U

V

W

Y

Z

For Product Safety Concerns and Information please contact our EU representative GPSR@taylorandfrancis.com
Taylor & Francis Verlag GmbH, Kaufingerstraße 24, 80331 München, Germany

www.ingramcontent.com/pod-product-compliance
Ingram Content Group UK Ltd.
Pitfield, Milton Keynes, MK11 3LW, UK
UKHW022223190926
475702UK00003B/484

* 9 7 8 1 0 3 2 8 0 3 4 4 9 *